THE

Fact Book

Other books in the FactsBook Series:

A. Neil Barclay, Albertus D. Beyers, Marian L. Birkeland, Marion H. Brown, Simon J. Davis, Chamorro Somoza, Alan F. Williams
The Leucocyte Antigen FactsBook

Rod Pigott and Christine Power
The Adhesion Molecule FactsBook

Robin E. Callard and Andy J. H. Gearing
The Cytokine FactsBook

Ed Conley
The Ion Channel FactsBook

Shirley Ayad, Ray Boot-Handford, Martin J. Humphries, Karl E. Kadler and Adrian Shuttleworth
The Extracellular Matrix FactsBook

THE
G-PROTEIN
LINKED
RECEPTOR
FactsBook

Steve Watson

University of Oxford,
Oxford, UK

Steve Arkinstall

Glaxo Institute for Molecular Biology,
Geneva, Switzerland

Academic Press
Harcourt Brace & Company, Publishers
LONDON   SAN DIEGO   NEW YORK   BOSTON
SYDNEY   TOKYO   TORONTO

This book is printed on acid-free paper

ACADEMIC PRESS LIMITED
24–28 Oval Road
LONDON NW1 7DX

United States Edition published by
ACADEMIC PRESS INC.
San Diego, CA 92101

A catalogue record for this book is available from the British Library

ISBN 0–12–738440–5

Designed by Eric Drewery and Adrian Singer
Typeset by Columns Design and Production Services Ltd, Reading
Printed and bound in Great Britain by
Mackays of Chatham PLC, Chatham, Kent

Contents

Section I SUPERFAMILY OF SEVEN TRANSMEMBRANE PROTEINS

Section II SUPERFAMILY OF HETEROTRIMERIC G-PROTEINS

Section III G-PROTEIN LINKED EFFECTOR AND SECOND MESSENGER SYSTEMS

Preface

We are grateful to Helen Brown, David Harrison, Joanne Wiley and Magali Leeman-Husler for help in the compilation of this book and to members of our respective research laboratories as well as our wives for their understanding and patience during preparation of the text. Drs J. Findlay (University of Leeds), A. Lyall (Glaxo, Greenford) and A. Chollet (Glaxo, Geneva) kindly provided initial help on the use of databases. S.P.W. acknowledges the Royal Society for a University Research Fellowship.

Every effort has been made to include all relevant material although by the time this book is in press additional cloning will mean that the lists are incomplete. The authors would appreciate feedback on the contents of the book with particular regard to omissions and inaccuracies which will be corrected in later editions.

Left: *Steve Watson*, Right: *Steve Arkinstall*

This book is dedicated to the memory of Philip Godfrey, a close friend and research colleague who introduced the authors to each other.

Abbreviations

AC	adenylyl cyclase
cyclicGMP-PDE	cyclic GMP-phosphodiesterase
α_{PDE}	cGMP-PDE α subunit
β_{PDE}	cGMP-PDE β subunit
γ_{PDE}	cGMP-PDE inhibitory γ subunit
CTx	Cholera toxin
DAG	1,2-diacylglycerol
EGF	epidermal growth factor
FAc	mixed fatty acylation sites
FGF	fibroblast growth factor
Gα	heterotrimeric G-protein α-subunit
Gβ	heterotrimeric G-protein β-subunit
Gγ	heterotrimeric G-protein γ-subunit
IP$_3$	inositol 1,4,5-trisphosphate
kDa	kilodalton
MAP kinase	mitogen activated protein kinase
Mys	myristoylation sites
NGF	nerve growth factor
PDGF	platelet derived growth factor
PI	phosphatidylinositol
PIP	phosphatidylinositol 4-phosphate
PIP$_2$	phosphatidylinositol 4,5-bisphosphate
PKC	protein kinase C
PKA	cyclicAMP-activated protein kinase A
PLA$_2$	phospholipase A$_2$
PLC	phospholipase C
PTx	Pertussius toxin
SH2	*Src* homology domain 2
SH3	*Src* homology domain 3
Amino acids	The one and three letter codes and derivations and their abbreviations are listed in Table 1, page xii.

Introduction

SCOPE OF THE BOOK

This book is a compendium of seven transmembrane receptors as well as associated heterotrimeric G-protein subunits and effector targets. It provides a general overview of each receptor family as well as primary amino acid sequence information and essential facts relating to structure, function, distribution and pathophysiological significance of each receptor belonging to that family. The information in this book was available in the published literature on 31st March, 1993. It is organized in three sections.

ORGANIZATION OF THE DATA

Section I

Section I describes the superfamily of seven transmembrane receptor proteins and an introduction describes in outline important structural features and general properties of these proteins. This is followed by individual entries organized primarily on receptor families, e.g. adrenoceptors, muscarinic receptors, etc. In a few instances, receptors have been grouped together, e.g. because of similarities in structure of their ligands (glycoprotein hormones; adenosine and adenine nucleotides) or because their endogenous ligand is not known. Where appropriate each entry is divided into three parts.

Endogenous ligands

A brief introduction to the endogenous ligands includes structures as well as short descriptions of their physiological role, distribution and synthesis. References to major reviews on the endogenous ligands are also given.

Receptor subtypes

This section contains a brief introduction to the receptor subtypes within each family. In most cases the receptor subtypes are seven transmembrane proteins or proteins forming intrinsic ion channels. The former are described in detail in the third part of each entry whilst similar information on the latter is presented in the Ion Channel FactsBook. Also included where appropriate are brief notes on the physiological and pathophysiological role of each receptor family as well as further potential receptor subtypes proposed on the basis of preliminary evidence. It is of note that in most cases there is no official scheme of classification of seven transmembrane receptor proteins and the nomenclature used in the book has followed that described in the 1993 *Trends in Pharmacological Science* Receptor Nomenclature Supplement *. This supplement is published annually and is compiled on the advice of experts in each receptor field and, where applicable, appropriate IUPHAR Receptor Committees. In view of the very rapid progress in the identification of new receptors and receptor ligands, the reader is referred to this publication for recent developments.

---

* An IUPHAR committee has been set up to establish an official classification of receptors (Vanhoutte, P. M. (1992) *Pharmacological Reviews* **44**, 349–340).

Individual receptor entries

Each of the molecularly characterized seven transmembrane receptors within a family are discussed in detail. A general key for presentation of each entry is given below.

Section II

Section II overviews the range of heterotrimeric G-protein subunits. An introduction describes major structural features as well as general functions within seven transmembrane receptor-linked signal transduction pathways. This is followed by entries on individual G-protein components separated into three parts for α, β and γ subunits. No rational nomenclature for G-protein subunits has yet been established and this section has adopted classifications used by major contributors to the field.

Section III

Section III describes G-protein-linked effector and second messenger systems and includes information on phospholipase C's, phospholipase A_2, adenylyl cyclases and retinal cyclicGMP-phosphodiesterases. The introduction overviews general catalytic properties of each class of effector molecule as well as describing the importance of second messenger formation to cell function. Individual entries give molecular details of each cloned effector protein together with information on their regulation by heterotrimeric G-protein subunits or other modulators. The nomenclature is based on major publications in the field. Ion channels are not included since they are covered in The Ion Channel FactsBook.

GENERAL KEY TO ORGANIZATION OF THE DATA

Endogenous receptor ligands

The relative potencies of endogenous receptor ligands are described. Where appropriate alternative names for each receptor, G-protein or effector component are also given.

Molecular representation

A molecular representation of receptors and adenylyl cyclases showing transmembrane domains and highlighting predicted sites of glycosylation, location of disulphide bridges and approximate lengths of extracellular and intracellular domains are shown.

Tissue Distribution

Major sites of distribution in cell lines, peripheral tissues as well as the central nervous system are described. This is based upon both RNA (Northern blots, in situ hybridization) and protein expression (Western blotting, immunocytochemistry). Note: in many cases detailed studies of distribution have not been performed and the omission of particular cell or tissue type does not indicate a clear demonstration of absence of expression.

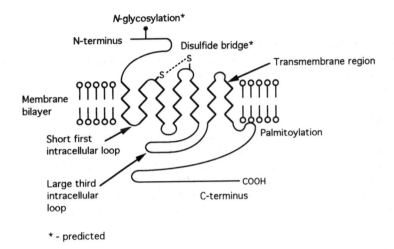

N-glycosylation*

N-terminus

Disulfide bridge*

Transmembrane region

Membrane bilayer

Short first intracellular loop

Palmitoylation

Large third intracellular loop

COOH

C-terminus

* - predicted

Receptor Pharmacology

Agonists: The most selective agonists and their selectivity relative to other receptors in the family is reported.

Antagonists: The most selective antagonists (pA_2 values in brackets) and an indication of their specificity relative to other receptors in the family are reported.

Radioligands: The most-selective ligands are described (K_d values in brackets). Examples of agonist and antagonist ligands, and tritiated and iodinated ligands, are given where possible.

Structures: Structures of receptor-selective agonists and antagonist molecules are shown.

Amino acid sequence (human if available)

The primary amino acid sequence of cloned receptors, G-protein subunits and effector molecules using single letter amino acid code are given (Table 1). Transmembrane domains are underlined and other likely sites of phosphorylation, fatty acylation, bacterial toxin ADP-ribosylation and isoprenylation are indicated using bold type and symbols summarized in Table 2. Potential sites of N-linked glycosylation are also shown based on the presence of the consensus amino acid sequence Asn,Xaa, Ser/Thr (Xaa=any amino acid except Pro). Primary sequences where applicable also show conserved domains of importance for protein–protein interaction or functional activities (Table 2). An asterisk (*) denotes predicted.

Note: Transmembrane domains correspond to regions of hydrophobicity predicted using PEPPLOT routine (based on hydrophobicity measures of Kyte and Doolittle and of Goldman, Engelman and Steiz) of the Wisconsin GCG sequence analysis software package version 7.

Molecular weights

Values given correspond both to theoretical size based on predicted primary amino acid sequence and also, where available, values derived experimentally using SDS–PAGE or gel filtration.

Database accession numbers

The PIR (Protein Identification Resource) and SWISSPROT databases for protein sequences and the GENBANK/EMBL databases for nucleic acid sequences are cited.

Gene structure

Where characterized the organization of genomic DNA is described. No diagram is shown in cases where the entire coding domain is contained within a single exon. Where identified, different lengths of mRNA are reported.

In the example shown below, the boxes represent exons with the numbers above each box indicating the amino acid located at the end of each exon and the number of the exon is shown below in bold. Solid lines connecting the boxes represent introns and hatched boxes untranslated regions. For receptors, solid bands within exons represent predicted transmembrane domains (TM). T1 is the site of translation initiation and STOP, indicates the predicted site of termination.

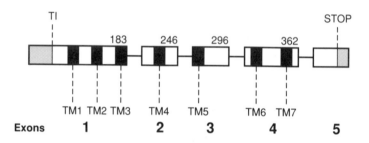

Gene localization and size

This is given where information is available.

Comments

Additional key points of interest are presented.

References

Key publications are cited and listed at the end of each entry. Reviews are indicated in bold print.

Table 1 *Amino acids and their derivatives: codes and abbreviations.*

Amino acid	Single letter code	Three letter code
Alanine	A	Ala
Arginine	R	Arg
Asparagine	N	Asn
Aspartic acid	D	Asp
Cysteine	C	Cys
Glutamine	Q	Gln
Glutamic acid	E	Glu
Glycine	G	Gly
Histidine	H	His
Isoleucine	I	Ile

Table 1 *Continued*

Amino acid	Single letter code	Three letter code
Leucine	L	Leu
Lysine	K	Lys
Methionine	M	Met
Phenylalanine	F	Phe
Proline	P	Pro
Serine	S	Ser
Threonine	T	Thr
Tryptophan	W	Trp
Tyrosine	Y	Tyr
Valine	V	Val

Abbreviation	Derivative
Aib	α-aminobutyric acid
d	deamino
d(CH$_2$)$_5$	1-(β-mercapto-β,β-cyclopentamethylenepropionic acid)
Hyp	4-hydroxyproline
Oic	(93as,7as)-octahydroindol-2-yl-carbonyl
Orn	ornithine
Phaa	phenylacetyl
Sar	sarcosine
Thi	(2-thienyl)-L-alanine
Tic	1,2,3,4-tetrahydroisoquinolin-2-yl-carbonyl
Tyr(Me)	O-methyl-tyrosine

Table 2 *Key to the sequences*

Transmembrane regions	: underlined
PLC catalytic X and Y regions	: boxed
Other domains (as indicated)	: shaded boxes
Carboxymethylation and isoprenylation/ geranylgeranylation	: $
ADP-ribosylation by cholera toxin	: CTx
ADP-ribosylation by pertussis toxin	: PTx
Myristoylation sites	: Mys
Mixed fatty acylation sites:	: FAc
Site of interaction with G-protein γ-subunits	: γ

SUPERFAMILY OF SEVEN TRANSMEMBRANE PROTEINS

Introduction: Seven Transmembrane Proteins

The superfamily of G-protein coupled receptors are integral membrane proteins characterized by amino acid sequences which contain seven hydrophobic domains, predicted to represent the transmembrane spanning regions of the proteins. They are found in a wide range of organisms and are involved in the transmission of signals to the interior of the cell through interaction with heterotrimeric G-proteins. They respond to a wide and diverse range of agents including lipid analogues, amino acid derivatives, small peptides and specialist stimuli such as light, taste and odour.

Brief notes are presented below on the main structural features and general properties of this protein superfamily. Extensive molecular and mechanistic research has been carried out on two receptor proteins, rhodopsin and the β_2 adrenoceptor, using a combination of biochemical and molecular biology techniques. In many cases, the generality of these results to other G-protein coupled receptors is uncertain.

A comprehensive treatise of the literature is beyond the scope of this brief introduction and the reader is referred to recent reviews for a more comprehensive discussion [1-13].

TRANSMEMBRANE DOMAINS

Hydropathicity plots suggest that all of the members of this family contain seven hydrophobic domains of 20–25 amino acids in length which are believed to represent transmembrane regions (Fig. 1). Based on structural similarities with the extensively characterized protein bacteriorhodopsin [14], which includes electron diffraction data, these regions are predicted to be α-helices and to be oriented to form a ligand binding pocket [2]. Protease digestion studies and detailed immunological mapping have provided supporting evidence for this model, including direct evidence that the N-terminal sequence is extracellular and the C-terminal sequence is intracellular [15,16]. Thus, each receptor is believed to have an extracellular N-terminal sequence which can vary in length from less than ten amino acids up to several hundred, followed by three sets of alternate intracellular and extracellular loops and a final intracellular C-terminal sequence. The majority of the intracellular and extracellular loops are predicted to be between 10 and 40 amino acids in length, although the third intracellular loop and the C-terminal sequence can have more than 150 residues.

LIGAND BINDING

The binding site for the diverse range of ligands which activate these proteins is contained, at least to some extent, in the transmembrane region of the receptor. The visual pigment rhodopsin is a special example of this in that a light signal induces isomerization of a covalently bound retinal chromophore located in the seventh transmembrane region.

For many small molecular weight ligands, e.g. adrenaline, the binding pocket is contained within the transmembrane regions, while for larger stimuli, e.g. small peptides [17] and glycolipid hormones [10], the extracellular region of the receptor also plays a role. Not surprisingly in light of the fact that the transmembrane regions are sites of agonist interaction, these regions are highly conserved between members of a receptor family and there is a far greater divergence in the extracellular regions. A notable exception here is the odorant family which has evolved to detect a wide range of odorous stimuli.

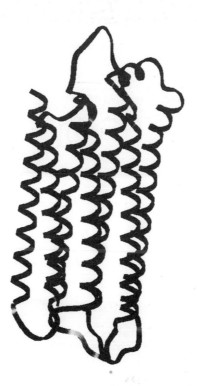

Figure 1 *Three dimensional model of the human NK₂ tachykinin receptor based on electron diffraction structure of bacteriorhodopsin [14]. Illustration shows only core seven transmembrane spanning domains with the extracellular side facing the top of the page. Note: the amino- and carboxyl-terminal portions of the receptor are omitted. This model was created by Professor J. Findlay et al. at the University of Leeds, UK.*

Within a given receptor family it is possible to detect conserved amino acids which have a key role in agonist binding. For example, in the adrenoceptor family, the carboxylate group on the conserved aspartate residue in the third transmembrane domain is believed to act as a counter-ion for the catecholamine nitrogen whilst two conserved serines in the fifth transmembrane domain interact with the meta- and para- hydroxyls on the catechol ring (Figure 2)[11]. These three residues are found in equivalent locations in all nine of the cloned adrenoceptors.

In many cases the binding sites for antagonists are distinct from those of the naturally occurring ligands. This is exemplified by the tachykinin NK₁ receptor in which the nonpeptide antagonist, CP 96345, binds to amino acids within the transmembrane region which do not participate in the binding of endogenous peptide ligands [17]. Moreover, CP 96345 also exhibits up to a 100-fold difference in affinity between species homologues of the NK₁ receptor in contrast to the similar affinities of the endogenous ligands; this difference is largely explained by a single, conservative amino acid substitution in a transmembrane domain [18,19]. Similar

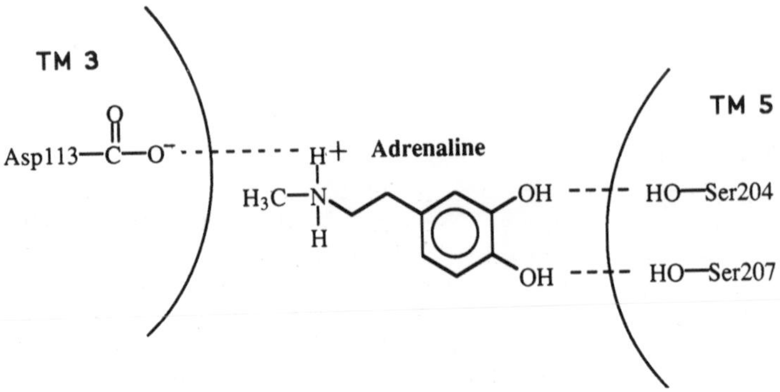

Figure 2   *Sites of interaction between the β2 adrenoceptor and adrenaline [11].*

differences in affinity of unnatural ligands for species homologues of equivalent receptors in other receptor families are beginning to emerge [3].

GLYCOSYLATION

The higher molecular weight of receptors than that predicted from their amino acid sequences suggests that many are glycosylated. This is supported by the observation that nearly all G-protein receptors have one or more asparagine residues in the extracellular domain, usually in the N-terminus, which are present in a glycosylation consensus sequence. Direct evidence for glycosylation has been shown for a number of receptors, including the purified β2 adrenoceptor in which endoglycosidase treatment decreases its molecular weight from ~ 65 kDa to ~ 49 kDa [20,21].

For most receptors it is believed that glycosylation has little role in agonist binding but that it may be of importance in determining the correct distribution of the receptor in the cell. For example, treatment of the β2 adrenoceptor with endoglycosidases or removal of the two asparagine residues predicted to undergo glycosylation by site-directed mutagenesis has little effect on agonist or antagonist binding or coupling of the receptor to adenylyl cyclase [8]. The potential importance of glycosylation, however, is exemplified by the finding that mutation of Thr17 to methionine in human rhodopsin, which destroys a consensus sequence for glycosylation, is one cause of the retinal degenerative disease, retinitis pig-mentosa [22].

CYSTEINE DISULFIDE BRIDGES

The presence of a cysteine residue towards the carboxyl side of the first extracellular loop and a second one towards the middle of the second extracellular loop is a constant finding in many G-protein coupled receptors. Direct chemical evidence has been obtained for a disulfide linkage between these two cysteine residues in rhodopsin and it has been shown that this is essential for achieving the correct tertiary structure [23]. Studies with the reducing agent dithiothreitol have provided indirect evidence that this disulfide linkage plays an important role in agonist and antagonist interactions in a large number of receptors and this has received direct support by site-directed mutagenesis of candidate cysteine residues in the β2 adrenoceptor [24,25]. Two further cysteine residues located next to each other in the third extracellular loop of the β2-adrenoceptor are also linked by a disulfide bond [9].

PALMITOYLATION

The β2 adrenoceptor and rhodopsin have one and two cysteine residues, respectively, in their C-terminal sequence which have been shown to bind palmitate via a thioester linkage [26,27]. It is believed that the covalently bound palmitic acid residues become intercalated in the membrane bilayer, thereby creating a fourth cytoplasmic loop. Site-directed mutagenesis of this cysteine residue in the β2 adrenoceptor uncouples the receptor from its associated G-protein, G_s [26]. The presence of one or two cysteines in a similar position in most other G-protein linked receptors suggests that this may be a common structural feature.

G-PROTEIN COUPLING

Site-directed mutagenesis, deletions and chimeric receptor studies have been used in an attempt to identify the region of the β2 adrenoceptor that couples with G_s. This work has highlighted a sequence of ~8 amino acids in the N-terminus and ~12 amino acids in the C-terminus of the third transmembrane loop as important determinants of this interaction [9]. However, it appears that additional regions of the receptor also participate in the binding to the G-protein, most notably in the second intracellular loop, and that it is the overall 3-dimensional structure of the receptor on the cytoplasmic side of the membrane that is important for the interaction with the G-protein. It has therefore not been possible to identify consensus amino acid sequences that confer G-protein specificity, and thus G-protein interactions cannot be predicted from the primary amino acid sequence.

REGULATION OF THE RECEPTOR BY PHOSPHORYLATION

Responses to G-protein coupled receptors undergo rapid desensitization during continued exposure to the appropriate stimulus and there is a great deal of evidence to support a role of protein phosphorylation in this set of events [5]. The intracellular portions of all G-protein coupled receptors are rich in serine and threonine residues, and many of these are found in consensus sequences for phosphorylation by protein kinases such as PKC and PKA. The functional importance of phosphorylation in desensitization has been studied extensively in rhodopsin and in the β2 adrenoceptor. Exposure of rhodopsin to light induces a conformational change which allows rhodospin kinase to phosphorylate a number of serine and threonine residues on its intracellular regions. These phosphorylations enable a 45 kDa protein, arrestin, to bind to the photoexcited rhodopsin and thus prevent activation of transducin [4]. β adrenergic receptor kinase (β-ARK) and β arrestin are believed to perform a similar function at the β2 adrenoceptor [4,9], and also possibly for other adenylyl cyclase-coupled receptors.

A number of other kinases, including PKA, also phosphorylate the β2 adrenoceptor, providing alternative pathways of desensitization.

GENE

Many G-protein coupled receptors are encoded by a single exon which lacks a cleavable N-terminal signal sequence. However, a number of proteins are encoded by multiple exons including members of the family of opsins (5 exons), tachykinin receptors (5 exons), bombesin receptors (3 exons), certain dopamine receptors (D_2, D_3 and D_4; 7 exons), endothelin receptors (7 exons) and the luteinizing hormone receptor (11 exons). Amongst this group, the luteinizing hormone receptor is unique in that the seven transmembrane region is encoded by a single exon, suggesting that it has been derived by addition of 10 exons to the N-terminus of an intronless gene [28].

References

1. Dohlman, H.G. et al. (1991) Model systems for the study of signal-transducing GTP-binding proteins Annu. Rev. Biochem. 60, 653–688.
2. Findley, J. et al. (1990) Three-dimensional modelling of G protein-linked receptors. Trends Pharmacol. Sci. 11, 492–499.
3. Hall, J. et al. (1993) Receptor subtypes or species homologues: relevance to drug discovery. Trends Pharmacol. Sci. in press.
4. Hargrave, P.A. and McDowell, J.H. (1992) Rhodopsin and phototransduction: a model system for G protein-linked receptors. FASEB J. 6, 2323–2331.
5. Hausdorff, W.P. et al. (1990) Turning off the signal: desensitisation of β-adrenergic receptor function. FASEB J. 4, 2881–2889.
6. Jackson, T.R. (1991) Structure and function of G protein coupled receptors. Pharmacol. Ther. 50, 425–442.
7. Kobilka, B. (1992) Adrenergic receptors as models for G protein-coupled receptors. Annu. Rev. Neurosci. 15, 87–114.
8. O'Dowd, B.F. et al. (1989) Structure of the adrenergic and related receptors. Annu. Rev. Neurosci. 12, 67–83.
9. Ostrowski, J. et al. (1992) Mutagenesis of the β_2-adrenergic receptor: how structure elucidates function. Annu. Rev. Pharmacol. Toxicol. 32, 167–185.
10. Reichert, L.E. et al. (1991) Structure-function relationships of the glycoprotein hormones and their receptors. Trends Pharmacol. Sci. 12, 199–203.
11. Strader, C.D. et al. (1989) Structural basis of β-adrenergic receptor function. FASEB J. 3, 1825–1832.
12. Strosberg. D. (1991) Structure/function relationship of proteins belonging to the family of receptors coupled to GTP-binding proteins. Eur. J. Biochem. 196, 1–10.
13. Venter, J.C. et al. (1989) Molecular biology of adrenergic and muscarinic cholinergic receptors. Biochem. Pharmacol. 38, 1197–1208.
14. Henderson, R. et al. (1990) J. Mol. Biol. 213, 899–929.
15. Wang, H. et al. (1989) J. Biol. Chem. 264, 14424–14431.
16. Dohlman, H.G. et al. (1987) J. Biol. Chem. 262, 14282–14288.
17. Fong, T.M. et al. (1993) Nature 362, 350–353.
18. Fong, T.M. et al. (1992) J. Biol. Chem. 267, 25668–25671.
19. Sachais, B.S. et al. (1993) J. Biol. Chem. 268, 2319–2323.
20. Benovic, J.L. et al. (1987) J. Rec. Res. 7, 257–281.
21. Rands, R. et al. (1990) J. Biol. Chem. 265, 10759–10764.
22. Sung, C.H. et al. (1991) Proc. Natl Acad. Sci. 88, 6481–6485.
23. Karnik, S.S. & Khorana, H.G. (1990) J. Biol. Chem. 265, 17520–17524.
24. Dohlman, H.G. et al. (1990) Biochemistry 29, 2335–2342.
25. Dixon, R.A.F. et al. (1987) EMBO J. 6, 3269–3275.
26. O'Dowd, B.F. et al. (1989) J. Biol. Chem. 264, 7564–7569.
27. Ovchinnikov, Y.A. et al. (1988) FEBS Lett. 230, 1–5.
28. Koo, Y.B. et al. (1991) Endocrinol. 128, 2279–2308.

Acetylcholine

INTRODUCTION

Acetylcholine is a neurotransmitter in the central and peripheral nervous systems and mediates its actions through two distinct receptor types, nicotinic and muscarinic [1-7]. Only the latter will be discussed in detail. Nicotinic receptors contain an intrinsic ion channel and are described in detail in the 'Ion Channel FactsBook' [8].

In the CNS, acetylcholine is a transmitter in interneurons e.g. caudate-putamen, nucleus accumbens, olfactory tubercle and islands of Calleja, and in projection neurons, notably the basal forebrain and pontomesencephalotegmental cholinergic systems. The physiological role of acetylcholine in the CNS, however, is poorly understood. Acetylcholine is implicated in higher functions of the brain, notably memory and cognition; consistent with this, there is a cholinergic deficiency in Alzheimer's disease, an illness associated with a severe impairment of cognitive function.

In the periphery acetylcholine is the major transmitter in autonomic ganglia, postganglionic parasympathetic nerve fibres and in the neuromuscular junction. It has an essential physiological role in stimulating contraction of skeletal and smooth muscles, and inhibition of cardiac muscle. It also regulates secretion in glandular tissue including the intestine and the parotid.

Muscarinic agonists are used clinically in the treatment of glaucoma where they are administered as eye drops. More minor uses include the suppression of atrial tachycardias, stimulation of intestinal motility and bladder emptying. They are of potential use in the treatment of Alzheimer's disease.

Muscarinic antagonists are used as a premedication in general anaesthesia to reduce bronchial and salivary secretions, and in the prevention of motion sickness. They are also used to a limited extent in the treatment of peptic ulcer, to induce pupillary vasodilatation to aid examination of the eye and in the treatment of certain inflammatory conditions. The use of muscarinic antagonists is limited by a frequent occurrence of unpleasant side-effects, e.g. dry mouth.

Acetylcholine

Distribution and synthesis

Acetylcholine is synthesized from acetyl CoA and choline in a reaction catalysed by choline acetyltransferase and the distribution of this enzyme is frequently used to reflect the distribution of acetylcholine. Choline acetyltransferase has a wide distribution within the CNS where it is found in nerve endings: the highest levels exist in the interpeduncular nucleus, caudate nucleus, retina and central spinal roots with only trace amounts present in dorsal spinal roots and cerebellum. In the periphery, choline acetyltransferase is found presynaptically in ganglia, postganglionic parasympathetic nerve terminals and in the neuromuscular junction.

Books/Supplement

1 Brown, D. (ed.) (1989) 'Muscarinic Receptors', Humana, New Jersey.
2 Levine, R. and Birdsall, N.J.M. (eds) (1989) Subtypes of muscarinic receptor IV. Trends Pharmacol. Sci.

Reviews

3 Bonner, T. I. (1989) The molecular basis of muscarinic receptor diversity. Trends Neurosci. **12**, 148–151
4 Caulfield, M. (1993) Muscarinic receptors: characterisation, function and coupling. Pharmacol. Ther. **58**, 319–381.
5 Hulme, E.C. et al. (1990) Muscarinic receptor subtypes. Annu. Rev. Pharmacol. Toxicol. **30**, 633–733
6 Mitchelson, F. (1988) Muscarinic receptor differentiation. Pharmacol. Ther. **37**, 357–423
7 Nathanson, N.N. (1987) Molecular properties of the muscarinic acetylcholine receptor. Annu. Rev. Neurosci. **10**, 195-236
8 Conley, E. (1994) The Ion Channel FactsBook, Academic Press, London, in press.

ACETYLCHOLINE RECEPTORS

Muscarinic M_1, M_2, M_3, M_4, M_5 receptors

These are all seven transmembrane spanning proteins. There are no selective agonists for the individual muscarinic receptor subtypes, but small differences in pharmacology exist between the five receptors [1-3]. No antagonist has a selectivity of more than tenfold [1-3]. The use of relative potencies of antagonists provides the most reliable means of defining a particular receptor subtype.

Muscarinic receptor subtypes are generally co-expressed with at least one other muscarinic receptor (the heart is a notable exception) making it difficult to map out their individual distributions. The distributions reported overleaf, therefore, are largely based on Northern analysis and may not correlate with protein expression.

[^{3}H]N-Methylscopolamine and [^{3}H](–)-3-quinuclidinylbenzilate are examples of nonselective antagonist radioligands with K_d values of less than 1 nM. Atropine (K_d 1 nM) is a nonselective antagonist.

Atropine

Nicotinic receptors

Each member of the nicotinic receptor family is made up of five subunits grouped together to form an intrinsic cation channel. The subunits are described in the 'The Ion Channel FactsBook' [4].

References
1 Buckley, N.J. et al, (1989) Mol. Pharmacol. **35**, 469–476.
2 Akiba, I. et al, (1988) FEBS Lett **235**, 257–261.
3 Peralta, E.G. et al, (1987) EMBO J. **6**, 3923–3929.
4 Conley, E. (1994) The Ion Channel FactsBook, Academic Press, London, in press.

THE M_1 RECEPTOR

The receptor was previously known as $M_{1\alpha}$ or the neural receptor.

Distribution

It is present in high levels and is distributed widely in neuronal cells in the CNS; it is particularly abundant in cerebral cortex and hippocampus. Its distribution is largely overlapping with that of M_3 and M_4 receptors. In the periphery M_1 receptors are found in autonomic ganglia and certain secretory glands, e.g. gastric. They are also found in cell lines, e.g. NB-OK1 cells.

Pharmacology

Agonists: No truly selective agonist has been described, although McNA343 is sometimes claimed to show selectivity for M_1 receptors.

Antagonists: Pirenzepine (pA$_2$ 8.0) and telenzepine (pA$_2$ 9.1) have selectivities of approximately tenfold.

Radioligands: [^{3}H]Pirenzepine (K_d 10 nM) and [^{3}H]telenzepine (K_d 2 nM) are weakly selective.

Structures:

Pirenzepine

Telenzepine

Predominant effector pathways

Activation of the phosphoinositide pathway via a pertussis-toxin-insensitive G-protein most likely of the G_q/G_{11} class.

Amino acid sequence [1]

```
                                TM 1
  1  MNTSAPPAVS PNITVLAPGK GPWQVAFIGI TTGLLSLATV TGNLLVLISF
           TM 2                                         TM 3
 51  KVNTELKTVN NYFLLSLACA DLIIGTFSMN LYTTYLLMGH WALGTLACDL
                                                   TM 4
101  WLALDYVASN ASVMNLLLIS FDRYFSVTRP LSYRAKRTPR RAALMIGLAW
                                      TM 5
151  LVSFVLWAPA ILFWQYLVGE RTVLAGQCYI QFLSQPIITF GTAMAAFYLP

201  VTVMCTLYWR IYRETENRAR ELAALQGSET PGKGGGSSSS SERSQPGAEG

251  SPETPPGRCC RCCRAPRLLQ AYSWKEEEEE DEGSMESLTS SEGEEPGSEV

301  VIKMPMVDPE AQAPTKQPPR SSPNTVKRPT KKGRDRAGKG QKPRGKEQLA
                                      TM 6                 TM 7
351  KRKTFSLVKE KKAARTLSAI LLAFILTWTP YNIMVLVSTF CKDCVPETLW

401  ELGYWLCYVN STINPMCYAL CNKAFRDTFR LLLLCRWDKR RWRKIPKRPG

451  SVHRTPSRQC
```

Residue 173 has been reported to be Met.

Amino acids	460
Molecular weight	51 420
Glycosylation	Asn2*, Asn12*
Disulfide bonds	Cys98 – Cys178*
Palmitoylation	Cys435*
Phosphorylation	Thr428*, Ser451*, Thr455*, Ser457*

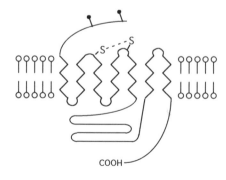

Database accession numbers

	PIR	SWISSPROT	EMBL/GENBANK	REFERENCES
Human	S09508	P11229	X15263	1
Mouse	A31897	P12657	J04192	2
Pig	A24325	P04761	X04413	3
Rat	A29514	P08482	M16406	4

Gene

The complete coding domain, 3'-untranslated sequence and a small portion of the 5'-untranslated sequence are contained in a single exon.

References

1 Peralta, E.G. et al. (1987) EMBO J. **6**, 3923–3929.
2 Shapiro, R.A. et al. (1989) J. Biol. Chem. **263**, 18397–18403; erratum, ibid. **264**, 6596.
3 Kubo, T. et al. (1986) Nature **323**, 411–416.
4 Bonner, T.I. et al. (1987) Science **237**, 527–532.

THE M2 RECEPTOR

The receptor was previously known as $M_{2\alpha}$ or cardiac M_2.

Distribution

It is found in low levels in the CNS where it has a restricted distribution. In contrast, M_2 receptors are expressed in high density in heart where they induce a decrease in inotropy and bradycardia; they are also found in smooth muscle, e.g. intestine, trachea and bladder.

Pharmacology

Agonists: No selective agonist has been described.

Antagonists: Methoctramine (pA2 7.9), himbacine (pA2 8.2) and AF-DX116 (pA2 7.3) have selectivities of less than tenfold.

Radioligands: [^{3}H]AF-DX116 and [^{3}H]AF-DX384 are weakly selective.

Structures:

Himbacine

AFDX116

Predominant effector pathways

Inhibition of adenylyl cyclase and L-type Ca^{2+} channels, and activation of K^+ channels through a pertussis-toxin-sensitive G-protein most likely of the G_i/G_o class.

Amino acid sequence [1,2]

```
                             TM 1
  1 MNNSTNSSNN SLALTSPYKT FEVVFIVLVA GSLSLVTIIG NILVMVSIKV
             TM 2                                      TM 3
 51 NRHLQTVNNY FLFSLACADL IIGVFSMNLY TLYTVIGYWP LGPVVCDLWL
                                          TM 4
101 ALDYVVSNAS VMNLLIISFD RYFCVTKPLT YPVKRTTKMA GMMIAAAWVL
                                   TM 5
151 SFILWAPAIL FWQFIVGVRT VEDGECYIQF FSNAAVTFGT AIAAFYLPVI

201 IMTVLYWHIS RASKSRIKKD KKEPVANQDP VSPSLVQGRI VKPNNNNMPS

251 SDDGLEHNKI QNGKAPRDPV TENCVQGEEK ESSNDSTSVS AVASNMRDDE

301 ITQDENTVST SLGHSKDENS KQTCIRIGTK TPKSDSCTPT NTTVEVVGSS
                                          TM 6
351 GQNGDEKQNI VARKIVKMTK QPAKKKPPPS REKKVTRTIL AILLAFIITW
              TM 7
401 APYNVMVLIN TFCAPCIPNT VWTIGYWLCY INSTINPACY ALCNATFKKT

451 FKHLLMCHYK NIGATR
```

Amino acids	466
Molecular weight	51 715
Glycosylation	Asn2*, Asn3*, Asn6*, Asn9*
Disulfide bonds	Cys96–Cys176*
Palmitoylation	Cys457*
Phosphorylation	Thr446*, Thr450*, Thr465*

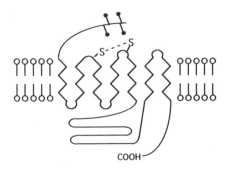

Database accession numbers

	PIR	SWISSPROT	EMBL/GENBANK	REFERENCES
Human	S10126	P08172	X15264; M16404	1,2
Pig	A27386	P06199	X04708	3,4
	A25656		M16331	
Rat	S10856	P10980	J03025	2,5

Gene

The complete coding domain, 3'-untranslated sequence and a small portion of the 5'-untranslated sequence are contained in a single exon.

Comment

The chick m2 sequence has been reported but the expressed receptor has high affinity for both AF-DX116 and pirenzepine [6].

References

[1] Peralta, E.G. et al. (1987) EMBO J. **6**, 3923–3929.
[2] Bonner, T.I. et al. (1987) Science **237**, 527–532.
[3] Kubo, T. et al. (1986) FEBS Lett. **209**, 367–372.
[4] Peralta, E.G. et al. (1987) Science **236**, 600–605.
[5] Gocayne, J. et al. (1987) Proc. Natl. Acad. Sci. USA **84**, 8298–8300.
[6] Nathanson, N.M. et al. (1991) J. Biol. Chem. **266**, 17382–17387.

THE M3 RECEPTOR

The receptor was previously known as $M_{2\beta}$ or glandular M_2. In one of the early cloning studies this receptor was named M_4 [1].

Distribution

It is present in high levels and is distributed widely in neuronal cells in the CNS. Its distribution is largely overlapping with that of M_1 and M_4 receptors. It is also found in peripheral ganglia, exocrine glands, e.g. parotid, lacrimal, smooth muscle, e.g intestine, bladder and trachea, and vascular endothelium. It is also found in cell lines, e.g. 132N1 cells.

Pharmacology

Agonists: No selective agonist has been described.

Antagonists: Hexahydrosiladifenidol (pA$_2$ 8.0) and *p*-fluorohexahydrosiladifenidol (pA$_2$ 7.8) have a selectivity of greater than tenfold relative to M_2, but little selectivity relative to M_1.

Radioligands: No selective ligand has been described.

Structures:

HO—Si—(CH₂)₃—N

Hexahydrosiladifenidol

F

HO—Si—(CH₂)₃—N

***p*-Fluoro-hexahydrosiladifenidol**

Predominant effector pathways

Activation of the phosphoinositide pathway via a pertussis-toxin-insensitive G-protein most likely of the G_q/G_{11} class.

Amino acid sequence [1]

```
  1 MTLHNNSTTS PLFPNISSSW IHSPSDAGLP PGTVTHFGSY NVSRAAGNFS
                    TM 1
 51 SPDGTTDDPL GGHTVWQVVF IAFLTGILAL VTIIGNILVI VSFKVNKQLK
          TM 2                                       TM 3
101 TVNNYFLLSL ACADLIIGVI SMNLFTTYII MNRWALGNLA CDLWLAIDYV
                                       TM 4
151 ASNASVMNLL VISFDRYFSI TRPLTYRAKR TTKRAGVMIG LAWVISFVLW
                                  TM 5
201 APAILFWQYF VGKRTVPPGE CFIQFLSEPT ITFGTAIAAF YMPVTIMTIL

251 YWRIYKETEK RTKELAGLQA SGTEAETENF VHPTGSSRSC SSYELQQQSM

301 KRSNRRKYGR CHFWFTTKSW KPSSEQMDQD HSSSDSWNNN DAAASLENSA

351 SSDEEDIGSE TRAIYSIVLK LPGHSTILNS TKLPSSDNLQ VPEEELGMVD

401 LERKADKLQA QKSVDDGGSF PKSFSKLPIQ LESAVDTAKT SDVNSSVGKS
                                                   TM 6
451 TATLPLSFKE ATLAKRFALK TRSQITKRKR MSLVKEKKAA QTLSAILLAF
                              TM 7
501 IITWTPYNIM VLVNTFCDSC IPKTFWNLGY WLCYINSTVN PVCYALCNKT

551 FRTTFKMLLL CQCDKKKRRK QQYQQRQSVI FHKRAPEQAL
```

Amino acids	590
Molecular weight	66 127
Glycosylation	Asn5*, Asn6*, Asn15*, Asn41*
Disulfide bonds	Cys141–Cys221*
Palmitoylation	Cys561*, Cys563*

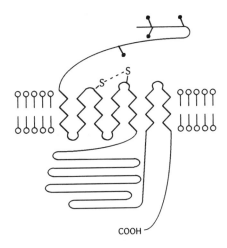

Database accession numbers

	PIR	SWISSPROT	EMBL/GENBANK	REFERENCES
Human	S10128	P20309	X15266	1
Pig	S01114	P11483	X12712	2
Rat	B29514; A29476	P08483	M16407; M62408 M18088; M62826	3,4,5

Gene

The complete coding domain, 3'-untranslated sequence and a small portion of the 5'-untranslated sequence are contained in a single exon.

References

1 Peralta, E.G. et al. (1987) EMBO J. **6**, 3923–3929.
2 Akiba, I. et al. (1988) FEBS Lett. **35**, 257–261.
3 Bonner, T.I. et al. (1987) Science **237**, 527–532.
4 Braun, T. et al. (1987) Biochem. Biophys. Res. Commun. **149**, 125–132.
5 Wess, J. et al. (1991) EMBO J. **10**, 3729–3734.

THE M4 RECEPTOR

In an early cloning study this receptor was named M3 [1].

Distribution

It is present in high levels and is distributed widely in neuronal cells in the CNS; its distribution is largely overlapping with M1 and M3 receptors. It is also found in cell lines, e.g. NG-108 cells.

Pharmacology

Agonists: No selective agonist has been described.

Antagonists: Tropicamide and secoverine have a modest selectivity [1].

Radioligands: No selective ligand has been described.

Predominant effector pathways

Inhibition of adenylyl cyclase and L-type Ca^{2+} channels, and activation of K^+ channels through a pertussis-toxin-sensitive G-protein most likely of the G_i/G_o class.

Amino acid sequence [1,2]

```
                                                  TM 1
     1 MANFTPVNGS SGNQSVRLVT SSSHNRYETV EMVFIATVTG SLSLVTVVGN
                                TM 2
    51 ILVMLSIKVN RQLQTVNNYF LFSLACADLI IGAFSMNLYT VYIIKGYWPL
          TM 3                                              TM 4
   101 GAVVCDLWLA LDYVVSNASV MNLLIISFDR YFCVTKPLTY PARRTTKMAG
                                                       TM 5
   151 LMIAAAWVLS FVLWAPAILF WQFVVGKRTV PDNQCFIQFL SNPAVTFGTA

   201 IAAFYLPVVI MTVLYIHISL ASRSRVHKHR PEGPKEKKAK TLAFLKSPLM

   251 KQSVKKPPPG EAAREELRNG KLEEAPPPAL PPPPRPVADK DTSNESSSGS

   301 ATQNTKERPA TELSTTEATT PAMPAPPLQP RALNPASRWS KIQIVTKQTG

   351 NECVTAIEIV PATPAGMRPA ANVARKFASI ARNQVRKKRQ MAARERKVTR
          TM 6                                      TM 7
   401 TIFAILLAFI LTWTPYNVMV LVNTFCQSCI PDTVWSIGYW LCYVNSTINP

   451 ACYALCNATF KKTFRHLLLC QYRNIGTAR
```

Amino acids	479
Molecular weight	53 058
Glycosylation	Asn3*, Asn8*, Asn13*
Disulfide bonds	Cys105–Cys185*
Palmitoylation	Cys470*

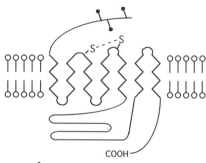

Database accession numbers

	PIR	SWISSPROT	EMBL/GENBANK	REFERENCES
Chick	A35546	P17200	J05212	3
Human	S10127	P08173	X15265; M16405	1,2
Rat	C29514	P08485	M16409	2

Gene

The complete coding domain, 3'-untranslated sequence and a small portion of the 5'-untranslated sequence are contained in a single exon.

References

1 Peralta, E.G. et al. (1987) EMBO J. **6**, 3923–3929.
2 Bonner, T.I. et al. (1987) Science **237**, 527–532.
3 Tietje, K.M. et al. (1990) J. Biol. Chem. **265**, 2828–2834.

THE M5 RECEPTOR

Although *m5* mRNA has been found in the CNS a translation product has not been identified. It is speculated that this receptor subtype may be important during development or may serve a unique function relative to the other receptor subtypes.

Pharmacology

Agonists: No selective agonist has been described.

Antagonists: No selective antagonist has been described; it has an intermediate affinity for pirenzepine and a low affinity for AF-DX116.

Radioligands: No selective ligand has been described.

Predominant effector pathways

Activation of the phosphoinositide pathway via a pertussis-toxin-insensitive G-protein most likely of the G_q/G_{11} class.

Amino acid sequence [1]

```
                                    TM 1
    1 MEGDSYHNAT TVNGTPVNHQ PLERHRLWEV ITIAAVTAVV SLITIVGNVL
                      TM 2
   51 VMISFKVNSQ LKTVNNYYLL SLACADLIIG IFSMNLYTTY ILMGRWALGS
          TM 3                                        TM 4
  101 LACDLWLALD YVASNASVMN LLVISFDRYF SITRPLTYRA KRTPKRAGIM
                                                      TM 5
  151 IGLAWLISFI LWAPAILCWQ YLVGKRTVPL DECQIQFLSE PTITFGTAIA

  201 AFYIPVSVMT ILYCRIYRET EKRTKDLADL QGSDSVbKAE KRKPAHRALF

  251 RSCLRCPRPT LAQRERNQAS WSSSRRSTST TGKPSQATGP SANWAKAEQL

  301 TTCSSYPSSE DEDKPATDPV LQVVYKSQGK ESPGEEFSAE ETEETFVKAE

  351 TEKSDYDTPN YLLSPAAAHR PKSQKCVAYK FRLVVKADGN QETNNGCHKV
                                                      TM 6
  401 KIMPCPFPVA KEPSTKGLNP NPSHQMTKRK RVVLVKERKA AQTLSAILLA
                          TM 7
  451 FIITWTPYNI MVLVSTFCDK CVPVTLWHLG YWLCYVNSTV NPICYALCNR

  501 TFRKTFKMLL LCRWKKKKVE EKLYWQGNSK LP
```

Amino acids	532
Molecular weight	60 186
Glycosylation	Asn8*, Asn13*
Palmitoylation	Cys511*
Disulfide bonds	Cys103–Cys183*

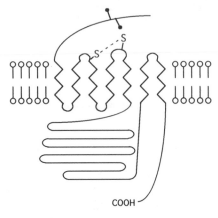

Database accession numbers

	PIR	SWISSPROT	EMBL/GENBANK	REFERENCES
Human	S09508	P08912	M80333; M35128; X52068; X15263	1,2
Rat	JT0531 A33354	P08911	M22926	1,3

Gene
The complete coding domain is contained in a single exon.

References
1 Bonner, T.I. et al. (1988) Neuron 1, 403–410.
2 Allard, W.J. et al. (1987) Nucleic Acids Res. 15, 10604.
3 Liao, C.F. et al. (1989) J. Biol. Chem. 264, 7328-7337.

Adenosine and adenine nucleotides

INTRODUCTION

In addition to their role in energy metabolism, purines, especially adenosine and adenine nucleotides, produce a wide range of pharmacological effects mediated by activation of cell surface receptors [1-15]. Studies of the latter, however, is complicated by rapid metabolism and uptake making it unclear which compound is exerting the effect.

ATP is an established co-transmitter in sympathetic nerves in the autonomic nervous system where it exerts an important physiological role in the regulation of smooth muscle activity, stimulating relaxation of intestinal smooth muscle and contraction of bladder. In addition, receptors for adenine nucleotides are involved in a number of other physiological pathways including the stimulation of platelet activation by ADP which is released from vascular endothelium following injury. ATP has excitatory effects in CNS.

Distinct receptors exist for adenosine. In the periphery, the main effects of adenosine include vasodilation, bronchoconstriction, immunosuppression, inhibition of platelet aggregation, cardiac depression, stimulation of nociceptive afferents, inhibition of neurotransmitter release and inhibition of the release of other factors, e.g. hormones. In CNS, adenosine exerts a pre- and post-synaptic depressant action, reducing motor activity, depressing respiration, inducing sleep and relieving anxiety. The physiological role of adenosine is thought to be to adjust energy demands in line with the oxygen supply. Many of the clinical actions of methylxanthines such as theophylline are thought to be mediated through antagonism of adenosine receptors, e.g. CNS stimulation, diuresis and possibly bronchodilation (inhibition of phosphodiesterase also involved).

Adenosine   ATP

Distribution

ATP and adenosine are ubiquitously distributed because of their essential role in metabolism.

Adenosine is produced by hydrolysis of AMP or by catabolism of *S*-homocysteine. Under conditions of stress, e.g. hypoxia, adenosine increases in the cell and is released, most likely by simple diffusion, where it can act on adenosine receptors.

ATP acts as a neurotransmitter, usually being released with other agents, e.g. noradrenaline. ATP and other nucleotides, e.g. ADP, are also released at the site of tissue injury where they can act on nearby cells, notably platelets.

Books

1 Stone, T.W. (ed.) (1985) Purines: Pharmacology and Physiological Roles, Macmillan, London.
2 Gerlach, E. and Backer, B.F. (eds) (1987) Topics and Perspectives in Adenosine Research, Springer-Verlag, Berlin.
3 Paton, D.M. (ed.) (1988) Adenosine and Adenine Nucleotides: Physiology and Pharmacology, Taylor & Francis, London.
4 Ribeiro, J.A. (ed.) (1989) Adenosine Receptors in the Nervous System, Taylor & Francis, London.
5 Jacobson, K.A. et al. (eds) (1990) Purines in Cellular Signalling, Springer-Verlag, Berlin
6 Williams, M. (ed.) (1990) Adenosine and Adenosine Receptors, Humana Press, New Jersey.

Reviews

7 Fredholm, B.B. and Dunwiddie, T.V. (1988) How does adenosine inhibit transmitter release? Trends Pharmacol. Sci. 9, 130–134.
8 Gordon, J.L. (1986) Extracellular ATP: effects, sources and fate. Biochem. J. 233, 309–319.
9 Kennedy, C. (1990) P_{1-} and P_{2-} purinoceptor subtypes – an update. Arch. Int. Pharmacodyn. Ther. 303, 30–50.
10 Linden, J. (1991) Structure and function of A_1 adenosine receptors. FASEB J. 5, 2668–2676
11 O'Connor, S.E. et al. (1991) Further subclassification of ATP receptors based on agonist studies. Trends Pharmacol. Sci. 12, 137–141.
12 Olsson, R.A. and Pearson, J.D. (1990) Cardiovascular purinoceptors. Physiol. Rev. 70, 761–845.
13 Rudolphi, K.A. et al. (1992) Neuroprotective role of adenosine in cerebral ischaemia. Trends Pharmacol. Sci. 13, 439–466.
14 Stiles, G. L. (1992) Adenosine receptors. J. Biol. Chem. 267, 6451–6454.
15 Stiles, G. L. and Olah, M.E. (1992) Adenosine receptors. Annu. Rev. Physiol. 54, 211–227.
16 Williams, M. (1987) Purine receptors in mammalian tissues: pharmacology and functional significance. Annu. Rev. Pharmacol. Ther. 27, 315–345.

ADENOSINE RECEPTORS AND P_2 PURINOCEPTORS

Purinoceptors were classified by Burnstock [1] as P_1 or P_2 depending on their preference for adenosine or adenine nucleotides, respectively. Adenosine receptors (P_1 purinoceptors) had been proposed earlier and are characterized by their affinity for adenosine and by the ability of methylxanthines, e.g. caffeine and theophylline, to act as antagonists (the latter are inactive on P_2 purinoceptors). The term adenosine receptor is preferred to P_1 purinoceptor. Adenosine has very low affinity for P_2 purinoceptors while many adenine nucleotides, particularly 5'-di- and triphosphates, are virtually inactive at adenosine receptors.

A_1, A_{2A}, A_{2B} and A_3 adenosine receptors

Four established subtypes of adenosine receptors have been identified and are described in the following pages. In addition, [^{3}H]2-phenylaminoadenosine (CV 1808) is reported to label a novel binding site that has been named an A_4 receptor [2].

The P_{2X} receptor

The P_{2X} receptor mediates the fast response of smooth muscle following sympathetic nerve stimulation. The receptor contains an intrinsic cation channel. α,β-MethyleneATP is a selective agonist and, in the absence of selective antagonists, can be used to block the receptor through desensitization. The P_{2X} receptor is described in further detail in the Ion Channel FactsBook [3].

The P_{2Y} receptor

This is a G-protein coupled receptor and is described in further detail on page 30.

The P_{2Z}-receptor

The P_{2Z} receptor has been demonstrated to be specific for ATP and has a limited distribution, being found primarily on mast cells and other immune cells. It is believed to contain an intrinsic cation channel. Activation of the receptor by very high concentrations of ATP in the absence of divalent cations appears to cause cell permeabilization.

The P_{2T} receptor

The P_{2T} receptor is believed to have an intrinsic cation channel. It has also been reported to inhibit adenylyl cyclase and to stimulate release of intracellular Ca^{2+} through an unknown mechanism. ADP is a potent agonist at this receptor while ATP and AMP are antagonists. The receptor has only been identified on platelets.

The nucleotide receptor

ATP and UTP stimulate phosphoinositide metabolism in some tissues, e.g. neutrophils, fibroblasts, PC12 cells, through a novel receptor, termed a nucleotide receptor or P_{2U} receptor, which may be coupled to a G-protein. UTP is the most potent nucleotide agonist whereas a number of selective P_{2X} and P_{2Y} agonists are inactive.

References
[1] Burnstock, G. (1978) In *Cell membrane receptors for Drugs and Hormones: a Multidisciplinary Approach* (Bolis, L. and Straub, R.W. eds) Raven Press, New York, pp. 107–118

[2] Cornfield, L.J. *et al.* (1992) *Mol. Pharmacol.* **42**, 552–561.
[3] Conley, E. (1994) The Ion channel FactsBook, Academic Press, London, in press.

THE A_1 ADENOSINE RECEPTOR

This receptor was previously named the R_i adenosine receptor in view of its ability to inhibit adenylyl cyclase.

Distribution

A_1 receptors are distributed widely in peripheral tissues, e.g. heart, adipose tissue, kidney, stomach and pancreas, where they are mainly inhibitory. They are also found in peripheral nerves, e.g. in intestine and vas deferens. They are present in high levels in CNS, notably cerebral cortex, hippocampus, cerebellum, thalamus and striatum, and in cell lines, e.g. DDT_1-MF2 cells, GH_3 cells.

Pharmacology

Agonists: 2-Chloro-N^6-cyclopentyladenosine *1* and 5'-chloro-N^6-adenosine are examples of selective, potent agonists. Until recently, A_1 receptors were identified by the rank order of potency of certain agonists: R-phenylisopropyladenosine > 2-chloroadenosine > 5'-N-ethylcarboxamidoadenosine.

Antagonists: 1,3-Dipropyl-8-cyclopentylxanthine (pA_2 8.3–9.3) *2* has a selectivity of more than 100-fold; 8-cyclopentyltheophylline (pA_2 7.4) is less selective (~tenfold).

Radioligands: [³H]1,3-Dipropyl-8-cyclopentylxanthine (K_d 0.4 nM) and [¹²⁵I]aminobenzyladenosine[3] (K_d 0.7 nM) are examples of selective antagonist and agonist radioligands, respectively.

Structures:

1,3-Dipropyl-8-cyclopentylxanthine

N^6-Cyclopentyladenosine

5'-Chloro-N^6-adenosine

Predominant effector pathways

Inhibition of adenylyl cyclase and voltage-dependent Ca^{2+} channels, and activation of K^+ channel through a pertussis-toxin-sensitive G-protein most likely of the G_i/G_o class.

A_1 receptors have also been reported to induce activation of phospholipase C and to potentiate the ability of other receptors to activate this pathway.

Amino acid sequence (canine) [4,5]

```
                  TM 1                                    TM 2
    1 MPPYISAFQA AYIGIEVLIA LVSVPGNVLV IWAVKVNQAL RDATFCFIVS
                        TM 3
   51 LAVADVAVGA LVIPLAILIN IGPQTYFHTC LMVACPVLIL TQSSILALLA
                        TM 4
  101 IAVDRYLRVK IPLRYKTVVT QRRAAVAIAG CWILSLVVGL TPMFGWNNLS
                        TM 5
  151 VVEQDWRANG SVGEPVIKCE FEKVISMEYM VYFNFFVWVL PPLLLMVLIY
                                        TM 6
  201 LEVFYLIRKQ LNKKVSASSG DPQKYYGKEL KIAKSLALIL FLFALSWLPL
                  TM 7
  251 HILNCITLFC PTCQKPSILI YIAIFLTHGN SAMNPIVYAF RIHKFRVTFL

  301 KIWNDHFRCQ PKPPIDEDLP EEKAED
```

Amino acids	326
Molecular weight	36 692
Glycosylation	Asn148*, Asn159*
Disulfide bonds	Cys80–Cys169*
Palmitoylation	Cys 309*

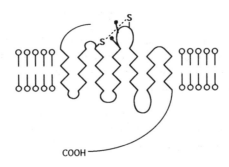

Database accession numbers

	PIR	SWISSPROT	EMBL/GENBANK	REFERENCES
Canine	C0341	P11616	X14051	*4,5*
	S12823			
Human				*6*
Rat	A40376	P25099	M64299	*7,8*
			M64299	
Bovine	S22213	P28190	M86261	*9,10*
	S23090		X65392	

Gene

The genomic sequence has not been determined. Two transcripts of 5.6 and 3.1 kb have been identified by Northern blot [7]. The gene for the canine A_1 receptor is RDC7 [4,5].

Comment

It is well established that the bovine A_1 adenosine receptor has a unique pharmacology exhibiting an ~ tenfold greater affinity for many agonists and antagonists, e.g. 1,3-dipropyl-8-cyclopentylxanthine. These features are also present on the expressed receptor suggesting that they reflect species variation in primary structure.

References

[1] Lohse, M.J. *et al.* (1988) *Naunyn-Schmiedeberg's Arch. Pharmacol.* **337** 687–689.

[2] Daly, J. *et al.* (1986) *J. Med. Chem.* **29** 1520–1524.

[3] Linden, J. *et al.* (1985) *Circ. Res.* **56** 279–284.

[4] Libert, F. *et al.* (1989) *Science* **244** 569–572.

[5] Libert, F. *et al.* (1990) *EMBO J.* **10** 1677–1682.

[6] Libert, F. *et al.* (1992) *Biochem. Biophys. Res. Commun.* **187** 919–926.

[7] Mahan, L.C. *et al.* (1991) *Mol. Pharmacol.* **40** 1–7.

[8] Reppert, S.M. *et al.* (1991) *Mol. Endocrinol.* **5** 1037–1048.

[9] Tucker, T. *et al.* (1992) *FEBS Lett.* **297** 107–111.

[10] Olah, M.E. *et al.* (1992) *J. Biol. Chem.* **267** 10764–10770.

THE A$_{2A}$ ADENOSINE RECEPTOR

This is also known as the high-affinity A$_2$ receptor subtype because it has an ~100-fold higher affinity for adenosine than the A$_{2B}$ (or low-affinity) subtype.

Distribution

The A$_{2A}$ receptor has a limited distribution in brain and is found in striatum, olfactory tubercle and nucleus accumbens. In the periphery, A$_2$ receptors mediate vasodilation, immunosuppression, inhibition of platelet aggregation and gluconeogenesis.

Pharmacology

Agonists: CGS 21680 [1] has a selectivity of more than 100-fold; CGS 22492 is also selective. Until recently A$_2$ receptors were identified by the rank order of potency of certain agonists: 5'-N-ethylcarboxamidoadenosine > 2-chloroadenosine > R-phenylisopropyladenosine.

Antagonists: CP 66713 (pA$_2$ 7.7) has a selectivity of approximately tenfold.

Radioligands: The agonist [³H]CGS 21680 (K_d 15 nM) [1] has a selectivity of more than 100-fold.

Structures:

CGS 21 680

CP66713

Predominant effector pathways
Activation of adenylyl cyclase through G$_s$.

Amino acid sequence [2]

```
                TM 1                                    TM 2
     1 MGSSVYITVE LAIAVLAILG NVLVCWAVWL NSNLQNVTNY FVVSLAAADI
                                         TM 3
    51 AVGVLAIPFA ITISTGFCAA CHGCLFIACF VLVLTQSSIF SLLAIAIDRY
                      TM 4
   101 IAIRIPLRYN GLVTGTRAKG IIAICWVLSF AIGLTPMLGW NNCGQPKEGK
```

```
                                 TM 5
   151 NHSQGCGEGQ  VACLFEDVVP  MNYMVYFNFF  ACVLVPLLLM  LGVYLRIFLA
                                                TM 6
   201 ARRQLKQMES  QPLPGERARS  TLQKEVHAAK  SLAIIVGLFA  LCWLPLHIIN
                            TM 7
   251 CFTFFCPDCS  HAPLWLMYLA  IVLSHTNSVV  NPFIYAYRIR  EFRQTFRKII

   301 RSHVLRQQEP  FKAAGTSARV  LAAHGSDGEQ  VSLRLNGHPP  GVWANGSAPH

   351 PERRPNGYAL  GLVSGGSAQE  SQGNTGLPDV  ELLSHELKGV  CPEPPGLDDP

   401 LAQDGAGVS
```

Amino acids	409
Molecular weight	44 365
Glycosylation	Asn151*
Disulfide bonds	Cys74–Cys163*
Palmitoylation	None

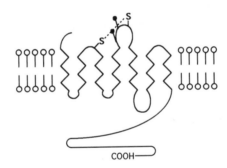

Database accession numbers

	PIR	SWISSPROT	EMBL/GENBANK	REFERENCES
Canine	D30341	P11617	X14052	3,4,5
Human		P29274	M97370	2
Rat				6

Gene

The genomic sequence has not been determined. The gene for the canine A_2 receptor is named RDC8.

References

1 Wan, W. et al. (1990) J. Neurochem. 55, 1763–1771.
2 Tiffany, H.L. and Murphy, P.M. (1993) submitted to EMBL/Genbank.
3 Libert, F. et al. (1989) Science 244, 569–572.
4 Maenhaut, C. et al. (1990) Biochem. Biophys. Res. Commun. 173, 1169–1178.
5 Libert, F. et al. (1990) Nucleic Acids Res. 18, 1914.
6 Chern, Y. et al. (1992) Biochem. Biophys. Res. Commun. 185, 304–309.

THE A2B ADENOSINE RECEPTOR

This is also known as the low-affinity A2 receptor subtype because it has an ~100-fold lower affinity for adenosine than the A2A (or high-affinity) subtype.

Distribution

The lack of receptor-selective ligands has prevented a complete description of the distribution and physiological role of the A2B receptor. Functional evidence suggests that it is widespread in human brain relative to the A2A receptor. However, in rat, its mRNA is found in low levels in brain and has a unique distribution in the periphery with high levels in intestine and urinary bladder.

Pharmacology

Agonists: No selective agonist has been described. It can be identified by the low potency of selective A1 and A2A ligands.

Antagonists: No selective agent has been described, although many xanthines are nonselective antagonists, e.g. theophylline.

Radioligands: [^{3}H]5'-N-Ethylcarboxamidoadenosine binds with low affinity relative to the A2A and A3 receptors.

Predominant effector pathways
Stimulation of cAMP through G$_s$.

Amino acid sequence [1]

```
               TM 1                                          TM 2
   1 MLLETQDALY VALELVIAAL SVAGNVLVCA AVGTANTLQT PTNYFLVSLA
                                      TM 3
  51 AADVAVGLFA IPFAITISLG FCTDFYGCLF LACFVLVLTQ SSIFSLLAVA
                                      TM 4
 101 VDRYLAICVP LRYKSLVTGT RARGVIAVLW VLAFGIGLTP FLGWNSKDSA
                                      TM 5
 151 TNNCTEPWDG TTNESCCLVK CLFENVVPMS YMVYFNFFGC VLPPLLIMLV
                                        TM 6
 201 IYIKIFLVAC RQLQRTELMD HSRTTLQREI HAAKSLAMIV GIFALCWLPV
                     TM 7
 251 HAVNCVTLFQ PAQGKNKPKW AMNMAILLSH ANSVVNPIVY AYRNRDFRYT

 301 FHKIISRYLL CQADVKSGNG QAGVQPALGV GL
```

Amino acids	332
Molecular weight	36 333
Glycosylation	Asn153*,Asn163*
Disulfide bonds	Cys78–Cys171*
Palmitoylation	Cys311*

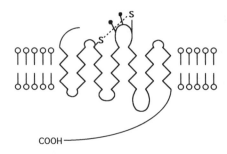

Database accession numbers

	PIR	SWISSPROT	EMBL/GENBANK	REFERENCES
Human		P29275	M97759	1
			M97370	
Rat		P29276	M91466	2

Gene

The genomic organization is not known.

Comment

There are differences in the pharmacology and distribution of the expressed rat and human A$_{2B}$ receptors and this is thought to be due to species variation rather than distinct receptor subtypes (the two proteins share 86% amino acid sequence homology).

References

1 Pierce, K.D. *et al.* (1992) *Biochem. Biophys. Res. Commun.* **187**, 86–93.
2 Stehle, J.H. *et al.* (1992) *Mol. Endocrinol.* **6**, 384–393.

THE A$_3$ ADENOSINE RECEPTOR

Distribution

mRNA is expressed in high levels in testis and in lower levels in lung, kidney and heart. It is also present in low levels in certain regions of the CNS including cerebral cortex, striatum and olfactory bulb. The high expression of the A$_3$ receptor in testis has led to the suggestion that it may have a role in reproduction.

Pharmacology

Agonists: No selective agonist has been described.

Antagonists: No xanthine antagonist has been described; a number of nonselective antagonists have been reported.

Radioligands: The agonist [^{125}I]N^6-2-(4-aminophenyl)ethyladenosine (K_d 15 nM) is selective to A$_1$ and A$_3$ receptors; [^{3}H]5'-N-ethylcarboxamidoadenosine (K_d 50 nM) is nonselective.

Predominant effector pathways

Inhibition of adenylyl cyclase through a pertussis-toxin-sensitive G-protein, most likely of the G_i/G_o class.

Amino acid sequence (rat) [1]

```
                     TM 1                                    TM 2
  1 MKANNTTTSA LWLQITYVTM EAAIGLCAVV GNMLVIWVVK LNRTLRTTTF
                                          TM 3
 51 YFIVSLALAD IAVGVLVIPL AIAVSLEVQM HFYACLFMSC VLLVFTHASI
                               TM 4
101 MSLLAIAVDR YLRVKLTVRY RTVTTQRRIW LFLGLCWLVS FLVGLTPMFG
                               TM 5
151 WNRKVTLELS QNSSTLSCHF RFVVGLDYMV FFSFITWILI PLVVMCIIYL
                                          TM 6
201 DIFYIIRNKL SQNLTGFRET RAFYGREFKT AKSLFLVLFL FALCWLPLSI
                TM 7
251 INFVSYFNVK IPEIAMCLGI LLSHANSMMN PIVYACKNKK VQRNHFVILR

301 ACRLCQTSDS LDSNLEQTTE
```

Amino acids	320
Molecular weight	36 644
Glycosylation	Asn4*, Asn5*
Disulfide bonds	None
Palmitoylation	Cys305*
Phosphorylation (PKC)	Thr44*, Thr117*, Thr115*

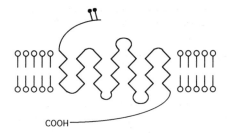

COOH

Database accession numbers

	PIR	SWISSPROT	EMBL/GENBANK	REFERENCES
Rat			M94152	[1]

The sequence of the rat A3 receptor is the same as that in the cDNA tgpcr1 [2]. The latter was found only in testis and its pharmacological profile was not examined.

Gene

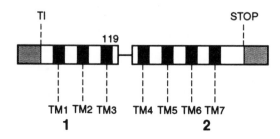

References
[1] Zhou, Q.-Y. *et al.* (1992) *Proc. Natl Acad. Sci.* USA **89**, 7432–7436.
[2] Meyerhof, W. *et al.* (1991) *FEBS Lett.* **284**, 155–160.

THE P$_{2Y}$ PURINOCEPTOR

ATP and ADP have similar affinities, while AMP has low affinity. It has not been cloned.

Distribution

The P$_{2Y}$ receptor is found in smooth muscle, e.g. taeni caeci, and in vascular tissue where it induces vasodilation through endothelium-dependent release of nitric oxide. It is also found in turkey erythrocytes.

Pharmacology

Agonists: The selective agonists 2-methylthio-ATP, homo-ATP and ATPαS have ~500, 75 and 50 times higher affinity than ATP at P$_{2Y}$ receptors. ADPβF is also selective and is resistant to degradation.

Antagonists: None have been described.

Radioligands: [^{35}S]ATPαS and [^{35}S]ADPβS.

Structures:

2-methylthio-ATP	$R_2 = SMe; X^3 = PO_4^{2-}$
homo-ATP	$X^1 = CH_2; X^3 = PO_4^{2-}$
ATPαS	$X^3 = S; X^3 = PO_4^{2-}$
ATPβF	$X^3 = F$

Predominant effector pathways

Activation of phosphoinositide metabolism through a pertussis-toxin-insensitive G-protein most likely of the G_i/G_o class.

Adrenaline and noradrenaline

INTRODUCTION

Noradrenaline is the major transmitter in postganglionic fibres of the sympathetic nervous system innervating smooth muscle, secretory glands and heart. Adrenaline is synthesized in chromaffin cells of the adrenal gland and is released into the plasma at times of stress or increased energy need where it stimulates glycogenolysis in liver and exerts potent actions on the cardiovascular system. Noradrenaline and, to a limited extent, adrenaline are also present in neurons in the CNS. Cell bodies of noradrenergic neurons are found in pons and medulla and send extensively branched axons to cerebral cortex, limbic system, hypothalamus, cerebellum and spinal cord [1-16].

In the periphery, the adrenergic system has an essential role in the regulation of the cardiovascular system. Increased sympathetic discharge to the heart leads to an increased rate and force of contraction mediated through β_1 receptors. Circulating adrenaline also acts on cardiac tissue but, in addition, acts on α_1 adrenoceptors in arterial smooth muscle stimulating vasoconstriction and on β_2 adrenoceptors in vascular beds of skeletal muscle stimulating vasodilation leading to increased blood flow. In the CNS, noradrenaline is thought to be involved in the regulation of mood and several psychoactive drugs alter noradrenergic function. There is also evidence that noradrenaline is involved in arousal and central regulation of blood pressure.

A large number of clinically important drugs exert their actions through adrenoceptors. β_2 selective agonists, e.g. salbutamol, are used in the acute treatment of asthma while α agonists prolong the action of local anaesthetics and act as nasal decongestants. Clonidine, an α_2-selective agonist, is used to treat hypertension through a central action. α_1 antagonists, e.g. prazosin, are also used to treat hypertension but have only limited therapeutic application. β antagonists, e.g. propranolol or atenolol, are the agents of choice in the treatment of hypertension and a number of other cardiovascular disorders, e.g. angina, certain cardiac dysrhythmias and cardiac infarction; they are also used in the treatment of anxiety and glaucoma. A combination of α and β blockers are used during surgical removal of phaeochromocytoma, a catecholamine-secreting tumour in the adrenal gland.

Noradrenaline   Adrenaline

Books

1 Limbird, L.E. (ed.) (1988) *The Alpha-2 Adrenergic Receptors*, Humana Press, New Jersey.

2 Perkins, J.P. (ed.) (1991) *The Beta-Adrenergic Receptors*, Humana Press, New Jersey.

3 Ruffolo, R.R. (ed.) (1988) *The Alpha-1 Adrenergic Receptors*, Humana Press, New Jersey.

Reviews

4 Brodde, O.-E. (1989) β-Adrenoceptors. In *Receptor Pharmacology and Function* (Williams, M. *et al.* eds), Marcel Dekker, New York, pp. 207–255.

5 Bylund, D.B. (1988) Subtypes of α2-adrenoceptors: pharmacological and molecular biological evidence converge. *Trends Pharmacol. Sci.* **9**, 356–361.

6 Bylund, D.B. *et al.* (1992) Subtypes of α1 and α2-adrenergic receptors. *FASEB. J.* **6**, 832–839.

7 Docherty, J.R. (1989) The pharmacology of α1 and α2-adrenoceptors: evidence for and against a further sub-division. *Pharmacol. Ther.* **44**, 241–284.

8 Harrison, J.K. *et al.* (1991) Molecular characterisation of α1 and α2-adrenoceptors. *Trends Pharmacol. Sci.* **12**, 62–67.

9 Kobilka, B.K. (1992) Adrenergic receptors as models for G protein-coupled receptors. *Annu. Rev. Neurosci.* **15**, 87–114.

10 McGrath, J.C. *et al.* (1989) Alpha-adrenoceptors: a critical review. *Med. Res. Rev.* **9**, 407–533.

11 Minneman, K.P. (1988) α1-Adrenergic receptor subtypes, inositol phosphates, and sources of cell Ca^{2+}. *Pharmacol. Rev.* **40**, 87–117.

12 O'Dowd, B.F. *et al.* (1989) Structure of the adrenergic and related receptors. *Annu. Rev. Neurosci.* **12**, 67–83.

13 Ostrowski, J. *et al.* (1992) Mutagenesis of the beta-2 adrenergic receptor: how structure elucidates function. *Annu. Rev. Pharmacol. Toxicol.* **32**, 167–185.

14 Ruffolo, R.R. *et al.* (1991) Structure and function of α-adrenoceptors. *Pharmacol. Rev.* **43**, 475–505.

15 Summers, R.J. *et al.* (1993) Adrenoceptors and their second messenger systems. *J. Neurochem.* **60**, 10–20.

16 Zaagsma, J. and Nahorski, S.F. (1990) Is the adipocyte β adrenoceptor a prototype for the recently cloned atypical "β3 adrenoceptor?", *Trends Pharmacol. Sci.* **11**, 3–7.

ADRENOCEPTORS

Adrenoceptors can be divided into three main types based on sequence information, receptor pharmacology and signalling mechanisms. Further subdivisions exist within each class. A large number of agonist and antagonists distinguish between the three main classes of adrenoceptor and examples of these are shown below. In contrast, only relatively small differences in affinity of agonists and antagonists exist in each class, especially within the α1 adrenoceptor and α2 adrenoceptor families.

α1 Adrenoceptors

There is evidence for four subtypes of α1 adrenoceptor all of which are G-protein coupled. The existence of the α1A receptor subtype was proposed to account for the 20-40-fold greater potency of the antagonist WB 4101 relative to the α1B receptor [1,2]. This subclassification was supported by the ability of chloroethylclonidine to selectively inactivate α1B receptors [3] and has since been extended to include a third subtype, the α1C receptor, which is also inhibited by chloroethylclonidine. Evidence for the existence of a fourth subtype of α_1 adrenoceptor has been suggested based on the affinity of antagonists, anatomical distribution and small differences in the pharmacology of cloned and endogenously expressed receptors [4]. All four receptors are described in the following pages.

Agonists which are selective to the α_1 adrenoceptor class include phenylephrine, methoxamine and cirazoline; examples of antagonists include prazosin (pA2 8.5–10.5) and corynanthine (pA2 6.5–7.5). [³H]Prazosin (K_d 0.1 nM) and [¹²⁵I]HEAT (K_d 0.1 nM) are radioligands selective to the α_1 class.

Phenylephrine

Methoxamine

Corynanthine

Prazosin

HEAT

α_2 Adrenoceptors

There are three subtypes of α_2 adrenoceptor all of which are G-protein coupled. The existence of two subtypes of α_2 receptors was first suggested by the different affinities of certain α_2 antagonists and has since been confirmed in cloning studies and extended to a third receptor subtype [5]. There is close overlap in the pharmacology of the α_{2B} and α_{2C} receptors. All three receptors are described in further detail in the following pages. In addition, there is evidence from binding studies for a fourth subtype of α_2 adrenoceptor which is present in bovine pineal gland and rat submaxillary gland [6]. This receptor has been tentatively named α_{2D} and is characterized by a low affinity for yohimbine and its structural analogue rauwolscine; however, it is believed to be a species homologue of α_{2A} [7].

Agonists which are selective to the α_2 adrenoceptor class include UK 14304, BHT 920 and clonidine; examples of antagonists include rauwolscine (pA2 9.0) and yohimbine (pA2 9.0). [³H]Yohimbine (K_d 1 nM) and [³H]rauwolscine (K_d 1 nM) are radioligands selective to the α_2 class.

UK 14304

BHT 920

Clonidine

Yohimbine

β Adrenoceptors

There are three subtypes of β adrenoceptor all of which are G-protein coupled. The existence of β_1 and β_2 receptor subtypes was established by the identification of selective agonists and antagonists. The existence of β_3 receptors was indicated by the low potencies of β antagonists and confirmed by the cloning of the receptor.

Isoprenaline is an example of an agonist selective to the β adrenoceptor class; propranolol (pA2 9.0) is an example of a nonselective β adrenoceptor antagonist.

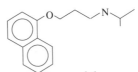

Isoprenaline

Propranolol

References
1 Battaglia, G. et al. (1983) *J. Neurochem.* **41**, 538–542.
2 Morrow, A.L. & Creese, I. (1986) *Mol. Pharmacol.* **29**, 321–330.
3 Han, C. et al. (1987) *Mol. Pharmacol.* **32**, 505–510.
4 Bylund, D.B. et al. (1992) *FASEB. J.* **6**, 832–839.
5 Bylund, D.B. (1988) *Trends Pharmacol. Sci.* **9**, 356–361.
6 Michel, A.D. et al. (1989) *Br. J. Pharmacol.* **98**, 890–897.
7 Link, A. et al. (1992) *Mol. Pharmacol.* **42**, 16–27.

THE α_{1A} RECEPTOR

The pharmacology of this receptor most closely resembles that of the cloned α_{1D} adrenoceptor. However, chloroethylclonidine is an irreversible antagonist at the latter and mRNA for the α_{1D} adrenoceptor has a distinct distribution to the α_{1A} adrenoceptor. Noradrenaline has a slightly greater potency than adrenaline at the α_{1A} adrenoceptor.

Distribution

The lack of selective ligands has prevented a detailed investigation of its distribution. It appears to be distributed widely in the rat with high levels in peripheral tissues, e.g. vas deferens, and in the CNS, e.g. hippocampus, cerebral cortex and brainstem.

Pharmacology

Agonists: Oxymetazoline, phenylephrine and methoxamine have weak selectivity for the α_{1A}, α_{1C} and α_{1D} relative to α_{1B}-adrenoceptors.

Antagonists: 5-Methylurapadil (pA$_2$ 10) has a similar affinity for α_{1A} and α_{1C} adrenoceptors but is 10–100 times less potent at α_{1B} and α_{1D} adrenoceptors. WB 4101 (pA$_2$ 9.2) is selective to α_{1A}, α_{1C} and α_{1D} adrenoceptors.

Structures:

WB 4101

Oxymetazoline

5-Methylurapadil

Predominant effector pathways

Coupled to the phosphoinositide pathway through a pertussis-toxin-insensitive G-protein most likely of the G_q/G_{11} class. It has also been suggested to stimulate direct entry of extracellular Ca^{2+}.

Amino acid sequence

It has not been cloned.

Comment

A report of the cloning of the α_{1A} receptor in human with sequence homology to the α_{1D} receptor has not been substantiated [1].

Reference
1 Bruno, J.F. *et al.* (1991) *Biochem. Biophys. Res. Commun.* **179**, 1485–1490.

THE α_{1B} RECEPTOR

Adrenaline and noradrenaline have similar affinities.

Distribution

Many tissues contain more than one α_1 adrenoceptor and identification of their distributions is hampered by the lack of selective ligands. α_{1B} Receptor mRNA is distributed widely with high levels in CNS, e.g. cerebral cortex and brainstem, and in peripheral tissues, e.g. kidney and lung. This corresponds reasonably well with data obtained from binding studies, although exceptions exist, e.g. rat aorta expresses predominantly α_{1B} receptors although no message has been detected.

Pharmacology

Agonists: No selective agonist has been described.

Antagonists: The irreversible antagonist chloroethylclonidine (100 μM; 20 min) inactivates α_{1B}, α_{1C} and α_{1D} adrenoceptors.

Predominant effector pathways

Stimulation of phosphoinositide metabolism through a pertussis-toxin-insensitive G-protein, most likely of the G_q/G_{11} class.

Amino acid sequence (rat) [1,2]

```
                                                            TM 1
    1 MNPDLDTGHN TSAPAHWGEL KDDNFTGPNQ TSSNSTLPQL DVTRAISVG
                                          TM 2
   51 VLGAFILFAI VGNILVILSV ACNRHLRTPT NYFIVNLAIA DLLLSFTVLP
                   TM 3
  101 FSATLEVLGY WVLGRIFCDI WAAVDVLCCT ASILSLCAIS IDRYIGVRYS
                   TM 4
  151 LQYPTLVTRR KAILALLSVW VLSTVISIGP LLGWKEPAPN DDKECGVTEE
          TM 5
  201 PFCALFCSLG SFYIPLAVIL VMYCRVYIVA KRTTKNLEAG VMKEMSNSKE
                                                            TM 6
  251 LTLRIHSKNF HEDTLSSTKA KGHNPRSSIA VKLFKFSREK KAAKTLGIVV
                        TM 7
  301 GMFILCWLPF FIALPLGSLF STLKPPDAVF KVVFWLGYFN SCLNPIIYPC

  351 SSKEFKRAFM RILGCQCRGG RRRRRRRRLG ACAYTYRPWT RGGSLERSQS

  401 RKDSLDDSGS CMSGQKRTLP SASPSPGYLG RGTQPPVELC AFPEWKPGAL

  451 LSLPEPPGRR GRLDSGPLFT FKLLGDPESP GTEATASNGG CDTTTDLANG

  501 QPGFKSNMPL GPGHF
```

Amino acids	515
Molecular weight	56 524
Chromosome	5
Glycosylation	Asn10*, Asn24*, Asn34*
Disulfide bonds	Cys118–Cys195*
Palmitoylation	Cys375*, Cys377*

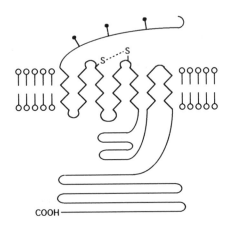

Database accession numbers

	PIR	SWISSPROT	EMBL/GENBANK	REFERENCES
Hamster		P18841	J04084	3
Rat	S08400	P15823	M60655	1,2
			X51585	

Gene

The gene contains introns in the coding domain [1], athough its organization is not known.

References

[1] Lomasney, J.W. et al. (1991) J. Biol. Chem. **266**, 6365–6369.
[2] Voigt, M.M. et al. (1990) Nucleic Acids Res. **18**, 1053.
[3] Cottechia, S. et al. (1988) Proc. Natl Acad. Sci. USA **85**, 7159–7163.

THE α_{1C} RECEPTOR

Adrenaline and noradrenaline have similar affinities.

Distribution

The α_{1C} receptor has not been detected in rat or bovine tissues by Northern analysis [1]. Expression of this receptor is therefore very low or highly specialized in tissue distribution or during development.

Pharmacology

Agonists: Oxymetazoline, phenylephrine and methoxamine have weak selectivity for α_{1A} and α_{1C} relative to α_{1B} adrenoeceptors.

Antagonists: WB 4101 (pA_2 9.2) and 5-methylurapadil (pA_2 9.2) have similar affinities for α_{1A} and α_{1C} receptors but are 10–100 times less potent at α_{1B} receptors. Chloroethylclonidine irreversibly inactivates α_{1C}, α_{1B} and α_{1D} adrenoceptors.

Predominant effector pathways
Stimulation of phosphoinositide metabolism through a pertussis-toxin-insensitive G-protein most likely of the G_q/G_{11} class.

Amino acid sequence (bovine) [1]

```
                                        TM 1
    1 MVFLSGNASD SSNCTHPPPP VNISKAILLG VILGGLILFG VLGNILVILS
                          TM 2
   51 VACHRHLHSV THYYIVNLAV ADLLLTSTVL PFSAIFEILG YWAFGRVFCN
      TM 3                                              TM 4
  101 VWAAVDVLCC TASIMGLCII SIDRYIGVSY PLRYPTIVTQ KRGLMALLCV
                                             TM 5
  151 WALSLVISIG PLFGWRQPAP EDETICQINE EPGYVLFSAL GSFYVPLTII

  201 LVMYCRVYVV AKRESRGLKS GLKTDKSDSE QVTLRIHRKN AQVGGSGVTS
                          TM 6
  251 AKNKTHFSVR LLKFSREKKA AKTLGIVVGC FVLCWLPFFL VMPIGSFFPD
          TM 7
  301 FRPSETVFKI AFWLGYLNSC INPIIYPCSS QEFKKAFQNV LRIQCLRRKQ

  351 SSKHTLGYTL HAPSHVLEGQ HKDLVRIPVG SAETFYKISK TDGVCEWKIF

  401 SSLPRGSARM AVARDPSACT TARVRSKSFL QVCCCLGPST PSHGENHQIP

  451 TIKIHTISLS ENGEEV
```

Amino acids	466
Molecular weight	51 466
Chromosome	8
Glycosylation	Asn7*, Asn13*, Asn22*
Disulfide bonds	Cys99–Cys176*

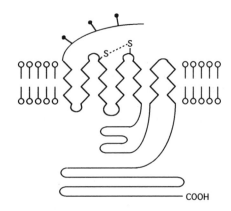

Database accession numbers

	PIR	SWISSPROT	EMBL/GENBANK	REFERENCES
Bovine	A35375	P18130	J05426	1

Gene

The genomic sequence contains at least one intron [2], but its organization is not known.

References

[1] Schwinn, D.A. et al. (1990) J. Biol. Chem. **265**, 8183–8189.
[2] Lomasney, J.W. et al. (1991) J. Biol. Chem. **266**, 6365–6369.

THE α_{1D} RECEPTOR

This was originally identified as the α_{1A} adrenoceptor [1]. However, chloro-ethylclonidine is an irreversible antagonist at this receptor and its mRNA has a distinct distribution to that of the α_{1A} adrenoceptor. Noradrenaline has a slightly greater potency than adrenaline.

Distribution

mRNA for the α_{1D} adrenoceptor is distributed widely in the rat with high levels in peripheral tissues, e.g. vas deferens, heart and spleen, and in CNS, e.g. hippocampus, cerebral cortex and brainstem.

Pharmacology

Agonists: There are no selective agonists.

Antagonists: WB 4101 (pA$_2$ 9.2) is selective relative to the α_{1B} adrenoceptor but not to the other α_1 adrenoceptors.

Predominant effector pathways

It has been shown to stimulate mobilization of Ca^{2+}, although the role of the phosphoinositide pathway is not established.

Amino acid sequence (rat) [1]

```
  1 MTFRDILSVT FEGPRSSSST GGSGAGGGAG TVGPEGGAVG GVPGATGGGA
                                                          TM 1
 51 VVGTGSGEDN QSSTGEPGAA ASGEVNGSAA VGGLVVSAQG VGVGVFLAAF
                                      TM 2
101 ILTAVAGNLL VILSVACNRH LQTVTNYFIV NLAVADLLLS AAVLPFSATM
             TM 3
151 EVLGFWAFGR TFCDVWAAVD VLCCTASILS LCTISVDRYV GVRHSLKYPA
            TM 4                                          TM 5
201 IMTERKAAAI LALLWAVALV VSVGPLLGWK EPVPPDERFC GITEEVGYAI

251 FSSVCSFYLP MAVIVVMYCR VYVVARSTTR SLEAGIKREP GKASEVVLRI
                                                          TM 6
301 HCRGAATSAK GYPGTQSSKG HTLRSSLSVR LLKFSREKKA AKTLAIVVGV
                                      TM 7
351 FVLCWFPFFF VLPLGSLFPQ LKPSEGVFKV IFWLGYFNSC VNPLIYPCSS

401 REFKRAFLRL LRCQCRRRRR RLWSLRPPLA SLDRRRAFRL RPQPSHRSPR

451 GPSSPHCTPG CGLGRHAGDA GFGLQQSKAS LRLREWRLLG PLQRPTTQLR

501 AKVSSLSHKI RSGARRAETA CALRSEVEAV SLNVPQDGAE AVICQAYEPG

551 DYSNLRETDI
```

Amino acids	560
Molecular weight	59 717
Chromosome	5
Glycosylation sites	Asn60*, Asn76*
Disulfide bonds	Cys163–Cys240*
Palmitoylation	Cys413*, Cys415*
Phosphorylation (PKA)	None

The sequence of ref. 2 differs in two introns, resulting in the presence of proline and arginine in positions 37 and 306, respectively.

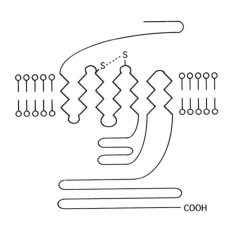

Database accession numbers

	PIR	SWISSPROT	EMBL/GENBANK	REFERENCES
Rat	A38731	P23944	M60654	1,2

Gene
The organization of the gene is unknown.

References
1 Lomasney, J.W. et al. (1991) J. Biol. Chem. **266**, 6365–6369.
2 Perez, D.M. et al. (1991) Mol. Pharmacol. **40**, 876–883.

THE α2A RECEPTOR
Adrenaline and noradrenaline have similar affinities.

Distribution

The distribution of the α2A receptor relative to other α2 receptor subtypes is not established because of the lack of selective ligands. Example prototypic tissues and cell lines expressing α2A receptors are human platelet and HT29 cells, respectively. Northern analysis has demonstrated a wide distribution of mRNA for the α2A receptor with high levels in rat CNS, e.g. brainstem, cerebral cortex, hippocampus, pituitary gland and cerebellum, and in peripheral tissues, e.g. kidney, aorta, skeletal muscle, spleen and lung.

Pharmacology

Agonists: The weak partial agonist, oxymetazoline, is 50–100 times selective relative to α2B and α2C receptors.

Antagonists: There are no selective antagonists.

Predominant effector pathways
Inhibition of adenylyl cyclase and L-type Ca^{2+} channels, and activation of K^+ channels through pertussis-toxin-sensitive G-proteins of the G_i/G_0 class.

Amino acid sequence [1,2]

```
                                        TM 1
  1 MGSLQPDAGN ASWNGTEAPG GGARATPYSL QVTLTLVCLA GLLMLLTVFG
                              TM 2
 51 NVLVIIAVFT SRALKAPQNL FLVSLASADI LVATLVIPFS LANEVMGYWY
      TM 3                                            TM 4
101 FGKAWCEIYL ALDVLFCTSS IVHLCAISLD RYWSITQAIE YNLKRTPRRI
                                                TM 5
151 KAIIITVWVI SAVISFPPLI SIEKKGGGGG PQPAEPRCEI NDQKWYVISS

201 CIGSFFAPCL IMILVYVRIY QIAKRRTRVP PSRRGPDAVA APPGGTERRP
```

```
251 NGLGPERSAG PGGAEAEPLP TQLNGAPGEP APAGPRDTDA LDLEESSSSD

301 HAERPPGPRR PERGPRGKGK ARASQVKPGD SLPRRGPGAT GPTPAAGPGE
                           TM 6
351 EERVGAAKAS RWRGRQNREK RFTFVLAVVI GVFVVCWFPF FFTYTLTAVG
         TM 7
401 CSVPRTLFKF FFWFGYCNSS LNPVIYTIFN HDFRRAFKKI LCRGDRKRIV
```

Amino acids	450
Molecular weight	49 016
Chromosome	10
Glycosylation	Asn10*, Asn14*
Disulfide bonds	Cys106–Cys188*
Palmitoylation	Cys442*

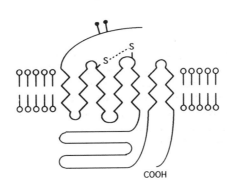

Database accession numbers

	PIR	SWISSPROT	EMBL/GENBANK	REFERENCES
Human	A34169	P08913	M18415	1,2
			M23533	
Mouse				3
Pig	A38316	P18871	J05652	4
Rat (rg20)	JH0190	P22909		5

Gene
The gene is intronless.

Comment
The rat and mouse receptors are characterized by a 20-fold lower affinity for yohimbine and rauwolscine. This has been attributed to the presence of serine in position 210 of the fifth transmembrane sequence instead of cysteine in the corresponding position in human and porcine [3].

References
1 Kobilka, B.K. *et al.* (1987) *Science* **238**, 650–656; *correction see ref. 4.*
2 Fraser, C.M. *et al.* (1989) *J. Biol. Chem.* **264**, 11754–11761.

[3] Link, R. *et al.* (1992) *Mol. Pharmacol.* **42**, 16–27.
[4] Guyer, C.A. *et al.* (1990) *J. Biol. Chem.* **265**, 17307–7317.
[5] Chalberg, S.C. *et al.* (1990) *Mol. Cell. Biochem.* **97**, 161–172.

THE α_{2B} RECEPTOR

Adrenaline and noradrenaline have similar affinities.

Distribution

Northern analysis has demonstrated the presence of mRNA for the α_{2B} adrenoceptor in only two peripheral tissues, liver and kidney, and absence of mRNA in brain. Examples of prototypic systems expressing α_{2B} adrenoceptors are neonatal rat lung and NG108 cells.

Pharmacology

Agonists: No selective agonist has been described.

Antagonists: Prazosin (pA2 7.5 nM) and ARC 239 (pA2 8.0 nM) have a selectivity of 10–100-fold for α_{2B} and α_{2C} relative to α_{2A} adrenoceptors.

Structures:

Prazosin

ARC 239

Predominant effector pathways

Inhibition of adenylyl cyclase and L-type Ca^{2+} channels through a pertussis-toxin-sensitive G-protein of the G_i/G_0 class.

Amino acid sequence [1-3]

```
                    TM 1                                              TM 2
     1 MDHQDPYSVQ ATAAIAAAIT FLILFTIFGN ALVILAVLTS RSLRAPQNLF
                                                       TM 3
    51 LVSLAAADIL VATLIIPFSL ANELLGYWYF RRTWCEVYLA LDVLFCTSSI
```

```
                                    TM 4
101 VHLCAISLDR YWAVSRALEY NSKRTPRRIK CIILTVWLIA AVISLPPLIY
                  TM 5
151 KGDQGPQPRG RPQCKLNQEA WYILASSIGS FFAPCLIMIL VYLRIYLIAK

201 RSNRRGPRAK GGPGQGESKQ PRPDHGGALA SAKLPALASV ASAREVNGHS

251 KSTGEKEEGE TPEDTGTRAL PPSWAALPNS GQGQKEGVCG ASPEDEAEEE

301 EEEEEEEEC EPQAVPVSPA SACSPPLQQP QGSRVLATLR GQVLLGRGVG
                                    TM 6
351 AIGGQWWRRR AHVTREKRFT FVLAVVIGVF VLCWFPFFFS YSLGAICPKH
        TM 7
401 CKVPHGLFQF FFWIGYCNSS LNPVIYTIFN QDFRRAFRRI LCRPWTQTAW
```

Amino acids	450
Molecular weight	49 948
Chromosome	2
Glycosylation	None
Disulfide bonds	Cys85–Cys164*
Palmitoylation	Cys442*

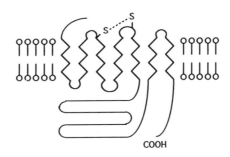

Database accession numbers

	PIR	SWISSPROT	EMBL/GENBANK	REFERENCES
Human	A36158	P18089	M34041	1–3
Mouse				4
Rat (rnga2)	A35642	P19328	M32061	5

Gene
The gene is intronless.

References
1 Lomasney, J.W. et al. (1990) Proc. Natl Acad. Sci. USA **87**, 5094–5098.
2 Weinshank, R.L. et al. (1990) Mol. Pharmacol. **38**, 681–688.
3 Chang, A.C. et al. (1990) Biochem. Biophys. Res. Commun. **172**, 817–823.
4 Chruscinski, A.J. et al. (1992) Biochem. Biophys. Res. Commun. **186**, 1280–1287.
5 Zeng, D. et al. (1990) Proc. Natl Acad. Sci. USA **87**, 3102–3106.

THE α_{2C} RECEPTOR

Adrenaline and noradrenaline have similar affinities.

Distribution

The distribution of the α_{2C} adrenoceptor relative to the other α_2 receptor subtypes is not established because of the lack of selective ligands. Northern analysis has demonstrated the presence of mRNA for the α_{2C} adrenoceptor in rat brain, including cerebral cortex, cerebellum, hippocampus and brainstem, but not in peripheral tissues. Examples of prototypic systems expressing α_{2C} adrenoceptors are the opossum kidney and OK cell line.

Pharmacology

Agonists: No selective agonist has been described.

Antagonists: Prazosin (pA$_2$ 7.5 nM) and ARC 239 (pA$_2$ 8.0 nM) have a selectivity of 10–100-fold for α_{2B} and α_{2C} relative to α_{2A} adrenoceptors.

Predominant effector pathways

Inhibition of adenylyl cyclase through a pertussis-toxin-sensitive G-protein of the G_i/G_0 class.

Amino acid sequence [1]

```
  1 MASPALAAAL AVAAAAGPNA SGAGERGSGG VANASGASWG PPRGQYSAG
       TM 1                                        TM 2
 51 VAGLAAVVGF LIVFTVVGNV LVVIAVLTSR ALRAPQNLFL VSLASADILV
                            TM 3
101 ATLVMPFSLA NELMAYWYFG QVWCGVYLAL DVLFCTSSIV HLCAISLDRY
                       TM 4
151 WSVTQAVEYN LKRTPRRVKA TIVAVWLISA VISFPPLVSL YRQPDGAAYP
         TM 5
201 QCGLNDETWY ILSSCIGSFF APCLIMGLVY ARIYRVAKRR TRTLSEKRAP

251 VGPDGASPTT ENGLGAAAGA GGENSHCAPP PADVEPDESS AAAERRRRGAL

301 RRGGRRRAGA EGGAGGADGQ GAGPGAAQSG ALTASRSPGP GGRLSRASSR
                                               TM 6
351 SVEFFLSRRR RARSSVCRRK VAQAREKRFT FVLAVVMGVF VLCWFPFFFI
                       TM 7
401 YSLYGICREA CQVPGPLFKF FFWIGYCNSS LNPVIYTVFN QDFRPSFKHI

451 LFRRRRRGFR Q
```

Amino acids	461
Molecular weight	49 513
Chromosome	4

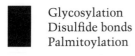

Glycosylation	Asn19*, Asn33*
Disulfide bonds	Cys124–Cys202*
Palmitoylation	None

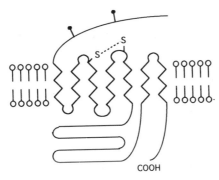

COOH

Database accession numbers

	PIR	SWISSPROT	EMBL/GENBANK	REFERENCES
Human	A31237	P18825	J03853	1
Mouse				2
Rat (rg10)	A37869	P22086	M58316	3,4
			M97516	

Gene
The gene is intronless.

References
1 Regan, J.W. et al. (1988) Proc. Natl Acad. Sci. USA **85**, 6301–6305; correction see Lomasney, J.W. et al. (1991) Biochem. Biophys. Acta **1095**, 127–139.
2 Link, R. et al. (1992) Mol. Pharmacol. **42**, 16–27.
3 Voigt, M.M. et al. (1991) FEBS Lett. **278**, 45–50.
4 Flordellis, C.S. et al. (1991) Proc. Natl Acad. Sci. USA **88**, 1019–1023.

THE β1 RECEPTOR
Noradrenaline has a slightly higher potency than adrenaline.

Distribution
Whereas the original subclassification suggested a high degree of tissue specificity, it is now clear that β1 and β2 receptors frequently coexist but with one subtype predominating. β1 Receptors predominate in cardiac tissue, where they mediate positive inotropic and chronotropic effects, and in the kidney where they enhance renin release.

Pharmacology
Agonists: Xamoterol has a selectivity of more than tenfold relative to β2 receptors. Noradrenaline is slightly less selective.

Antagonists: CGP 20 712A (pA$_2$ 8.5–9.3) has a selectivity of more than three orders of magnitude; atenolol and betaxolol are less selective (10–100 times).

Radioligands: [^{3}H]Bisoprolol (K_d 10 nM) has a selectivity of 50-fold.

Structures:

CGP 20712A

Atenolol

Betaxolol

Xamoterol

Predominant effector pathways

Activation of adenylyl cyclase through Gs; direct activation of voltage-operated Ca^{2+} channels has been reported in heart but is of uncertain physiological significance.

Amino acid sequence [1]

```
  1 MGAGVLVLGA SEPGNLSSAA PLPDGAATAA RLLVPASPPA SLLPPASESP
              TM 1                                         TM 2
 51 EPLSQQWTAG MGLLMALIVL LIVAGNVLVI VAIAKTPRLQ TLTNLFIMSL
                                                  TM 3
101 ASADLVMGLL VVPFGATIVV WGRWEYGSFF CELWTSVDVL CVTASIETLC
                                       TM 4
151 VIALDRYLAI TSPFRYQSLL TRARARGLVC TVWAISALVS FLPILMHWWR
                          TM 5
201 AESDEARRCY NDPKCCDFVT NRAYAIASSV VSFYVPLCIM AFVYLRVFRE

251 AQKQVKKIDS CERRFLGGPA RPPSPSPSPV PAPAPPPGPP RPAAAAATAP
```

```
                              TM 6
301 LANGRAGKRR PSRLVALREQ KALKTLGIIM GVFTLCWLPF FLANVVKAFH
         TM 7
351 RELVPDRLFV FFNWLGYANS AFNPIIYCRS PDFRKAFQGL LCCARRAARR

401 RHATHGDRPR ASGCLARPGP PPSPGAASDD DDDDVVGATP PARLLEPWAG

451 CNGGAAADSD SSLDEPCRPG FASESKV
```

Amino acids	477
Molecular weight	51 223
Glycosylation	Asn15*
Disulfide bonds	Cys131–Cys209*
Palmitoylation	Cys392*

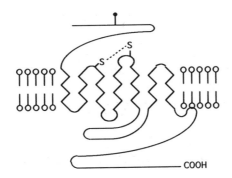

Database accession numbers

	PIR	SWISSPROT	EMBL/GENBANK	REFERENCES
Human		P08588	J03019	1
Rat	A36618	P18090	J05561	2,3
	S12591		D00634	
Turkey	P25896	P07700	M14379	4

Gene
The gene is intronless.

Comment
Although the sequence of the turkey adrenoceptor has closest homology with the β_1 adrenoceptor, it has a distinct pharmacological specificity [5].

References
[1] Frielle, T. et al. (1987) Proc. Natl Acad. Sci. USA 84, 7920–7924.
[2] Machida, C.A. et al. (1990) J. Biol. Chem. 265, 12960–12965.
[3] Shimomura, H. et al. (1990) Nucleic Acids Res. 18, 4591.
[4] Yarden, Y. (1986) Proc. Natl Acad. Sci. USA 83, 6795–6799.
[5] Minneman, K. et al. (1980) Mol. Pharmacol. 17, 1–7.

THE β_2 RECEPTOR

Adrenaline has about tenfold higher affinity than noradrenaline.

Distribution

Whereas the original subclassification suggested a high degree of tissue specificity, it is now clear that β_1 and β_2 receptors usually coexist but with one subtype predominating. β_2 Receptors mediate relaxation of smooth muscle including vascular beds, bronchus, intestine and uterus. They mediate glycogenlysis and glucogenesis in liver and regulate cell metabolism in skeletal muscle. They inhibit the activity of leukocytes and other blood cells and are also found in heart where their physiological role is uncertain. β_2 Receptors are located presynaptically in nerves where they facilitate neurotransmitter release, and in brain where they regulate a variety of physiological processes.

Pharmacology

Agonists: Procaterol is ~100-fold selective to β_2 receptors relative to β_1 receptors.

Antagonists: ICI 118 551 (pA$_2$ 8.3–9.2) has a selectivity of ~ 100 fold. Butoxamine (pA$_2$ 6.2) and α-methylpropranolol (pA$_2$ 8.5) are of lower selectivity.

Radioligands: [^{3}H]ICI 118 551 (K_d 10 nM) has a selectivity of ~ 100 fold.

Structures:

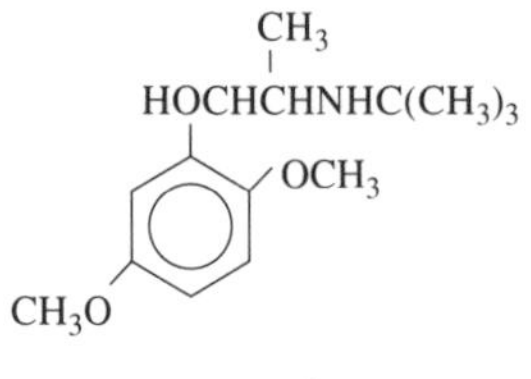

Butoxamine

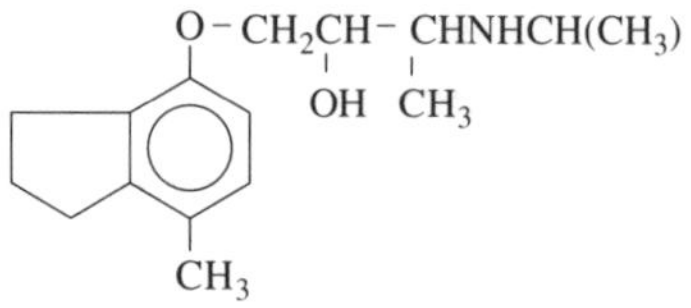

ICI 118551

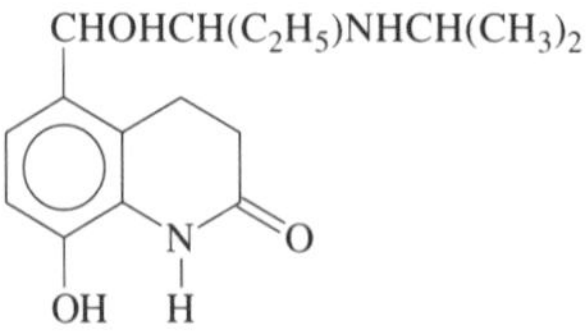

Procaterol

Predominant effector pathways

Activation of adenylyl cyclase through G_s.

Amino acid sequence [1-5]

```
                                                     TM 1
  1 MGQPGNGSAF LLAPNRSHAP DHDVTQQRDE VWVVGMGIVM SLIVLAIVFG
                          TM 2
 51 NVLVITAIAK FERLQTVTNY FITSLACADL VMGLAVVPFG AAHILMKMWT
       TM 3
101 FGNFWCEFWT SIDVLCVTAS IETLCVIAVD RYFAITSPFK YQSLLTKNKA
    TM 4                                                  TM 5
151 RVIILMVWIV SGLTSFLPIQ MHWYRATHQE AINCYANETC CDFFTNQAYA

201 IASSIVSFYV PLVIMVFVYS RVFQEAKRQL QKIDKSEGRF HVQNLSQVEQ
                          TM 6
251 DGRTGHGLRR SSKFCLKEHK ALKTLGIIMG TFTLCWLPFF IVNIVHVIQD
       TM 7
301 NLIRKEVYIL LNWIGYVNSG FNPLIYCRSP DFRIAFQELL CLRRSSLKAY

351 GNGYSSNGNT GEQSGYHVEQ EKENKLLCED LPGTEDFVGH QGTVPSDNID

401 SQGRNCSTND SLL
```

Amino acids	413
Molecular weight	46 557
Chromosome	5
Glycosylation	Asn6*, Asn15*
Disulfide bonds	Cys106–Cys184*
Palmitoylation	Cys341

Palmitoylation of Cys341 has been confirmed by site-directed mutagenesis and is thought to have a crucial role in coupling to G_s [6].

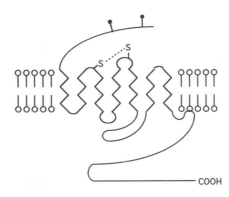

Database accession numbers

	PIR	SWISSPROT	EMBL/GENBANK	REFERENCES
Hamster	A03159	P04274	X03804	7
Human	A27525	P07550	Y00106	1–5
	A28405		X04827	
	A29061		M15169	
	A29026		J02960	
	A29574			
Mouse	S00260	P18762	X15643	8,9
Rat	S08227	P10608	J03024	10
	S10855		X17607	

Gene

The gene is intronless.

References

1 Schofield, P.R. *et al.* (1987) *Nucleic Acids Res.* **15**, 3636.
2 Kobilka, B.K. *et al.* (1987) *Proc. Natl Acad. Sci. USA* **84**, 46–50.
3 Chung, F.Z. *et al.* (1987) *FEBS Lett.* **211**, 200–206.
4 Emorine, L.J. *et al.* (1987) *Proc. Natl Acad. Sci. USA* **84**, 6995–6999.
5 Kobilka, B.K. *et al.* (1987) *J. Biol. Chem.* **262**, 7321–7327.
6 O'Dowd,B.F. *et al.* (1989) *J. Biol. Chem.* **264**, 7564–7569.
7 Dixon, R.A.F. *et al.* (1986) *Nature* **321**, 75–79.
8 Allen, J.M. *et al.* (1988) *EMBO J.* **7**, 133–138.
9 Nakada, M.T. *et al.* (1989) *Biochem. J.* **260**, 53–59.
10 Gocayne, J. *et al.* (1987) *Proc. Natl Acad. Sci. USA* **84**, 8296–8300.

THE β_3 RECEPTOR

The β_3 adrenoceptor was proposed to explain the low potency of β antagonists in inhibiting lipolysis and gut motility. It has also been named the atypical β adrenoceptor. Noradrenaline has a slightly greater potency than adrenaline.

Distribution

β_3 Receptors have a very limited distribution and are present in low levels. They are found in adipose tissue and gastrointestinal tract where they stimulate lipolysis and increased gut motility. They appear to coexist with β_2 receptors in skeletal muscle, with the latter predominating.

Pharmacology

Agonists: BRL 37 344 is selective relative to β_2 receptors and almost inactive at β_1 receptors.

Antagonists: No selective antagonist has been described although most of the classical β-antagonists have low affinity.

Radioligands: No selective ligand has been described. [^{125}I]Iodocyanopindolol (K_d 500 pM) has ~tenfold greater affinity relative to β_1 and β_2 receptors.

Structure:

OH   CH₃

–CHCH₂NHCHCH₂– –OCH₂COOH

Cl

BRL 37344

Predominant effector pathways
Activation of adenylyl cyclase through G_s.

Amino acid sequence [1]

```
                                                      TM 1
   1 MAPWPHENSS LAPWPDLPTL APNTANTSGL PGVPWEAALA GALLALAVLA
                                  TM 2
  51 TVGGNLLVIV AIAWTPRLQT MTNVFVTSLA AADLVMGLLV VPPAATLALT
              TM 3
 101 GHWPLGATGC ELWTSVDVLC VTASIETLCA LAVDRYLAVT NPLRYGALVT
          TM 4
 151 KRCARTAVVL VWVVSAAVSF APIMSQWWRV GADAEAQRCH SNPRCCAFAS
        TM 5
 201 NMPYVLLSSS VSFYLPLLVM LFVYARVFVV ATRQLRLLRG ELGRFPPEES
                                                      TM 6
 251 PPAPSRSLAP APVGTCAPPE GVPACGRRPA RLLPLREHRA LCTLGLIMGT
                                  TM 7
 301 FTLCWLPFFL ANVLRALGGP SLVPGPAFLA LNWLGYANSA FNPLIYCRSP

 351 DFRSAFRRLL CRCGRRLPPE PCAAARPALF PSGVPAARSS PAQPRLCQRL

 401 DG
```

Amino acids	402
Molecular weight	42 931
Glycosylation	Asn8*, Asn27*
Disulfide bonds	Cys110–Cys189*

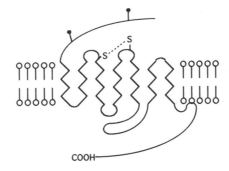

COOH

Database accession numbers

	PIR	SWISSPROT	EMBL/GENBANK	REFERENCES
Human		P13945	M29932	1
Mouse		P25962	X60438	2
Rat	A41679	P26255	M74716	3

Gene
The gene is intronless.

References

[1] Emorine, L.J. *et al.* (1989) *Science* **245**, 1118–1121.
[2] Nahmias, C. *et al.* (1991) *EMBO J.* **10**, 3721–3727.
[3] Muzzin, P. *et al.* (1991) *J. Biol. Chem.* **266**, 24053–24058.

Angiotensin

INTRODUCTION

Angiotensin II is the principal mediator of the renin-angiotensin system and circulates in the bloodstream stimulating vasoconstriction and retention of salt and water [1-4]. It also stimulates increased fluid intake and regulates the neuroendocrine system. Many of its actions are mediated by release of hormones from endocrine glands, e.g. vasopressin, catecholamines, aldosterone, growth hormone, etc.

Angiotensin is probably involved in most forms of hypertension, including conditions where the levels of angiotensin and renin are normal. Angiotensin converting enzyme (ACE) inhibitors, e.g. captopril, reduce plasma levels of angiotensin II and cause a lowering of blood pressure; ACE inhibitors are widely prescribed in the treatment of hypertension. Angiotensin antagonists represent potential novel antihypertensive agents.

Structure of angiotensins

Angiotensin I	Asp–Arg–Val–Tyr–Ile–His–Pro–Phe–His–Leu
Angiotensin II	Asp–Arg–Val–Tyr–Ile–His–Pro–Phe
Angiotensin III	Arg–Val–Tyr–Ile–His–Pro–Phe

The majority of the biological actions are mediated by angiotensin II and, to a limited extent, angiotensin III.

Distribution and synthesis

Angiotensinogen, a large glycoprotein (~58 kDa) secreted by the liver, is metabolized in the blood to angiotensin I by renin. Angiotensin I is cleaved by ACE located on endothelial cells in the pulmonary circulation to angiotensin II, which, in turn, is converted to angiotensin III by the aminopeptidase angiotensinase A.

Reviews

[1] Mendelsohn, F.A.O. (1985) Localization and properties of angiotensin receptors. *J. Hypertens.* **3**, 307–316.

[2] Smith, R.D. *et al.* (1992) Pharmacology of nonpeptide angiotensin II receptor antagonists. *Annu. Rev. Pharmacol. Toxicol.* **32**, 135–167.

[3] Steckelings, U.M. *et al.* (1992) Angiotensin receptor subtypes in the brain. *Trends Pharmacol. Sci.* **13**, 365–368.

[4] Timmermans, P.B. *et al.* (1991) Nonpeptide angiotensin II antagonists. *Trends Pharmacol. Sci.* **12**, 55–62.

ANGIOTENSIN RECEPTORS

AT$_{1A}$ and AT$_{1B}$ receptors

Molecular cloning studies have identified two highly homologous (~95%) AT$_1$ receptors in a single species. Both are seven transmembrane spanning proteins and are described below.

The AT$_2$ receptor

A second binding site for angiotensin II has been identified for which the nonpeptide PD 123177 (EXP 655) and the peptide [*p*-aminoPhe^6]angiotensin II are highly selective; losartan has low affinity [1]. The site is abundant in adrenal medulla, uterus and brain.

Recent evidence suggests that this is a functional receptor with a novel mode of signal transduction. Activation of AT$_2$ receptors has been reported to decrease basal cGMP levels and to inhibit atrial natriuretic peptide stimulated formation of cGMP, possibly through activation of a protein tyrosine phosphatase [2,3]. The receptor does not appear to be coupled to a G-protein [4].

The receptor was previously named AII$_2$, AII$_\beta$ or AII type A.

Other receptors

Although in the majority of systems angiotensin II is more potent than angiotensin III, in some tissues this potency order is reversed. This may reflect the existence of a further subtype of angiotensin receptor.

It has been claimed that the *mas* oncogene encodes for an angiotensin receptor, although this has not been substantiated (see Orphan receptors, page 229).

References
1 Timmermans, P.B. *et al.* (1991) *Trends Pharmacol. Sci.* **12**, 55–62.
2 Sumners, C. *et al.* (1991) *Proc. Natl Acad. Sci. USA* **88**, 7567–7571.
3 Bottari, S.P. *et al.* (1992) *Biochem. Biophys. Res. Commun.* **183**, 206–211.
4 Speth, R.C. and Kim, K.H. (1990) *Biochem. Biophys. Res. Commun.* **169**, 997–1006.

THE AT$_1$ RECEPTORS

This is the major subclass of angiotensin receptor. Angiotensin II is slightly more potent than angiotensin III and induces activation in low nanomolar concentrations; angiotensin I has low affinity [1]. It was previously named the AII$_1$, AII$_\alpha$ or AII type B receptor.

Two highly homologous forms of the AT$_1$ receptor have been identified, named AT$_{1A}$ and AT$_{1B}$ (see below). They exhibit a similar pharmacological profile.

Distribution

AT$_1$ receptors are found on blood vessels, other smooth muscles, e.g. uterus and bladder, and endocrine glands, e.g. adrenal. Receptors are also present in kidney, liver and on presynaptic sympathetic nerve terminals where they potentiate release of noradrenaline. They are distributed widely in CNS and are present in high levels in hypothalamus and anterior pituitary where they stimulate release of vasopressin and ACTH, respectively.

Pharmacology

Agonists: There are no selective agonists.

Antagonists: Losartan (DuP 753; pA$_2$ 8.6 nM) is the prototypic nonpeptide antagonist; it is highly selective to the AT_1 receptor type [2]. Many other nonpeptide antagonists, structurally related to losartan, also exist, e.g. L 158809 [3] (pA$_2$ 10.5 nM). Peptide antagonists have weak partial agonist activity and are nonselective, e.g. saralasin (pA$_2$ 8.7).

Radioligands: The antagonists [^{3}H]losartan (K_d 0.5 nM) and [^{125}I]-[Sar^1,Ile^8]angiotensin II (K_d 0.5 nM) bind to a single site. The agonist [^{125}I]angiotensin II binds to a high-affinity site (K_d 0.1–1 nM) and, in some cases, a second, low-affinity site (K_d ~50 nM).

Losartan

Predominant effector pathways
Activation of phosphoinositide metabolism through a pertussis-toxin-insensitive G-protein most likely of the G_q/G_{11} class and inhibition of adenylyl cyclase.

Amino acid sequence [4-6]

```
                              TM 1
   1 MILNSSTEDG IKRIQDDCPK AGRHNYIFVM IPTLYSIIFV VGIFGNSLVV
                    TM 2
  51 IVIYFYMKLK TVASVFLLNL ALADLCFLLT LPLWAVYTAM EYRWPFGNYL
     TM 3                                              TM 4
 101 CKIASASVSF NLYASVFLLT CLSIDRYLAI VHPMKSRLRR TMLVAKVTCI
                                                     TM 5
 151 IIWLLAGLAS LPAIIHRNVF FIENTNITVC AFHYESQNST LPIGLGLTKN
                                                     TM 6
 201 ILGFLFPFLI ILTSYTLIWK ALKKAYEIQK NKPRNDDIFK IIMAIVLFFF
                                        TM 7
 251 FSWIPHQIFT FLDVLIQLGI IRDCRIADIV DTAMPITICI AYFNNCLNPL

 301 FYGFLGKKFK RYFLQLLKYI PPKAKSHSNL STKMSTLSYR PSDNVSSSTK

 351 KPAPCFEVE
```

Amino acids	359
Molecular weight	41 060
Glycosylation	Asn4*, Asn176*, Asn188*
Disulfide bonds	Cys101–Cys180*
Palmitoylation	None

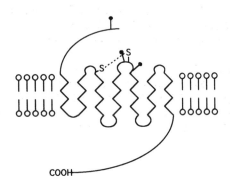

Database accession numbers

	PIR	SWISSPROT	EMBL/GENBANK	REFERENCES
Bovine	S15403	P25104	X62294	7
Human			Z11162	4–6
			M91464	
Mouse AT$_{1A}$				8
Mouse AT$_{1B}$				8
Rat AT$_{1A}$	JQ1055	P25095	M74054	9–15
	S15404		X64052	
	S20424			
Rat AT$_{1B}$	S20423	P29089	X64052	11–15

Gene

The human AT$_1$ receptor gene is contained in a single exon indicating an intronless stucture of the coding region. In rat, the AT$_{1A}$ receptor has three exons, two of which encode parts of the 5' untranslated regions while the third encompasses the entire coding sequence.

The AT$_{1B}$ receptor

The AT$_{1B}$ receptor has been identified in rat [11-15] and mouse [8] and there is evidence for its presence in man [4]. It shares ~95% homology in its amino acid sequence with the AT$_{1A}$ receptor from the same species and has also been named the AT$_3$ receptor. AT$_{1A}$ and AT$_{1B}$ receptors have similar tissue distributions and only small differences in the binding of angiotensin II or its analogues have been observed. The gene encoding the rat AT$_{1B}$ receptor also lacks introns in the coding region.

References

1 Bumpus, F.M. *et al.* (1990) *Hypertens.* **17**, 720–721.
2 Timmermans, P.B. *et al.* (1991) *Trends Pharmacol. Sci.* **12**, 55–62.

[3] Chang, R.S.L *et al.* (1992) *J. Pharmacol. Ther.* **262**, 133–138.

[4] Mauzy, C.A. *et al.* (1992) *Biochem. Biophys. Res. Commun.* **186**, 277–284.

[5] Takayanagi, R. *et al.* (1992) *Biochem. Biophys. Res. Commun.* **183**, 910–916.

[6] Furuta, H. *et al.* (1992) *Biochem. Biophys. Res. Commun.* **183**, 8–13.

[7] Sasaki, K. *et al.* (1991) *Nature* **351**, 230–233.

[8] Sasamura, H. *et al.* (1992) *Biochem. Biophys. Res. Commun.* **185**, 253–259.

[9] Murphy, T.J. *et al.* (1991) *Nature* **351**, 233–236.

[10] Iwai, N. and Inagami, T. (1991) *Biochem. Biophys. Res. Commun.* **177**, 299–304.

[11] Langford, K. *et al.* (1992) *Biochem. Biophys. Res. Commun.* **183**, 1025–1032.

[12] Sandberg, K. *et al.* (1992) *J. Biol. Chem.* **267**, 9455–9458.

[13] Iwai, N. and Inagami, T. (1992) *FEBS Lett.* **298**, 257–260.

[14] Elton, T.S. *et al.* (1992) *Biochem. Biophys. Res. Commun.* **184**, 1067–1073.

[15] Ye, M.Q. and Healy, T.P. (1992) *Biochem. Biophys. Res. Commun.* **185**, 204–210.

Bombesin

Bombesins are a family of peptide neurotransmitters which share a common C-terminal amino acid sequence Trp–Ala–X–Gly–His–X–Met–NH$_2$ [1-4]. Full biological activity resides in this region, with the N-terminal sequence having a modulatory role. Bombesin is not found in mammalian tissue.

In the periphery, bombesin-related peptides stimulate smooth muscle contraction, e.g. intestine, uterus, bladder, etc., and glandular secretion, e.g. pancreas, anterior pituitary, etc. In brain, these peptides are believed to play a role in thermoregulation, homeostasis, metabolism and behaviour and have been reported to elicit analgesia and excessive grooming together with central regulation of a variety of peripheral effects, e.g. increase in sympathetic activity. The peptides are mitogens in a number of cell lines including small-cell lung carcinomas where they may have an autocrine growth function.

Structures

Bombesin   Pyr–Gln–Arg–Leu–Gly–Asn–Gln–Trp–Ala–Val–Gly–His–Leu–Met–NH$_2$

GRP$_{1-27}$ H–Ala–Pro–Val–Ser–Val–Gly–Gly–Gly–Thr–Val–Leu–Ala–Lys–Met–Tyr–Pro–Arg–Gly–Asn–His–Trp–Ala–Val–Gly–His–Leu–Met–NH$_2$

GRP$_{18-27}$ H–Gly–Asn–His–Trp–Ala–Val–Gly–His–Leu–Met–NH$_2$

Neuromedin B   Gly–Asn–Leu–Trp–Ala–Thr–Gly–His–Phe–Met–NH$_2$

GRP$_{18-27}$ (gastrin releasing peptide) was previously known as neuromedin C. Additional members of the family exist in lower organisms, notably amphibia, e.g. bombesin, ranatensin, litorin and phyllolitorin.

Distribution and synthesis

Bombesin-like immunoreactivity is present in high amounts in brain, intestine and certain other tissues, e.g. adrenal gland, lung and pituitary. Most of the immunoreactivity in brain is associated with neurons.

Two genes encode for mammalian bombesins. The preproGRP gene transcript encodes for a precursor of ~147 amino acids which gives GRP and GRP$_{18-27}$. The preproNMB gene transcript encodes for a precursor of ~117 amino acids which is metabolized to neuromedin B. The distribution of cells expressing mRNAs for GRP and neuromedin B in brain is distinct.

Book
1 Bombesin-like peptides in health and disease. *Ann. N.Y. Acad. Sci.* (1988) **547**.

Reviews
2 Battey, J. and Wada, E. (1991) Two distinct receptor subtypes for mammalian bombesin-like peptides. *Trends Neurosci.* **14**, 524–528.
3 Jensen, R. and Coy, D. (1991) Progress in the development of potent bombesin receptor antagonists. *Trends Pharmacol. Sci.* **12**, 13–18.

4 Lebacq-Verheyden, A. M. *et al.* (1990) Peptide growth factors and their receptors II. *Handbook Exp. Pharmacol.* **95**, 71–124.

THE BB$_1$ RECEPTOR

Neuromedin B is the most potent mammalian peptide, inducing activation in low nanomolar concentrations. GRP has a 20–100-fold lower potency. This receptor has also been named the neuromedin B-preferring receptor.

Distribution

The BB$_1$ receptor has been characterized in rat oesphagus and rat urinary bladder [1,2]. It is widespread within the CNS and is found in high levels in olfactory nucleus and thalamic regions and in lower levels in frontal cortex, dentate gyrus, amygdala and dorsal raphe.

Pharmacology

Agonists: Neuromedin B has a selectivity of more than 20-fold [1].

Antagonists: No antagonist has been described.

Radioligands: [^{125}I]BH-Neuromedin B (K_d 3 nM) is weakly selective; [^{125}I]-[Tyr^4]bombesin (K_d 2nM) is nonselective.

Predominant effector pathways
Activation of the phosphoinositide pathway through a pertussis-toxin-insensitive G-protein, most likely of the G$_q$/G$_{11}$ class.

Amino acid sequence [3]

```
                                                      TM 1
   1 MPSKSLSNLS VTTGANESGS VPEGWERDFL PASDGTTTEL VIRCVIPSLY
                                         TM 2
  51 LLIITVGLLG NIMLVKIFIT NSAMRSVPNI FISNLAAGDL LLLLTCVPVD
                                 TM 3
 101 ASRYFFDEWM FGKVGCKLIP VIQLTSVGVS VFTLTALSAD RYRAIVNPMD
            TM 4
 151 MQTSGALLRT CVKAMGIWVV SVLLAVPEAV FSEVARISSL DNSSFTACIP
                 TM 5
 201 YPQTDELHPK IHSVLIFLVY FLIPLAIISI YYYHIAKTLI KSAHNLPGEY
                      TM 6
 251 NEHTKKQMET RKRLAKIVLV FVGCFIFCWF PNHILYMYRS FNYNEIDPSL
        TM 7
 301 GHMIVTLVAR VLSFGNSCVN PFALYLLSES FRRHFNSQLC CGRKSYQERG

 351 TSYLLSSSAV RMTSLKSNAK NMVTNSVLLN GHSMKQEMAM
```

Amino acids	390
Molecular weight	43 453
Glycosylation	Asn8*, Asn16*, Asn192*
Disulfide bonds	Cys116–Cys198*
Palmitoylation	Cys341*
Phosphorylation	Thr260*, Ser364*

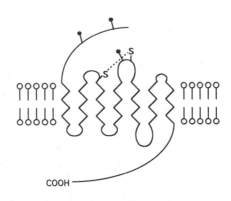

Database accession numbers

	PIR	SWISSPROT	EMBL/GENBANK	REFERENCES
Human	B41007	P28336	M73482	3
Rat	JH0374	P24053		4

Gene (human) [3]

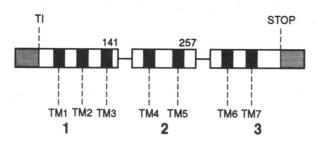

Two mRNA species of 3.2 and 2.7 kb have been identified and are thought to be transcripts from the same gene [4].

References
[1] Rouissi, N. et al. (1991) Br. J. Pharmacol. 103, 1141–1147.
[2] von Schrenck, T. et al. (1990) Am. J. Physiol. 259, G468–473.
[3] Corjay, M.H. et al. (1991) J. Biol. Chem. 266, 18771–18779.
[4] Wada, K. et al. (1991) Neuron 6, 421–430.

THE BB₂ RECEPTOR

GRP is the most potent mammalian peptide, inducing activation in low nanomolar concentrations. Bombesin is slightly less potent and neuromedin B is ~two orders of magnitude less potent [1]. It has also been named the GRP-preferring receptor.

Distribution

The BB₂ receptor has a widespread distribution in peripheral tissue. High levels are found in smooth muscle, e.g. intestine, stomach and bladder, and in secretory glands, e.g. pancreas. In brain it is found in high levels in hypothalamus and is present in other areas in lower levels, e.g. olfactory tract, dendate gyrus, cortex. It is found in a number of cell lines, e.g. Swiss 3T3 fibroblasts and small-cell lung carcinomas.

Pharmacology

Agonists: The receptor can be identified by the rank order of potency of naturally occurring agonists: GRP > bombesin >> neuromedin B [2,3].

Antagonists: [DPhe⁶]Bombesin₆₋₁₄ ethyl ester (K_d 2–10 nM), [DPhe⁶,Cpa¹⁴, ψ¹³⁻¹⁴]bombesin₆₋₁₄ (K_d 10–40 nM) and AcGRP₂₀₋₂₆ ethyl ester (pA₂ 8.0) are highly selective [4]. A number of antagonists are, in fact, weak partial agonists, e.g. [DPhe⁶,Cpa³,Ψ¹³⁻¹⁴]bombesin.

Radioligands: The antagonist [¹²⁵I]-[DTyr⁶]bombesin₆₋₁₃ methyl ester is selective [5]. [¹²⁵I]-[Tyr⁴]bombesin (K_d 0.2–2 nM) and [¹²⁵I]GRP (K_d 0.2–2 nM) are nonselective.

Predominant effector pathways

Activation of the phosphoinositide pathway through a pertussis-toxin-insensitive G-protein most likely of the G_q/G_{11} class.

Amino acid sequence [6]

```
                                              TM 1
  1 MALNDCFLLN LEVDHFMHCN ISSHSADLPV NDDWSHPGIL YVIPAVYGVI
                                TM 2
 51 ILIGLIGNIT LIKIFCTVKS MRNVPNLFIS SLALGDLLLL ITCAPVDASR
                     TM 3
101 YLADRWLFGR IGCKLIPFIQ LTSVGVSVFT LTALSADRYK AIVRPMDIQA
              TM 4
151 SHALMKICLK AAFIWIISML LAIPEAVFSD LHPFHEESTN QTFISCAPYP
                          TM 5
201 HSNELHPKIH SMASFLVFYV IPLSIISVYY YFIAKNLIQS AYNLPVEGNI
                          TM 6
251 HVKKQIESRK RLAKTVLVFV GLFAFCWLPN HVIYLYRSYH YSEVDTSMLH
                TM 7
301 FVTSICARLL AFTNSCVNPF ALYLLSKSFR KQFNTQLLCC QPGLIIRSHS

351 TGRSTTCMTS LKSTNPSVAT FSLINGNICH ERYV
```

Amino acids	384
Molecular weight	43 198
Glycosylation	Asn20*
Disulfide bonds	Cys113–Cys196*
Palmitoylation	Cys390*
Phosphorylation (PKC)	Ser258*, Ser360*

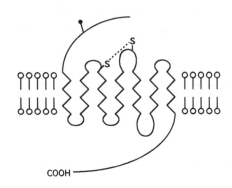

Database accession numbers

	PIR	SWISSPROT	EMBL/GENBANK	REFERENCES
Human			M73481	6
Mouse		P21729	M57922	7,8
			M61000	

Gene (human)

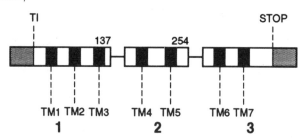

In mouse two different mRNAs have been identified in Northern blots of 3.1 and 7.2 kb which are identical in the protein-encoding domain. They are thought to be derived from the same gene with the difference in size being due to the presence of a long 3' untranslated domain [7].

References
1 Rouissi, N. *et al.* (1991) *Br. J. Pharmacol.* **103**, 1141–1147.
2 von Schrenck, T. *et al.* (1990) *Am. J. Physiol.* **259**, G468–473.
3 Feldman, R. I. *et al.* (1990) *J. Biol. Chem.* **265**, 17364–17372.
4 Jensen, R. and Coy, D. (1991) *Trends Pharmacol. Sci.* **12**, 13–18.
5 Mantey, S. *et al.* (1993) *Mol. Pharmacol.* in press.
6 Corjay, M.H. *et al.* (1991) *J. Biol. Chem.* **266**, 18771–18779.

[7] Battey, J. et al. (1991) Proc. Natl Acad. Sci. USA **88**, 395–399.
[8] Spindel, E.R. et al. (1990) Mol. Endocrinol. **43**, 1950–1963.

THE BB3 RECEPTOR

This receptor has only recently been identified through cloning studies. GRP is approximately two orders of magnitude more potent than neuromedin B. However, the K_i (~ 300 nM) of GRP for this receptor is relatively high, suggesting that there may be an alternative endogenous ligand with higher affinity [1].

Distribution

mRNA expression has been detected in secondary spermatocytes in testis and uteri of pregnant animals. It is also present in a variety of lung carcinoma cell lines.

Pharmacology

The receptor can be labelled with [^{125}I]-[Tyr^4]bombesin; however, an extensive pharmacological characterization has not been determined.

Predominant effector pathways
Not known.

Amino acid sequence (guinea-pig) [1]

```
                                                              TM 1
  1 MSQKQPQSPN QTLISITNDT ESSSSVVSND TTNKGWTGDN SPGIEALCAI
                                         TM 2
 51 YITYAVIISV GILGNAILIK VFFKTKSMQT VPNIFITSLA LGDLLLLLTC
                          TM 3
101 VPVDATHYLA EGWLFGRIGC KVLSFIRLTS VGVSVFTLTI LSADRYKAVV
                     TM 4
151 KPLERQPSNA ILKTCAKAGC IWIMSMIFAL PEAIFSNVHT LRDPNKNMTS
                     TM 5
201 EWCAFYPVSE KLLQEIHALL SFLVFYIIPL SIISVYYSLI ARTLYKSTLN
                          TM 6
251 IPTEEQSHAR KQVESRKRIA KTVLVLVALF ALCWLPNHLL NLYHSFTHKA
              TM 7
301 YEDSSAIHFI VTIFSRVLAF SNSCVNPFAL YWLSKTFQKQ FKAQLFCCKG

351 ELPEPPLAAT PLNSLAVMGR VSGTENTHIS EIGVASFIGR PMKKEENRV
```

Amino acids	399
Molecular weight	44 341
Chromosome	X
Glycosylation	Asn10*, Asn18*, Asn29*
Disulfide bonds	Cys120–Cys203*
Palmitoylation	Cys347*, Cys348*
Phosphorylation (PKC)	Ser265*

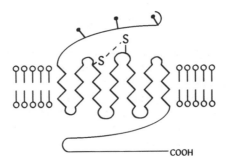

Database accession numbers

	PIR	SWISSPROT	EMBL/GENBANK	REFERENCES
Guinea-pig			X67126	1
Human				2

References
1 Gorbulev, V. *et al.* (1992) *Eur. J. Biochem.* **208**, 405–410.
2 Fathi, Z. *et al.* (1993) *J. Biol. Chem.* **268**, 5979–5984.

Bradykinin

INTRODUCTION

Bradykinins are a family of structurally very similar peptides of 8–11 amino acids in length derived from large precursors (kininogens) by the action of a group of serine proteases, the kallikreins [1-10]. They activate sensory fibres, contract venous smooth muscle, stimulate release of cytokines, induce connective tissue proliferation and mediate endothelium-dependent vasodilation.

Bradykinin (BK) antagonists are of potential use in the treatment of inflammation, asthma, mild pain and endotoxic shock.

Structures

BK	Arg–Pro–Pro–Gly–Phe–Ser–Pro–Phe–Arg
[des-Arg^9]BK	Arg–Pro–Pro–Gly–Phe–Ser–Pro–Phe
[Hyp^3]BK	Arg–Pro–Hyp–Gly–Phe–Ser–Pro–Phe–Arg
Kallidin	Lys–Arg–Pro–Pro–Gly–Phe–Ser–Pro–Phe–Arg
Ornitho-kinin	Arg–Pro–Pro–Gly–Phe–Thr–Pro–Leu–Arg

T-kinin (Ile-Ser-BK) is found in rat.

Distribution and synthesis

Two major kinin precursor proteins, high molecular weight kininogen and low molecular weight kininogen, are synthesized in liver and circulate in plasma. High molecular weight kininogen is cleaved by plasma kallikrein to BK or by tissue kallikrein to kallidin. Low molecular weight kininogen is cleaved to kallidin by tissue kallikrein.

Reviews

1 Bathon, J.M. and Proud, D. (1991) Bradykinin antagonists. *Annu. Rev. Pharmacol. Toxicol.* **31**, 129–162.
2 Bhoola, K.D. *et al.* (1992) Bioregulation of kinins: kallikreins, kininogens and kininases. *Pharmacol. Rev.* **44**, 1–81.
3 Burch, R.M. *et al.* (1990) Bradykinin receptor antagonists. *Med. Res. Rev.* **10**, 237–269.
4 Burch, R.M. and Farmer, S.G. (1992) Biochemical and molecular pharmacology of kinin receptors. *Annu. Rev. Parmacol. Toxicol.* **32**, 511–536.
5 Dray, A. and Perkins, M. (1993) Bradykinin and inflammation. *Trends Pharmacol. Sci.* **16**, 99–103.
6 Hall, J.M. (1992) Bradykinin receptors. *Pharmacol. Ther.* **56**, 131–190.
7 Regoli, D. and Barabé, J. (1980) Pharmacology of bradykinin and related kinins. *Pharmacol. Rev.* **32**, 1–46.
8 Regoli, D. *et al.* (1990) New selective bradykinin receptor antagonists and bradykinin B2 receptor characterisation. *Trends Pharmacol. Sci.* **11**, 156–161.
9 Steranka, L. R. (1989) Antagonists of B2 bradykinin receptors. *FASEB J.* **3**, 2019–2025.

10 Schachter, M. (1979) Kallikreins (kininogenases) – A group of serine proteases with bioregulatory actions. *Pharmacol. Rev.* **31**, 1–17.

BRADYKININ RECEPTORS

The B₁ receptor

The B_1 receptor is identified by the activity of [des-Arg^9]BK (or BK$_{1-8}$) which is active at nanomolar concentrations (1–100 nM) but virtually inactive on B_2 receptors. The B_1 receptor has a very limited distribution in the periphery but has not been described in the CNS. It stimulates contraction of rabbit aorta and rat duodenum, and mediates relaxation of rabbit mesenteric and coeliac arteries.

In rabbit vascular smooth muscle, the expression of B_1 receptors is increased following various pathological insults *in vitro* or *in vivo*. This is particularly marked in rabbit anterior mesenteric vein and aorta *in vitro* where responsiveness to [des-Arg^1]BK, initially nonexistent, increases with the time of incubation reaching a maximum after 6 h [1].

The B₂ receptor

This is a seven transmembrane spanning receptor and is described below.

The B₃ receptor

The existence of B_3 receptors in guinea-pig trachea has been proposed [2,3]. [DPhe^7]-substituted analogues of BK are antagonists at B_1 and B_2 receptors but are inactive at B_3 receptors [2,3]. The B_3 receptor may be the guinea-pig homologue of the B_2 receptor.

References
1 Regoli, D. and Barabé, J. (1980) *Pharmacol. Rev.* **32**, 1–46.
2 Farmer, S.G. *et al.* (1989) *Mol. Pharmacol.* **36**, 1–8.
3 Farmer, S.G. *et al.* (1991) *Br. J. Pharmacol.* **102**, 785–787.

THE B₂ BRADYKININ RECEPTOR

This is the predominant subtype of BK receptor. BK and kallidin induce activation in nanomolar concentrations and [des-Arg^9]BK is virtually inactive.

Distribution

The receptor has a widespread distribution in peripheral tissues. B_2 Receptors mediate slow contraction of a number of smooth muscles, e.g. veins, intestine, uterus, trachea and lung, induce endothelium-dependent relaxation of arteries and arterioles, and stimulate natriuresis/diuresis in kidney. BK also induces hyperalgesia through activation of B_2 receptors in sensory nerve fibres and dorsal root ganglion neurons. B_2 Receptors are found on cell lines, e.g. N1E-115 neuroblastoma, Swiss 3T3 fibroblasts.

Pharmacology

Agonists: [Phe^8,ψ(CH$_2$-NH)Arg^9]BK and [Hyp^3,Tyr(Me)^8]BK are examples of potent agonists.

Antagonists: Peptide antagonists include: HOE-140 [1] (DArg[Hyp^3,Thi^5,DTic^7, Oic^8]BK; pA$_2$ 7.5–10.5) and DArg[Hyp^3,Thi^5,HypE(*trans*-propyl)^7,Oic^8]BK[2,3] (pA$_2$ 8.5–9.6); HOE-140 is not active at B$_1$ receptors. Earlier antagonists were [DPhe^7]-substituted analogues of BK with pA$_2$ values between 5 and 8; many of these show partial agonist activity.

Radioligands: The agonists [^{125}I]Tyr-BK and [^{3}H]BK are nonselective unless used in the presence of [des-Arg^9]BK; [^{3}H]BK has been reported to bind to either one or two sites with K_d values in the range 10-100 pM and 0.3-10 nM [4,5].

Predominant effector pathways

In the majority of tissues BK has been reported to induce activation of phosphoinositide metabolism through a pertussis-toxin-insensitive G-protein [4,6], most likely of the G$_q$/G$_{11}$ class. BK has also been reported to inhibit adenylyl cyclase through a pertussis-toxin-sensitive G-protein [3], and to raise levels of cGMP and open ion channels[6]. It also stimulates PLA$_2$, possibly by direct activation of a G protein [6].

Amino acid sequence [7,8]

```
                                      TM 1
    1 MLNVTLQGPT LNGTFAQSKC PQVEWLGWLN TIQPPFLWVL FVLATLENIF
                    TM 2
   51 VLSVFCLHKS SCTVAEIYLG NLAAADLILA CGLPFWAITI SNNFDWLFGE
        TM 3                                              TM 4
  101 TLCRVVNAII SMNLYSSICF LMLVSIDRYL ALVKTMSMGR MRGVRWAKLY
                                                      TM 5
  151 SLVIWGCTLL LSSPMLVFRT MKEYSDEGHN VTACVISYPS LIWEVFTNML
                                                      TM 6
  201 LNVVGFLLPL SVITFCTMQI MQVLRNNEMQ KFKEIQTERR ATVLVLVVLL
                                              TM 7
  251 LFIICWLPFQ ISTFLDTLHR LGILSSCQDE RIIDVITQIA SFMAYSNSCL

  301 NPLVYVIVGK RFRKKSWEVY QGVCQKGGCR SEPIQMENSM GTLRTSISVE

  351 RQIHKLQDWA GSRQ
```

Amino acids	364
Molecular weight	41 140
Glycosylation	Asn3*, Asn12*, Asn180*
Disulfide bonds	Cys103–Cys184*
Phosphorylation	Cys334*, Cys339*
Phosphorylation (PKA)	Ser316*
Phosphorylation (PKC)	Thr170*, Thr237*, Thr342*

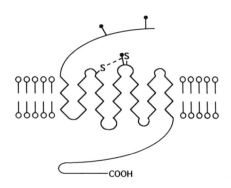

Database accession numbers

	PIR	*SWISSPROT*	*EMBL/GENBANK*	*REFERENCES*
Human			M88714	*7,8*
Rat		P25023	M59967	*9*

Gene

mRNAs of 4.0, 5.7 and 6.5 kb have been observed, although Southern blots have identified only a single gene in rat [7]. The encoding region of the gene is contained in a single exon although there is an intron in the 5' untranslated region [8].

Comment

There is a wide variation in pA_2 values of bradykinin antagonists and in certain tissues agonist activity has been reported. This may reflect the existence of B_2 receptor subtypes or, more likely, species differences in receptor structure as demonstrated by comparison of the cloned B_2 receptors in human and in rat [8].

References

[1] Hock, F.J. *et al.* (1991) *Br. J. Pharmacol.* **102**, 769–773.
[2] Kyle, D.J. *et al.* (1991) *J. Med. Chem.* **34**, 2649–2653.
[3] Farmer, S.G. *et al.* (1991) *Br. J. Pharmacol.* **102**, 785–787.
[4] Leeb-Lundberg, L.M.F. *et al.* (1990) *J. Biol. Chem.* **265**, 9621–9627.
[5] Seguin, L. *et al.* (1992) *J. Neurochem.* **59**, 2125–2133.
[6] Farmer, S.G. *et al.* (1992) *Annu. Rev. Pharmacol. Toxicol.* **32**, 511–536.
[7] Hess, J.F. *et al.* (1992) *Biochem. Biophys. Res. Commun.* **184**, 260–268.
[8] Eggerickx, T. *et al.* (1992) *Biochem. Biophys. Res. Commun.* **187**, 1306–1313.
[9] McEachern, A.E. *et al.* (1991) *Proc. Natl Acad. Sci. USA* **88**, 7724–7728.

C5a anaphylatoxin

The accumulation of phagocytic cells at the site of injury or infection is regulated by substances which stimulate chemotaxis, granule secretion, superoxide generation and upregulation of cell surface adhesion molecules e.g. MAC-1, CR1, in cells of the immune system [1-2]. The list of chemoattractant substances includes C5a, N-formylmethionyl-containing peptides, interleukin 8, leukotriene B4 and platelet activating factor. Many of these agents participate in anaphylactoid and septic shock.

Structure

Human C5a: TLQKKIEEIA AKYKHSVVKK CCYDGACVNN DETCEQRAAR
ISLGPRCIKA FTECCVVASQ LRANISHKDM QLGR

Synthesis and distribution

C5 is one of 13 plasma proteins making up the complement system, a defence mechanism responsible for clearing foreign particles or organisms from the bloodstream. When the complement system is activated, C5 is proteolytically cleaved to C5a and C5b.

Reviews

[1] Hugli, T.E. *et al.* (1981) The structural basis for anaphlatoxins and chemotactic functions of C3a, C4a and C5a. *CRC Crit. Rev. Immunol.* **1**, 321–366.
[2] Oppenheim, P. *et al* (1991) Properties of the novel proinflammatory supergene "intercrine" cytokine family. *Annu. Rev. Immunol.* **9**, 617–648.

THE C5a RECEPTOR

Distribution

The receptor is present on cells of the immune system, e.g. neutrophils, macrophages, mast cells and related cell lines, e.g. U937 and HL60. It is also found in smooth muscle.

Pharmacology

Agonists: C5a induces activation in nanomolar concentrations. C-terminal peptide analogues of C5a, 6–8 amino acids in length, are agonists at micromolar concentrations [1].

Antagonists: None have been described.

Radioligands: The receptor can be radiolabelled with [^{125}I]recombinant C5a (K_d 1 nM); two binding sites have been observed on the expressed receptor [2].

Effector pathway

Activation of phosphoinositide metabolism through a pertussis-toxin-insensitive G-protein most likely of the G_q/G_{11} class.

Amino acid sequence [3,4]

```
                                        TM 1
  1 MNSFNYTTPD YGHYDDKDTL DLNTPVDKTS NTLRVPDILA LVIFAVVFLV
                                TM 2
 51 GVLGNALVVW VTAFEAKRTI NAIWFLNLAV ADFLSCLALP ILFTSIVQHH
                     TM 3
101 HWPFGGAACA ILPSLILLNM YASILLLATI SADRFLLVFK PIWCQNFRGA
               TM 4
151 GLAWIACAVA WGLALLLTIP SFLYRVVREE YFPPKVLCGV DYSHDKRRER
       TM 5                                          TM 6
201 AVAIVRLVLG FLWPLLTLTI CYTFILLRTW SRRATRSTKT LKVVVAVVAS
                                            TM 7
251 FFIFWLPYQV TGIMMSFLEP SSPTFLLLNK LDSLCVSFAY INCCINPIIY

301 VVAGQGFQGR LRKSLPSLLR NVLTEESVVR ESKSFTRSTV DTMAQKTQAV
```

Amino acids	350
Molecular weight (polypeptide)	39 320
Molecular weight (SDS PAGE)	52–55 kDa
Chromosome	19
Glycosylation	Asn5*
Disulfide bonds	Cys109–Cys188*
Palmitoylation	None

The C5a receptor is unusual in having six aspartic acid residues in the N-terminal sequence conferring acidic properties which may bind to the basic nature of C5a.

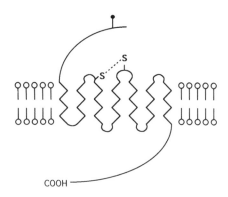

Database accession numbers

	PIR	SWISSPROT	EMBL/GENBANK	REFERENCES
Human		P21730	J05327	2,3
			X58674	
Mouse			S46665	4

Gene
Southern analysis of genomic DNA indicates a simple restriction pattern for the C5a receptor consistent with a single gene [1]. Isolation and sequence analysis of genomic clones suggests that there are no introns within the coding sequence [1].

Reference
[1] Drapeau, G. *et al.* (1993) *Biochem. Pharmacol.* **45**, 1289–1299.
[2] Boulay, F. *et al.* (1991) *Biochemistry* **30**, 2993–2999.
[3] Gerard, N.P. and Gerard, C. (1991) *Nature* **349**, 614–617.
[4] Gerard, N.P. *et al.* (1992) *J. Immunol.* **149**, 2600–2606.

Calcitonin

Although calcitonin and calcitonin-gene related peptide (CGRP) are derived by alternative splicing of the same gene, they share limited structural homology and have distinct physiological roles mediated by distinct receptors [1].

The major physiological role of calcitonin is to inhibit bone resorption thereby leading to a reduction in plasma Ca^{2+}. In addition, it enhances excretion of ions in kidney, prevents absorption of ions in intestine and inhibits secretion in endocrine cells, e.g. pancreas and pituitary. In CNS, calcitonin has been reported to be analgesic and to suppress feeding and gastric acid secretion. The physiological relevance of the latter effects is uncertain.

Calcitonin is used in the treatment of Paget's disease of the bone.

Structure (human)

```
            S────────────────────S
Calcitonin  Cys–Gly–Asn–Leu–Ser–Thr–Cys–Met–Leu–Gly–Thr–Tyr–Thr–Glu–Asp–Phe–Asn–
            –Lys–Phe–His–Thr–Phe–Pro–Glu–The–Ala–Ile–Gly–Val–Gly–Ala–Pro–NH₂
```

There is a high degree of species heterogeneity although all forms of calcitonin have cysteine residues at position 1 and 7 linked by a disulfide bond and identical amino acids at positions 4, 5, 6, 9, 28 and 32. There is evidence for additional forms of calcitonin in human.

Distribution and synthesis

In higher vertebrates calcitonin is produced in C cells of the thyroid gland by proteolytic cleavage of a large precursor, preprocalcitonin (molecular weight 17 500). In brain, however, the same gene transcript is alternatively spliced to produce an mRNA which encodes for CGRP. Calcitonin-like immunoreactivity has been described in posterior hypothalamus, median eminence and pituitary (the synthesis of calcitonin in brain has not been demonstrated).

Review

[1] Epand, R.M. and Caulfield, M.P. (1990) Calcitonin and parathyroid hormone receptors. In Comprehensive Medicinal Chemistry, vol. 3 (Membranes and Receptors) (Hansch, C., Sammes, B.G. and Taylor, J.B. eds), Pergamon Press, Oxford, pp. 1023–1045.

THE CALCITONIN RECEPTOR

Distribution

The receptor is found predominantly on osteoclasts or on immortal cell lines derived from these cells (~1-3 million receptors per cell). It is found in lower amounts in brain, e.g. hypothalamus and pituitary, and in peripheral tissues, e.g. testes, kidney, liver and lymphocytes. It has also been described in lung and breast cancer cell lines.

Pharmacology

Agonists: Human and salmon calcitonin are commonly used agonists. The receptor can tolerate a wide degree of amino acid substitutions in the structure of calcitonin. CGRP has ~1000 times lower affinity relative to calcitonin.

Antagonists: None have been described.

Radioligands: The majority of studies have used [^{125}I]salmon calcitonin (K_d 0.1–1.0 nM); human calcitonin has low affinity for these sites.

Predominant effector pathways
Activation of adenylyl cyclase through G_s is the predominant signalling pathway, although calcitonin has also been described to have both stimulatory and inhibitory actions on the phosphoinositide pathway.

Amino acid sequence (porcine) [1]

```
  1 MRFTLTRWCL TLFIFLNRPL PVLPDSADGA HTPTLEPEPF LYILGKQRML

 51 EAQHRCYDRM QKLPPYQGEG LYCNRTWDGW SCWDDTPAGV LAEQYCPDYF
                                                        TM 1
101 PDFDAAEKVT KYCGEDGDWY RHPESNISWS NYTMCNAFTP DKLQNAYILY
                                                        TM 2
151 YLAIVGHSLS ILTLLISLGI FMFLRSISCQ RVTLHKNMFL TYVLNSIIII
                                                        TM 3
201 VHLVVIVPNG ELVKRDPPIC KVLHFFHQYM MSCNYFWMLC EGVYLHTLIV
          TM 4
251 VSVFAEGQRL WWYHVLGWGF PLIPTTAHAI TRAVLFNDNC WLSVDTNLLY
       TM 5                                            TM 6
301 IIHGPVMAAL VVNFFFLLNI LRVLVKKLKE SQEAESHMYL KAVRATLILV
                                     TM 7
351 PLLGVQFVVL PWRPSTPLLG KIYDYVVHSL IHFQGFFVAI IYCFCNHEVQ

401 GALKRQWNQY QAQRWAGRRS TRAANAAAAT AAAAAALAET VEIPVYICHQ

451 EPREEPAGEE PVVEVEGVEV IAMEVLEQET SA
```

Amino acids	482
Molecular weight	55 259
Glycosylation	Asn74*, Asn126*, Asn131*
Disulfide	Cys237–Cys290*
Palmitoylation	None

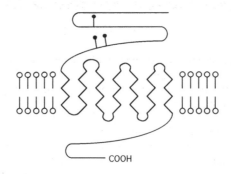

Database accession numbers

	PIR	SWISSPROT	EMBL/GENBANK	REFERENCE
Pig	A32985	P25117	M74420	1

Comment

A binding site in certain nuclei of rat brain, e.g. nucleus accumbens, with similar affinities for calcitonin and CGRP has been described [2]. This site has been termed a C_3 receptor although it has no known functional correlate.

References

[1] Lin, H.Y. *et al.* (1991) *Science* **254**, 1022–1024.
[2] Sexton, D. *et al.* (1988) *Neurochem. Int.* **12**, 323–335.

Calcitonin-gene related peptide (CGRP) and amylin

INTRODUCTION

Although calcitonin-gene related peptide (CGRP) and calcitonin are derived by alternative splicing of the same gene they share only limited structural homology and have distinct physiological roles mediated by distinct receptors. CGRP shares ~50% homology with amylin [1-4].

CGRP is believed to have important physiological roles in the CNS and in the cardiovascular system. Intracerebroventricular injection of CGRP in rats reduces food intake and growth hormone secretion but stimulates gastric acid secretion. In the periphery, CGRP increases heart rate and force, and causes vasodilation. CGRP has also been reported to have longer term actions on neuronal differentiation.

Amylin (previously known as islet amyloid polypeptide) was identified in plaques of patients suffering from type II diabetes. Many of its actions are opposite to those of insulin, e.g. it inhibits insulin-stimulated glycogen synthesis and glucose uptake. It has been proposed that amylin has a role in the onset of type II diabetes.

Structure (human)

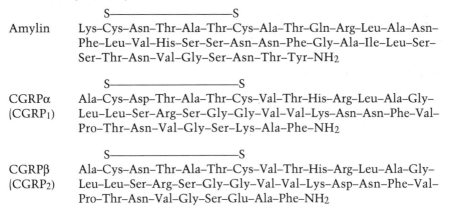

Amylin
S——————————S
Lys–Cys–Asn–Thr–Ala–Thr–Cys–Ala–Thr–Gln–Arg–Leu–Ala–Asn–
Phe–Leu–Val–His–Ser–Ser–Asn–Asn–Phe–Gly–Ala–Ile–Leu–Ser–
Ser–Thr–Asn–Val–Gly–Ser–Asn–Thr–Tyr–NH$_2$

CGRPα
(CGRP$_1$)
S——————————S
Ala–Cys–Asp–Thr–Ala–Thr–Cys–Val–Thr–His–Arg–Leu–Ala–Gly–
Leu–Leu–Ser–Arg–Ser–Gly–Gly–Val–Val–Lys–Asn–Asn–Phe–Val–
Pro–Thr–Asn–Val–Gly–Ser–Lys–Ala–Phe–NH$_2$

CGRPβ
(CGRP$_2$)
S——————————S
Ala–Cys–Asn–Thr–Ala–Thr–Cys–Val–Thr–His–Arg–Leu–Ala–Gly–
Leu–Leu–Ser–Arg–Ser–Gly–Gly–Val–Val–Lys–Asp–Asn–Phe–Val–
Pro–Thr–Asn–Val–Gly–Ser–Glu–Ala–Phe–NH$_2$

There are small variations in structure of CGRP and amylin between species.

Distribution and synthesis

CGRP is distributed throughout the CNS and in peripheral nerves associated with the cardiovascular system; it is found in particularly high levels in sensory nerves. In brain, the primary RNA transcript is processed to give mRNA which encodes for CGRP; in thyroid, the same gene transcript is alternatively spliced to produce mRNA which encodes for calcitonin. A second gene also exists which gives rise to mRNA which encodes for CGRP but not calcitonin.

Amylin is found in secretory granules in β cells of the pancreas and is most likely co-secreted with insulin; it is derived from a precursor of 89 amino acids. Very low levels of amylin have also been detected in gastrointestinal tract, lung and dorsal root ganglia.

Reviews

1 Breimer, L.H. *et al.* (1988) Peptide from the calcitonin genes: molecular genetics, structure and function. *Biochem. J.* **255**, 377–390.
2 Nishi, M. *et al.* (1990) Islet amyloid polypeptide. *J. Biol. Chem.* **265**, 4173–4176.
3 Poyner, D.R. (1992) Calcitonin gene-related peptide: multiple actions, multiple receptors. *Pharmacol. Ther.* **56**, 23–53.
4 Yamamoto, A.I. and Tohyama, M. (1989) Calcitonin gene-related peptide in the nervous system. *Prog. Neurobiol.* **33**, 335–386.

CGRP RECEPTORS

CGRP$_1$ and CGRP$_2$ receptors

These are described below. A binding site in certain nuclei of rat brain, e.g. nucleus accumbens, with similar affinities for calcitonin and CGRP has been described [1]. This site has been termed a C_3 receptor although it has no known functional correlate.

Amylin receptor

It is believed that amylin produces its metabolic effects through a high-affinity receptor which has low affinity for CGRP. Evidence has been presented that at least some of the actions of amylin are unrelated to cAMP [2].

References

1 Sexton, P.M. *et al.* (1988) *Neurochem. Int.* **12**, 323–335.
2 Deems, R.O. *et al.* (1991) *Biochem. Biophys. Res. Commun.* **174**, 716–720.

THE CGRP$_1$ AND CGRP$_2$ RECEPTORS

Two subtypes of CGRP receptor have been proposed based on small differences in potency of analogues and fragments of CGRP. CGRP stimulates both receptors in nanomolar concentration while amylin and calcitonin have ~50 and 1000 times lower affinities, respectively.

Distribution

CGRP receptors are distributed widely in the CNS with the highest levels in cerebellum and dorsal spinal cord. In the periphery, CGRP receptors are distributed throughout the cardiovascular system, in secretory tissues, e.g. adrenals, pituitary, exocrine pancreas, in smooth muscles, e.g. vas deferens, spleen, kidney, and in bone.

CGRP$_1$ receptors are present in rat atria and ileum, and CGRP$_2$ receptors in rat vas deferens and spleen. Both receptors are present in brain. A more complete description of the distribution of each receptor will require the development of selective agents with high affinities.

Pharmacology

Agonists: Shorter fragments of CGRP or the reduced form of the molecule show greatly reduced potency suggesting that receptor recognition requires the full length sequence with an intact disulfide link. [Cys(ACM)^{2,7}]CGRP is selective to the CGRP$_2$ receptor [1].

Antagonists: The CGRP fragments, CGRP$_{8-37}$ and CGRP$_{12-37}$, are ~tenfold selective to the CGRP$_1$ receptor (pA$_2$ 6.0–8.0 at the CGRP$_1$ receptor) [2].

Radioligands: [^{125}I-iodohistidyl^{10}]HumanCGRPα and [^{125}I-iodohistidyl^{10}]ratCGRPα bind to high-affinity (Kd 0.2–2.0 nM) and low-affinity (Kd 5–20 nM) sites.

Predominant effector pathways

Activation of adenylyl cyclase by stimulation of G$_s$.

Amino acid sequence

Neither receptor has been cloned.

References
1 Dennis, T. *et al.* (1989) *J. Pharmacol. Ther.* **251**, 718–725.
2 Dennis, T. *et al.* (1989) *J. Pharmacol. Ther.* **254**, 123–128.

Cannabinoid

The ability of marijuana to activate the cannabinoid receptor provides a molecular explanation for its psychoactive effects and other CNS actions which include hallucinations, memory deficits, altered time and space perception, CNS depression and appetite stimulation. The endogenous ligand at the cannabinoid receptor is not known, although there is evidence that it may be a derivative of arachidonic acid, anandamide [1].

Agonists at the cannabinoid receptor have potential therapeutic use as analgesics and anti-emetic agents.

Anandamide

Reference
1 Devane, W.A. (1992) *Science* **258**, 146–1949.

THE CANNABINOID RECEPTOR

Anandamide induces activation in nanomolar concentrations.

Distribution

The cannabinoid receptor is widespread throughout the CNS with high levels in dendate gyrus, hippocampus and cerebral cortex and more moderate levels in hypothalamus and amygdala. The receptor is also present in a variety of neural cell lines, e.g. NG108-15. In the periphery, it is present in testis and in vas deferens.

Pharmacology

Agonists: (–)-Δ^9Tetrahydrocannabinol, levonantrado, cannibinol, CP 55940, nabilone and WIN 55212-2 are active in high nanomolar concentrations.

Antagonists: No antagonist has been described.

Radioligands: [^{3}H]CP 55940 (K_d 1 nM), [^{3}H]WIN 55212 (K_d 2 nM) and [^{3}H]HU-243 (K_d 0.4 nM) bind to a single site.

Structures:

Δ^9Tetrahydrocannabinol

Effector pathway

Inhibition of adenylyl cyclase through a pertussis-toxin-sensitive G-protein most likely of the G_i/G_o class.

Amino acid sequence [1]

```
  1  MKSILDGLAD TTFRTITTDL LYVGSNDIQY EDIKGDMASK LGYFPQKFPL

 51  TSFRGSPFQE KMTAGDNPQL VPADQVNITE FYNKSLSSFK ENEENIQCGE
                      TM 1
101  NFMDIECFMV LNPSQQLAIA VLSLTLGTFT VLENLLVLCV ILHSRSLRCR
          TM 2                                      TM 3
151  PSYHFIGSLA VADLLGSVIF VYSFIDFHVF HRKDSRNVFL FKLGGVTASF
                                                     TM 4
201  TASVGSLFLT AIDRYISIHR PLAYKRIVTR PKAVVAFCLM WTIAIVIAVL
                                 TM 5
251  PLLGWNCEKL QSVCSDIFPH IDETYLMFWI GVTSVLLLFI VYAYMYILWK
                                                        TM 6
301  AHSHAVRMIQ RGTQKSIIIH TSEDGKVQVT RPDQARMDIR LAKTLVLILV
                                         TM 7
351  VLIICWGPLL AIMVYDVFGK MNKLIKTVFA FCSMLCLLNS TVNPIIYALR

401  SKDLRHAFRS MFPSCEGTAQ PLDNSMGDSD CLHKHANNAA SVHRAAESCI

451  KSTVKIAKVT MSVSTDTSAE AL
```

Amino acids	472
Molecular weight	52 857
Glycosylation	Asn78*, Asn84*, Asn113*
Disulfide bonds	None
Palmitoylation	Cys415*, Cys431*, Cys449*

The first 116 amino acids are identical in rat and human receptors.

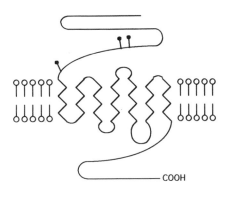

COOH

Database accession numbers

	PIR	SWISSPROT	EMBL/GENBANK	REFERENCES
Human	S13668	P21554	X54937	1
Rat	A33117	P202722	X55812	2

References

[1] Gerard, C. *et al.* (1990) *Biochem. J.* **279**, 129–134.
[2] Matsuda, L.A. *et al.* (1990) *Nature* **346**, 561–564.

Chemokines

INTRODUCTION

Chemokines are proteins with important physiological and pathophysiological roles in a wide range of acute and chronic inflammatory processes [1,2]. They share sequence homology, characterized by the conservation of a four cysteine motif, and the family can be divided according to whether the first two cysteines are adjacent, the C–C family, or separated by an intervening residue, the C–x–C family. In general, C–x–C chemokines are potent chemoattractants for neutrophils but not for monocytes while, in contrast, C–C chemokines are chemoattractant for monocytes but not for neutrophils. Certain chemokines are also chemoattractant for lymphocytes and induce activation of other blood leukocytes.

Many of these substances participate in pathophysiological conditions, e.g. anaphylactoid and septic shock.

Structure (human)

IL-8   AVLPRSAKEL RCQCIKTYSK PFHPKFIKEL RVIESGPHCA
            NTEIIVKLSD GRELCLDPKE NWVQRVVEKF LKRAENS

The C–x–C chemokine family includes interleukin 8 (IL-8), melanoma growth-stimulatory activity (MGSA or Groα) and neutrophil activating peptide-2 (NAP-2). Human IL-8 is produced by a variety of cell types including monocytes, alveolar macrophages, endothelial cells, fibroblasts, epithelial cells and hepatoma cells; it is composed of 77 amino acids in endothelial cells but the first five amino acids are absent in the other cell types. IL-8 has also been known as monocyte-derived neutrophil chemotactic factor (MDNCF), neutrophil activating protein (NAP-1), monocyte-derived neutrophil activating peptide (MONAP) or granulocyte chemotactic protein (GCP).

The C–C family includes human monocyte chemotactic protein-1 (MCP-1), RANTES and macrophage inflammatory proteins (MIP-1α, MIP-1β).

Reviews
1 Baggliolini, M. *et al.* (1989) Neutrophil-activating peptide-1/interleukin8, a novel cytokine that activates neutrophils. *J. Clin. Invest.* **84**, 1045–1049.
2 Oppenheim, J.J. *et al.* (1991) Properties of the novel proinflammatory supergene "intercrine" cytokine family. *Annu. Rev. Immunol.* **9**, 617–648.

CHEMOKINE RECEPTORS

There is little or no cross-competition between members of the C–C or C–x–C families for their individual receptors.

C–C chemokine receptors

One receptor has been identified in cloning studies; it is a G-protein-coupled receptor and is described below.

C–x–C chemokine receptors

Two receptors have been identified in cloning studies and named IL8R$_A$ and IL8R$_B$. Both receptors are G-protein coupled and are described below.

THE C–C CHEMOKINE RECEPTOR

C-C chemokines induce activation in low nanomolar concentrations; C–x–C chemokines are inactive.

Distribution

It is found in monocytes, lymphocytes, basophils and eosinophils. mRNA is also found in premonocytic cell lines, e.g. U937 and HL60 cells.

Pharmacology

Agonists: MCP-1 and MIP-1α induce activation in low nanomolar concentrations and are highly selective relative to C–x–C chemokine receptors.

Antagonists: None have been described.

Radioligands: [^{125}I]MIP-1α (K_d 5 nM) binds to a single site in monocytes and in cells expressing the recombinant C–C receptor.

Predominant effector pathways

Mobilization of Ca^{2+} has been demonstrated in monocytes and in cells expressing the recombinant C–C receptor through an uncharacterized G-protein; pertussis toxin inhibits several of its actions.

Amino acid sequence [1]

```
                                        TM 1
  1 METPNTTEDY DTTTEFDYGD ATPCQKVNER AFGAQLLPPL YSLVFVIGLV
                         TM 2
 51 GNILVVLVLV QYKRLKNMTS IYLLNLAISD LLFLFTLPFW IDYKLKDDWV
         TM 3                                        TM 4
101 FGDAMCKILS GFYYTGLYSE IFFIILLTID RYLAIVHAVF ALRARTVTFG

151 VITSIIIWAL AILASMPGLY FSKTQWEFTH HTCSLHFPHE SLREWKLFQA
         TM 5                                    TM 6
201 LKLNLFGLVL PLLVMIICYT GIIKILLRRP NEKKSKAVRL IFVIMIIFFL
                                        TM 7
251 FWTPYNLTIL ISVFQDFLFT HECEQSRHLD LAVQVTEVIA YTHCCVNPVI

301 YAFVGERFRK YLRQLFHRRV AVHLVKWLPF LSVDRLERVS STSPSTGEHE

351 LSAGF
```

Amino acids	355
Molecular weight	41 172
Glycosylation	Asn5*
Disulfide bonds	Cys107–Cys183*
Palmitoylation	None
Phosphorylation (PKC)	None

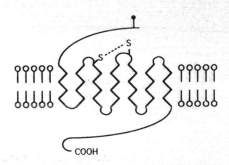

Database accession number

	PIR	SWISSPROT	EMBL/GENBANK	REFERENCE
Human			L09230	1

Gene

The relatively simple hybridization pattern observed in Southern blots suggests that the gene may be intronless. Low stringency washing conditions suggest that there may be a second gene [1].

Comment

C–C chemokines also bind in nanomolar concentrations to a viral gene product, CMV US28, which shares homology with the C–C receptor suggesting a role for C–C chemokines in viral immunity.

References

[1] Neote, K. et al. (1993) Cell **72**, 415–425.

THE IL8R$_A$ RECEPTOR

The IL8R$_A$ receptor shares ~77% sequence homology with the IL8R$_B$ receptor. It was previously named the high-affinity receptor [1], although IL-8 is now believed to have similar affinity at both receptors inducing activation in low nanomolar concentrations [2]. It has also been named IL8R$_B$ [3].

Distribution

It is found in high density in neutrophils, monocytes, basophils, melanoma cells and in related cell lines e.g. HL60, and in lower density in T cells. There are similar levels of IL8R$_A$ and IL8R$_B$ receptors on neutrophils [2].

Pharmacology

Agonists: IL-8 induces activation in nanomolar concentrations; MGSA and NAP-2 are active at micromolar concentrations.

Antagonists: None have been described.

Radioligands: [^{125}I]IL-8 (K_d 1–4 nM) binds to a single site but is nonselective.

Predominant effector pathways

IL-8 has been reported to stimulate the phosphoinositide pathway through an uncharacterized G-protein; pertussis toxin also inhibits several of its actions.

Amino acid sequence [1]

```
                                               TM 1
  1 MSNITDPQMW DFDDLNFTGM PPADEDYSPC MLETETLNKY VVIIAYALVF
                                    TM 2
 51 LLSLLGNSLV MLVILYSRVG RSVTDVYLLN LALADLLFAL TLPIWAASKV
                TM 3
101 NGWIFGTFLC KVVSLLKEVN FYSGILLLAC ISVDRYLAIV HATRTLTQKR
         TM 4
151 HLVKFVCLGC WGLSMNLSLP FFLFRQAYHP NNSSPVCYEV LGNDTAKWRM
         TM 5                                   TM 6
201 VLRILPHTFG FIVPLFVMLF CYGFTLRTLF KAHMGQKHRA MRVIFAVVLI
                                                TM 7
251 FLLCWLPYNL VLLADTLMRT QVIQETCERR NNIGRALDAT EILGFLHSCL

301 NPIIYAFIGQ NFRHGFLKIL AMHGLVSKEF LARHRVTSYT SSSVNVSSNL
```

Amino acids	350
Molecular weight (polypeptide)	39 806
Molecular weight (protein)	58–67 kDa
Chromosome	2
Glycosylation	Asn3*, Asn16*, Asn181*, 182*, Asn193*
Disulfide bond	Cys110–Cys187*
Palmitoylation	None

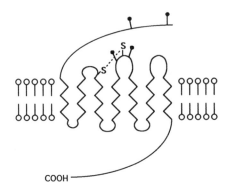

Database accession numbers

	PIR	SWISSPROT	EMBL/GENBANK	REFERENCES
Human IL8R$_A$			M73969	1,2
			M68932	
Rabbit IL8R$_A$			J05705	4
			M58021	

Gene
The coding sequence is contained in a single exon [3].

References
[1] Holmes, W.E. et al. (1991) Science **253**, 1278–1280.
[2] Lee, J. et al. (1992) J. Biol. Chem. **267**, 16283–16287.
[3] Ahuja, S.K. et al. (1992) Nature Genet. **2**, 31–36.
[4] Thomas, K.M. et al. (1990) J. Biol. Chem. **265**, 20061-20064; correction ibid. **267**, 13780.

THE IL8R$_B$ RECEPTOR

The IL8R$_B$ receptor shares ~77% sequence homology with the IL8R$_A$ receptor. It was previously named the low-affinity receptor [1], although IL-8 is now believed to have similar affinity at both receptors inducing activation in low nanomolar concentrations [2]. It has also been named IL8R$_A$ [3].

Distribution

It is found in high density in neutrophils, monocytes, basophils, melanoma cells and in related cell lines, e.g. HL60, and in lower density in T cells. There are similar levels of IL8R$_A$ and IL8R$_B$ receptors on neutrophils [2].

Pharmacology

Agonists: MGSA and NAP-2 have a selectivity of more than 100-fold relative to the IL8R$_A$ receptor [2]; they are about equipotent with IL-8.

Antagonists: None have been described.

Radioligands: [^{125}I]MGSA (K_d 1 nM) is selective. [^{125}I]IL-8 (K_d 1–4 nM) binds to a single site but is nonselective.

Predominant effector pathways
IL-8 has been reported to stimulate the phosphoinositide pathway through an uncharacterized G-protein; pertussis toxin also inhibits several of its actions.

Amino acid sequence [1]

```
                                                      TM 1
  1 MEDFNMESDS FEDFWKGEDL SNYSYSSTLP PFLLDAAPCE PESLEINKYF
                                     TM 2
 51 VVIIYALVFL LSLLGNSLVM LVILYSRVGR SVTDVYLLNL ALADLLFALT
```

```
                                 TM 3
101 LPIWAASKVN GWIFGTFLCK VVSLLKEVNF YSGILLLACI SVDRYLAIVH
              TM 4
151 ATRTLTQKR YLVKFICLSIW GLSLLLALPV LLFRRTVYSS NVSPACYEDM
          TM 5
201 GNNTANWRML LRILPQSFGF IVPLLIMLFC YGFTLRTLFK AHMGQKHRAM
       TM 6                                        TM 7
251 RVIFAVVLIF LLCWLPYNLV LLADTLMRTQ VIQETCERRN HIDRALDATE

301 ILGILHSCLN PLIYAFIGQK FRHGLLKILA IHGLISKDSL PKDSRPSFVG

351 SSSGHTSTTL
```

Amino acids	360
Molecular weight	40 123
Chromosome	2
Glycosylation	Asn22*, Asn191*, Asn202*
Disulfide bonds	Cys119–Cys196*
Palmitoylation	None

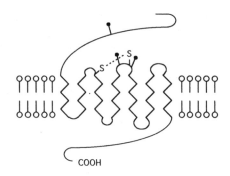

Database accession numbers

	PIR	SWISSPROT	EMBL/GENBANK	REFERENCES
Human IL8R_B	A39445	P25024	M73969	1,2

Gene
The coding sequence is contained in a single exon [3].

Comment
A pseudogene has been identified, IL8RAP (EMBL/GenBank M98335) [3].

References
[1] Murphy, P.M. et al. (1991) Science 253, 1280–1283.
[2] Lee, J. et al. (1992) J. Biol. Chem. 267, 16283–16287.
[3] Ahuja, S.K. et al. (1992) Nature Genet. 2, 31–36.

Cholecystokinin (CCK) and gastrin

CCKs and gastrins are naturally occurring peptides which share a common C-terminal amino acid sequence, Gly–Trp–Met–Asp–Phe–NH$_2$ [1-3]. Full biological activity resides in this region.

The major physiological actions of CCK in the periphery include gall bladder contraction, pancreatic enzyme secretion and regulation of secretion/absorption in the gastrointestinal tract. In the CNS, CCK induces analgesia, satiety and a decrease in exploratory behaviour. CCK coexists with dopamine in mesolimbic and mesocortical neurons and has been claimed to be anxiolytic.

The major physiological role of gastrin is to stimulate acid secretion in the stomach. In addition, gastrin has trophic effects on the gastric mucosa.

Structures

CCK-8   Asp–Tyr(SO$_3$H)–Met–Gly–Trp–Met–Asp–Phe–NH$_2$

CCK-33   Lys–Ala–Pro–Ser–Gly–Arg–Val–Ser–Met–Ile–Lys–Asn–Leu–Gln–Ser–Leu–Asp–Pro
         Ser–His–Arg–Ile–Ser–Asp–Arg–Asp–Tyr(SO$_3$H)–Gly–Met–Gly–Trp–Met–Asp–Phe–NH$_2$

G-34   Pyr–Leu–Gly–Pro–Gln–Gly–Pro–Pro–His–Leu–Val–Ala–Asp–Pro–Ser–Lys–Lys–Gln–
         Pro–Gly–Trp–Leu–Glu–Glu–Glu–Glu–Glu–Ala–Tyr*–Gly–Trp–Met–Asp–Phe–NH$_2$

The major forms of CCK are CCK-33, CCK-8 and CCK-4 (or G-4); CCK-8 is the predominant species. Other identified molecular forms include CCK-58, CCK-39 and CCK-22. There is species variation in the amino acid sequence of CCK. Caerulein is an octapeptide from frog skin which shares seven of the last eight C-terminal amino acids in CCK.

There are approximately equal amounts of sulfated and nonsulfated forms of gastrin. The major forms are G-34 (big gastrin), G-17 (little gastrin), G-14 (minigastrin) and G-4 (or CCK-4).

Distribution and synthesis

CCK is produced from a single amino acid precursor of 115 amino acid (human) which encodes for a single copy of CCK-33. CCK is present throughout the digestive tract with high concentrations in duodenum and jejunum. It is also found in peripheral nerves to other smooth muscles, e.g. bladder and uterus, and to secretory glands, e.g. exocrine pancreas. It is one of the most abundant peptides in brain with a widespread distribution; it is found in high levels in cerebral cortex but is absent from cerebellum and pineal body.

Gastrin is produced from a single gene transcript and is found predominantly in stomach and intestine, but is also present in vagal nerves. It is absent from brain.

Reviews

1 Jensen, R. *et al.* (1989) Interaction of CCK with pancreatic acinar cells. *Trends Pharmacol. Sci.* **10**, 418–423.
2 Lindefors, N. *et al.* (1993) CCK peptides and mRNA in the human brain. *Prog. Neurobiol.* **40**, 671–691.
3 Woodruff, G. and Hughes, J. (1991) Cholecystokinin antagonists. *Annu. Rev. Pharmacol. Toxicol.* **31**, 469–501.

THE CCK$_A$ RECEPTOR

Sulfated forms of CCK induce activation in nanomolar concentrations and are more potent than nonsulfated forms. In contrast, sulfated and nonsulfated forms of gastrin have similar potencies. The receptor has been called the CCK$_1$ receptor.

Distribution

The highest levels of the CCK$_A$ receptor are found in peripheral tissues, notably pancreas, stomach, intestine and gall bladder. It has a limited distribution in brain, which includes the area postrema, nucleus tractus solitaris and interpeduncular nucleus.

The receptor has been implicated in the pathogenesis of schizophrenia, Parkinson's disease, drug addiction and feeding disorders.

Pharmacology

Agonists: The CCK-4 analogue, A-71623, has a selectivity of greater than 100 times and is active in low nanomolar concentrations [1]. The receptor can also be identified by the higher potency (10–100 fold) of sulfated CCK-8 relative to desulfated CCK-8 or CCK-4.

Antagonists: Devazepide [2] (pA$_2$ 9.8; previously named L-364718), lorglumide [3] (pA$_2$ 7.2) and 2-NAP[4] (pA$_2$ = 7.0) are competitive antagonists with a selectivity of > 100-fold.

Radioligands: [^{3}H]Devazepide (K_d 0.2 nM) is highly selective; [^{125}I]BH-CCK-8 (K_d 0.1–1.0 nM) is nonselective unless used in the presence of an excess of a CCK$_B$-selective ligand.

Structures:

Lorglumide   2-NAP

A-71623

Devazepide

Predominant effector pathways

Activation of the phosphoinositide pathway through a pertussis-toxin-insensitive G-protein most likely of the G_q/G_{11} class.

Amino acid sequence (rat) [5]

```
  1 MSHSPARQHL VESSRMDVVD SLLMNGSNIT PPCELGLENE TLFCLDQPQP
           TM 1                                        TM 2
 51 SKEWQSALQI LLYSIIFLLS VLGNTLVITV LIRNKRMRTV TNIFLLSLAV
                                          TM 3
101 SDLMLCLFCM PFNLIPNLLK DFIFGSAVCK TTTYFMGTSV SVSTFNLVAI
                                 TM 4
151 SLERYGAICR PLQSRVWQTK SHALKVIAAT WCLSFTIMTP YPIYSNLVPF
                                 TM 5
201 TKNNNQTANM CRFLLPSDAM QQSWQTFLLL ILFLLPGIVM VVAYGLISLE

251 LYQGIKFDAS QKKSAKEKKP STGSSTRYED SDGCYLQKSR PPRKLELQQL
                                          TM 6
301 SSGSGGGSRLN RIRSSSSAAN LIAKKRVIRM LIVIVVLFFL CWMPIFSANA
          TM 7
351 WRAYDTVSAE KHLSGTPISF ILLLSYTSSC VNPIIYCFMN KRFRLGFMAT

401 FPCCPNPGPP GVRGEVGEEE DGRTIRALLS RYSYSHMSTS APPP
```

Amino acids	444
Molecular weight (polypeptide)	49 657
Molecular weight (actual)	85–95 kDa
Glycosylation	Asn25*. Asn28*, Asn39*, Asn206*
Disulfide bonds	Cys129–Cys211*
Palmitoylation	Cys403*, Cys404*
Phosphorylation (PKC)	Ser260*, Ser264*, Ser275*, Thr424*

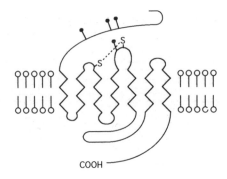

Database accession numbers

	PIR	SWISSPROT	EMBL/GENBANK	REFERENCE
Rat			M88096	5

Gene

Genomic organization is unknown.

References

1 Lin, C.W. *et al*, (1991) *Mol. Pharmacol.* **39**, 346–351.
2 Chang, R.S.L. and Lotti, V.J. (1986) *Proc. Natl Acad. Sci. USA* **83**, 4923–4926.
3 Makovec, F. *et al.* (1985) *Arzneimittelforsch* **35**, 1048–1051.
4 Hull, R.A.D. *et al.* (1993) *Br. J. Pharmacol.* **108**, 734–740.
5 Wank, S.A. *et al.* (1992) *Proc. Natl Acad. Sci. USA* **89**, 3125–3129.

THE CCK$_B$ RECEPTOR

Sulfated and desulfated forms of CCK have similar potencies and induce activation in nanomolar concentrations. Shorter C-terminal fragments, e.g. CCK-4, have slightly lower potencies. The receptor is also known as the gastrin receptor and was previously called the CCK$_2$ receptor.

Distribution

The receptor has a widespread distribution in the CNS and has been implicated in the pathogenesis of panic-anxiety attacks caused by CCK-related peptides. It has a more limited distribution in the periphery where it is found in smooth muscle, e.g. oesphageal sphincter, gall bladder, and secretory glands, e.g. pancreas.

Pharmacology

Agonists: pBC 264 [1] has a selectivity of more than two orders of magnitude; desulfated CCK-8 and CCK-5 (pentagastrin) have similar potencies to CCK-8 and are weakly selective.

Antagonists: The nonpeptide antagonists CI 988 (pA$_2$ 8.1) [2], L-365260 (pA$_2$ 8.1) [3], LY 262691 (K_i 30 nM) have selectivities of more than three orders of magnitude.

Radioligands: [³H]L-365260 (K_d 2 nM) and [³H]pBC 264 (K_d 0.2 nM) *4* are highly selective; [³H] or [¹²⁵I]gastrin are less selective. [³H]CCK-8 (K_d 0.2 nM) is nonselective.

Structures:

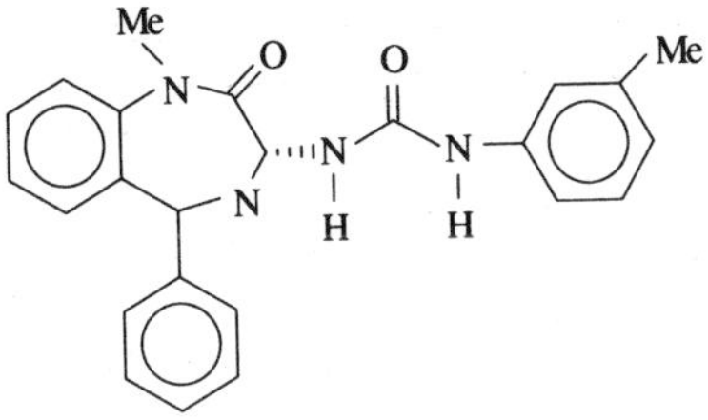

L-365260

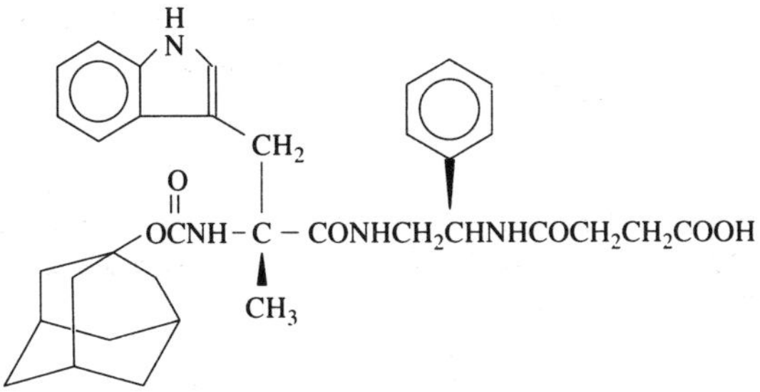

CI 988

pBC 264: Propionyl–Tyr(SO₃H)–gNle–Gly–Trp–(NMe)–Nle–Asp–Phe–NH₂

Predominant effector pathways

Activation of phosphoinositide pathway through a pertussis-toxin-insensitive G-protein most likely of the G_q/G_{11} class.

Amino acid sequence [5]

```
  1 MELLKLNRSV QGTGPGPGAS LCRPGAPLLN SSSVGNLSCE PPRIRGAGTR
      TM 1                                        TM 2
 51 ELELAIRITL YAVIFLMSVG GNMLIIVVLG LSRRLRTVTN AFLLSLAVSD
                                     TM 3
101 LLLAVACMPF TLLPNLMGTF IFGTVICKAV SYLMGVSVSV STLSLVAIAL
                         TM 4
151 ERYSAICRPL QARVWQTRSH AARVIVATWL LSGLLMVPYP VYTVVQPVGP
                         TM 5
201 RVLQCVHRWP SARVRQTWSV LLLLLLFFIP GVVMAVAYGL ISRELYLGLR

251 FDGDSDSDSQ SRVRNQGGLP GAVHQNGRCR PETGAVGEDS DGCYVQLPRS
```

```
                                              TM 6
301 RPALELTALT APGPGSGSRP TQAKLLAKKR VVRMLLVIVV LFFLCWLPVY
                           TM 7
351 SANTWRAFDG PGAHRALSGA PISFIHLLSY ASACVNPLVY CFMHRRFRQA

401 CLETCARCCP RPPRARPRAL PDEDPPTPSI ASLSRLSYTT ISTLGPG
```

Amino acids	447
Molecular weight (polypeptide)	48 419
Molecular weight (protein)	75 kDa
Chromosome	11
Glycosylation	Asn7*, Asn30*, Asn36*
Disulfide bonds	Cys127–Cys205*
Palmitoylation	Cys408*
Phosphorylation (PKC)	Ser82*, Ser300*
Phosphorylation (PKA)	Ser154*, Thr321*, Ser442*

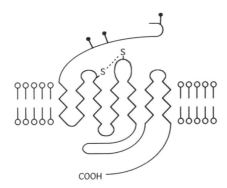

COOH

Database accession numbers

	PIR	SWISSPROT	EMBL/GENBANK	REFERENCES
Canine			M87834	6
Human			L04473	5
Rat				7

Gene

The genomic organization is not known.

Comment

Canine and human CCK$_B$ receptors can be distinguished by the affinity of two nonpeptide antagonists, L-365260 and L-364718. Replacement of Val 319 in the sixth transmembrane region of the human receptor with the corresponding amino acid in the canine receptor, Leu 355, alters their affinity to the values seen for the canine receptor [8]. Complementary results are obtained when Leu 355 in the canine receptor is replaced with valine. These results demonstrate that the differences in antagonist affinity reflects a species difference.

References

1 Charpentier, B. *et al.* (1988) *Peptides* **9**, 835–841.
2 Hughes, J. *et al.* (1988) *Proc. Natl Acad. Sci. USA* **87**, 6728–6732.
3 Bock, M.G. *et al.* (1989) *J. Med. Chem.* **32**, 13–16.
4 Durieux, C. *et al.* (1992) *Mol. Pharmacol.* **41**, 1089–1095.
5 Pisegna, J.R.. *et al.* (1992) *Biochem. Biophys. Res. Commun.* **189**, 296–303.
6 Kopin, A.S. *et al.* (1992) *Proc. Natl Acad. Sci. USA* **89**, 3605–3609.
7 Wank, S.A. *et al.* (1992) *Proc. Natl Acad. Sci. USA* **89**, 8691–8695.
8 Beinborn, M. *et al.* (1993) *Nature* **362**, 348–350.

Dopamine

INTRODUCTION

Dopamine is an intermediate in the biosynthetic pathway of noradrenaline and adrenaline, and is a major neurotransmitter in its own right in the CNS and in the peripheral nervous system [1-9]. It has an essential role in a number of physiological functions and pathophysiological states.

Dopamine

In the CNS, dopamine is involved in the regulation of movement. Activation of the nigrostriatal pathway, or stimulation of dopamine receptors in the striatum, leads to "stereotyped" behaviour in rodents. A lesion in the nigrostriatal pathway in humans underlies Parkinson's disease, a motor disorder associated with rigidity, tremor and hypokinesia, while Huntington's disease, a syndrome associated with severe voluntary movements, is also a disorder of striatonigral degeneration and is linked with excess dopamine transmission. Dopamine is also a neurotransmitter in the mesolimbic system where it is involved in cognitive and emotional responses, and may be associated with schizophrenia, although there is little evidence for an increase in dopamine transmission in this disorder. Dopamine is also believed to have physiological roles in inhibiting prolactin secretion in the anterior pituitary and in initiating nausea and vomiting through an action in the medulla. In the periphery, dopamine mediates vasodilation in the kidney and increased contractility in myocardium.

Clinically, L-dopa is the agent of choice in Parkinson's disease and is believed to mediate its action by conversion to dopamine; several agents which mimic, release or enhance the action of dopamine are also effective, e.g. bromocriptine. Dopamine antagonists, e.g. chlorpromazine, are the agents of choice in the treatment of schizophrenia and are used in the early stages of treatment of Huntington's disease; several are also anti-emetic. Many of the actions of amphetamine are mediated through the dopamine system.

Distribution and synthesis

Dopamine has a much more restricted distribution in brain relative to noradrenaline. Approximately 75% of dopamine in the brain is found in the nigrostriatal pathway, the axons of which terminate in the striatum. Dopamine is also found in cell bodies which originate in the midbrain and project to parts of the limbic system, especially the nucleus accumbens. It is present in a group of short neurons from the hypothalamus to the median eminence and pituitary gland; local dopaminergic interneurons are also found in the olfactory bulb and in the retina. Dopamine is present in some sympathetic neurons in the periphery, notably in kidney.

Dopamine is synthesized through the same route as noradrenaline, namely conversion of tyrosine to dopa followed by the action of dopa decarboxylase. Dopamine neurons lack dopamine β-hydroxylase and therefore do not produce noradrenaline.

Books
[1] Creese, I. *et al.* (eds) (1987) Dopamine Receptors, Liss, New York.

Reviews
[2] Andersen, P.H. *et al.* (1990) Dopamine receptor subtypes: beyond the D_1/D_2 classification. *Trends Pharmacol. Sci.* **11**, 231–236.
[3] Grandy, D.K. and Civelli, O. (1992) G protein-coupled receptors: the new dopamine receptor subtypes. *Current Opinions Neurobiol.* **2**, 275–281.
[4] Horn, A. (1990) Dopamine receptors, In Comprehensive Medicinal Chemistry, vol. 3, Membranes and Receptors (Hansch, C., Sammes, P.G. and Taylor, J.B. eds) Pergamon Press, Oxford, pp. 229–290.
[5] O'Dowd, B.F. (1993) Structures of dopamine receptors. *J. Neurochem.* **60**, 804–816.
[6] Sibley, D.R. and Monsma, F.J. (1992) Molecular biology of dopamine receptors *Trends Pharmacol. Sci.* **13**, 61–69.
[7] Sokoloff, P. *et al.* (1992) The third dopamine receptor (D_3) as a novel target for antipsychotics *Biochem. Pharmacol.* **43**, 659–666.
[8] Strange, P. (1991) Interesting times for dopamine receptors. *Trends Neurosci.* **14**, 43–45.
[9] Vadasz, C. *et al.* (1992) Genetic aspects of dopamine receptor binding in the mouse and rat brain: an overview. *J. Neurochem.* **59**, 793–803.

DOPAMINE RECEPTORS

D_1 and D_5 receptors (D_1-LIKE)
These two receptors are seven transmembrane proteins with close sequence homology and similar pharmacological profiles. They are described below.

D_2, D_3 and D_4 receptors (D_2-LIKE)
These three receptors are seven transmembrane proteins with close sequence homology and similar pharmacological profiles. They are described below.

Common structural features of cloned dopamine receptors
Several amino acids are conserved between the five subtypes of dopamine receptor, some of which are implicated in the binding of dopamine. The conserved aspartate residue in the third transmembrane region may provide the counterion to the protonated amine while the two conserved serine residues in the fifth transmembrane region are thought to form hydrogen bonds with the catechol hydroxyl groups.

THE D$_1$ RECEPTOR

The D$_1$ receptor has a similar pharmacology to the D$_5$ receptor. It is also called the D$_{1A}$ receptor.

Distribution

The absence of ligands which exhibit high selectivity to D$_1$ or D$_5$ receptors prevents a full description of their respective distributions. In general, localization of mRNA for the D$_1$ receptor is in agreement with the distribution of [^{3}H]SCH 23390 binding sites. Highest levels are found in caudate-putamen, nucleus accumbens and olfactory tubercle, with lower levels in frontal cortex, habenula, amygdala, hypothalamus and thalamus. In the periphery, binding sites are found in kidney, liver, heart and parathyroid gland, although mRNA encoding for the D$_1$ receptor has only been found in parathyroid gland.

Pharmacology

Agonists: SKF 38393 (partial agonist) and fenoldopam have a selectivity of more than tenfold at D$_1$ and D$_5$ receptors and induce activation in subnanomolar concentrations.

Antagonists: SCH 23390 (K_d 0.1–0.5 nM) has a selectivity of more than 1000-fold relative to D$_{2\text{-LIKE}}$ receptors; several other selective D$_1$ antagonists have been described, e.g. SKF 83566.

Radioligands: The antagonists [^{3}H]SCH 23390 (K_d 0.4 nM) and [^{125}I]SCH 23982 (K_d 0.5nM), and the agonist [^{3}H]SKF 38393 have high selectivity at D$_1$ and D$_5$ receptors.

Structures:

SKF 38393

Fenodopam

SCH 23390

SKF 83566

Predominant effector pathways

Stimulation of adenylyl cyclase via G_s. Several reports suggest that D_1 receptors are also able to stimulate phosphoinositide metabolism [1].

Amino acid sequence [2-5]

```
                              TM 1
  1 MRTLNTSAMD GTGLVVERDF SVRILTACFL SLLILSTLLG NTLVCAAVIR
               TM 2                                     TM 3
 51 FRHLRSKVTN FFVISLAVSD LLVAVLVMPW KAVAEIAGFW PFGSFCNIWV
                                              TM 4
101 AFDIMCSTAS ILNLCVISVD RYWAISSPFR YERKMTPKAA FILISVAWTL
                                                        TM 5
151 SVLISFIPVQ LSWHKAKPTS PSDGNATSLA ETIDNCDSSL SRTYAISSSV

201 ISFYIPVAIM IVTYTRIYRI AQKQIRRIAA LERAAVHAKN CQTTTGNGKP
                              TM 6
251 VECSQPESSF KMSFKRETKV LKTLSVIMGV FVCCWLPFFI LNCILPFCGS
               TM 7
301 GETQPFCIDS NTFDVFVWFG WANSSLNPII YAFNADFRKA FSTLLGCYRL

351 CPATNNAIET VSINNNGAAM FSSHHEPRGS ISKECNLVYL IPHAVGSSED

401 LKKEEAAGIA RPLEKLSPAL SVILDYDTDV SLEKIQPITQ NGQHPT
```

Amino acids	446
Molecular weight	49 293
Chromosome	5
Glycosylation	Asn5*, Asn175*
Disulfide bonds	Cys96 - Cys186*
Palmitoylation	Cys348*
Phosphorylation (PKC)	Thr186*, Ser259*, Ser263*
Phosphorylation (PKA)	Thr186*, Thr268*

There are two potential initiator methionines in the human gene sequence and three in rat. There is evidence that the first initiator in rat is not functional, and protein structures have therefore been predicted using the first initiator methionine in human and the second in rat.

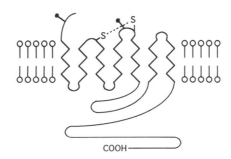

Database accession numbers

	PIR	SWISSPROT	EMBL/GENBANK	REFERENCES
Human		P21728	X58987	2–5
			X59308	
			X55760	
Rat		P18901	M35077	4,5
Rhesus Macaque				6

Gene
The gene is intronless.

References
[1] Mahan, L.C. et al. (1990) Proc. Natl Acad. Sci. USA **87**, 2196–2200.
[2] Dearry, A. et al. (1990) Nature **347**, 72–76.
[3] Sunahara, R.K. et al. (1990) Nature **347**, 80–83.
[4] Zhou, Q.-Y. et al. (1990) Nature **347**, 76-80
[5] Monsma, F.J. et al. (1990) Proc. Natl Acad. Sci. USA **87**, 6723–6727.
[6] Machida, C.A. et al. (1992) Mol. Pharmacol. **41**, 652–659.

THE D_2 RECEPTOR

D_2 receptors have a similar pharmacology to D_3 and D_4 receptors and the lack of highly selective ligands has prevented a complete description of their individual distributions. It is also called the D_{2A} receptor.

Two forms of the D_2 receptor exist, D_{2S} and D_{2L} (also been called D_{2A-S} and D_{2A-L}), which are derived from the same gene by alternative splicing. The long form of the receptor contains an additional 29 amino acids in the third cytoplasmic loop. Both forms are present in all tissues and regions where D_2 receptors are expressed; the long form usually predominates in rat but not in human. Small pharmacological differences exist between the long and short forms.

Distribution

The distribution of mRNA for this receptor is in reasonable agreement with that mapped in radioligand binding studies using D_2-selective ligands. High levels are present in the major dopamine projection areas including the caudate-putamen, nucleus accumbens and olfactory tubercle; it is also found in cell bodies of dopaminergic neurons in substantia nigra and ventral tegmental area. In the periphery, it is found in pituitary, heart and blood vessels.

Pharmacology

Agonists: N-0437 (aminotetralin) and bromocriptine have a selectivity of between 10- and 100-fold relative to D_{1-LIKE} receptors and induce activation in subnanomolar concentrations.

Antagonists: (–)-Sulpiride (K_d 0.1 nM), YM 09151-2 (K_d 0.1 nM) and domperidone (K_d 1 nM) have a selectivity of more than three orders of magnitude to D_{2-LIKE} receptors. Many other, less selective, antagonists have also been described.

Radioligands: Several D$_{2\text{-LIKE}}$ selective radioligands have been described including the antagonists [^{3}H]YM 09151-2 (K_d 0.1 nM), [^{3}H]domperidone (K_d 1 nM) and [^{125}I]iodosulpiride (K_d 0.5 nM); the agonist [^{3}H]N-0347 is also selective. Spiperone (K_d 0.1 nM) is the most commonly used ligand, but also binds to some 5-HT receptor subtypes and α_1-adrenoceptors.

Structures:

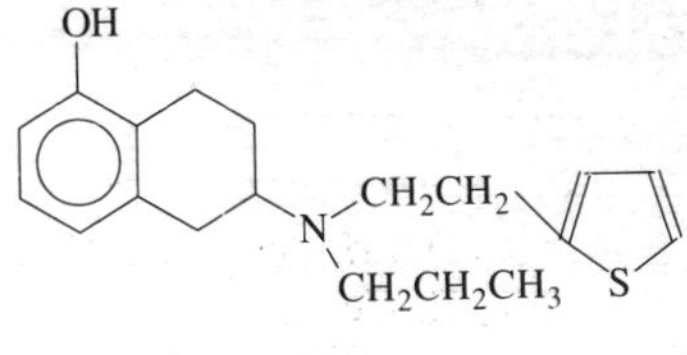

N-0437

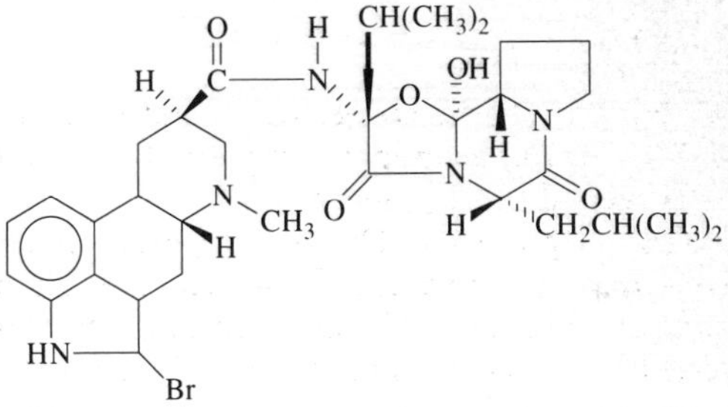

Bromocriptine

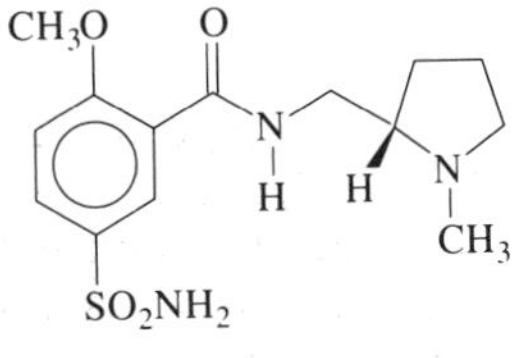

(–)-Sulpiride

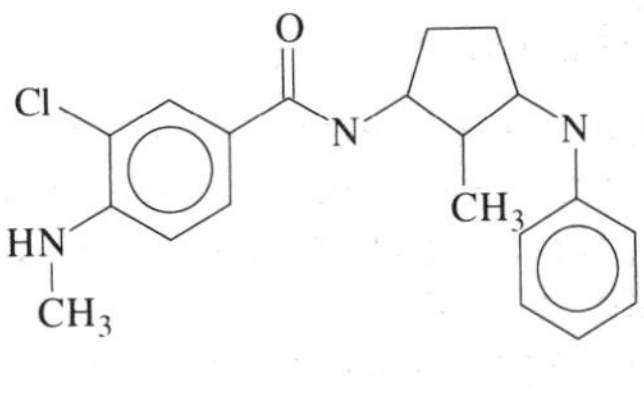

YM 09151-2

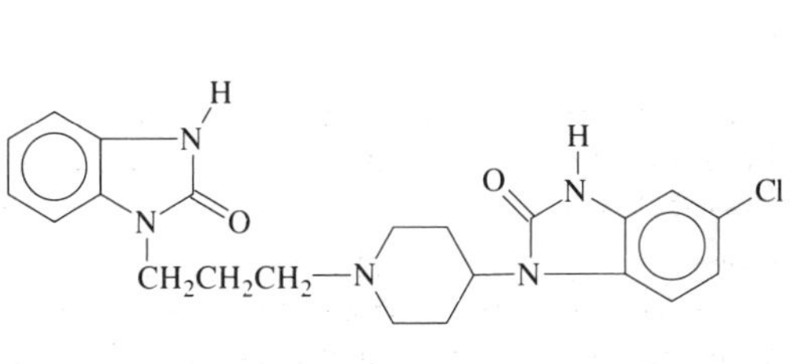

Domperidone

Spiperone

Predominant effector pathways

Inhibition of adenylyl cyclase and Ca^{2+} channels, and activation of K^+ channels through a pertussis-toxin-sensitive G-protein most likely of the G_i/G_o class. Activation and inhibition of phosphoinositide metabolism, and potentiation of arachidonic acid release has been described but it is not established whether these are direct actions.

Amino acid sequence [1-4]

```
                                                TM 1
  1 MDPLNLSWYD DDLERQNWSR PFNGSDGKAD RPHYNYYATL LTLLIAVIVF
                          TM 2
 51 GNVLVCMAVS REKALQTTTN YLIVSLAVAD LLVATLVMPW VVYLEVVGEW
            TM 3
101 KFSRIHCDIF VTLDVMMCTA SILNLCAISI DRYTAVAMPM LYNTRYSSKR
    TM 4                                       TM 5
151 RVTVMISIVW VLSFTISCPL LFGLNNADQN ECIIANPAFV VYSSIVSFYV

201 PFIVTLLVYI KIYIVLRRRR KRVNTKRSSR AFRAHLRAPL KGNCTHPEDM
                                                         ^
251 KLCTVIMKSN GSFPVNRRRV EAARRAQELE MEMLSSTSPP ERTRYSPIPP
                     ^
301 SHHQLTLPDP SHHGLHSTPD SPAKPEKNGH AKDHPKIAKI FEIQTMPNGK
                                     TM 6
351 TRTSLKTMSR RKLSQQKEKK ATQMLAIVLG VFIICWLPFF ITHILNIHCD
            TM 7
401 CNIPPVLYSA FTWLGYVNSA VNPIIYTTFN IEFRKAFLKI LHC
```

The alternative splice sequence (amino acids 187–270) is shown (^).

Amino acids (long)	443
Amino acids (short)	414
Molecular weight (long)	50 619
Chromosome	11
Glycosylation	Asn5*, Asn17*, Asn23*
Disulfide bonds	Cys107–Cys182*
Palmitoylation	None

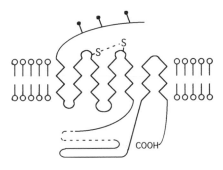

The dashed region in the third intracellular loop is absent in the D$_{2S}$ receptor.

Database accession numbers

	PIR	SWISSPROT	EMBL/GENBANK	REFERENCES
Bovine	S08163	P20288		5
Human	A33392	P14416	M30625	1–4
	A34502		M29066	
	S08417		X51646	
Mouse			X55674	6
Rat	A34046;	P13953	X14028	5, 7–10,
	A34555		X55674	11, 12
	S07791;		X17458	
	S08145			
	S01846			
	S09097			
	S09040			
	S13921			
Xenopus	S14827	P24628		13

Gene (human)

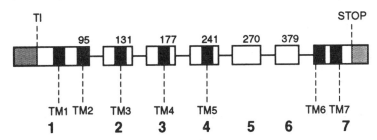

One exon (5) encodes for the 29 amino acid insert sequence present in the D$_{2L}$ receptor. The presence of a large intron(s) in the 5' flanking region of the gene has precluded its further characterization.

Comment

A splice variant in the 5' noncoding region has been described (EMBL/Genbank: X51362)[14].

References

[1] Grandy, D.K. *et al.* (1989) *Proc. Natl Acad. Sci. USA* **86**, 9762–9766.
[2] Dal Toso, R. *et al.* (1989) *EMBO J.* **8**, 4025–4034.
[3] Selbie, L.A. *et al.* (1989) *DNA* **8**, 683–689.
[4] Stormann, T.M. *et al.* (1990) *Mol. Pharmacol.* **37**, 1–6.
[5] Chio, C.L. *et al.* (1990) *Nature* **343**, 266–269.
[6] Montmayer, J.P. *et al.* (1991) *FEBS Lett.* **278**, 239–243.
[7] Bunzow, J.R. *et al.* (1988) *Nature* **336**, 783–787.
[8] Giros, B. *et al.* (1989) *Nature* **342**, 923–926.
[9] Monsma, F.J. *et al.* (1989) *Nature* **342**, 926–929.
[10] Eidne, K.A. *et al.* (1989) *Nature* **342**, 865.
[11] Miller, J.C. *et al.* (1990) *Biochem. Biophys. Res. Commun.* **166**, 109–112.
[12] O'Dowd, B.F. *et al.* (1990) *FEBS Lett.* **262**, 8–12.
[13] Martens, G.J.M. *et al.* (1991) *FEBS Lett.* **281**, 85–89.
[14] Robakis, N.K. *et al.* (1990) *Nucleic Acids Res.* **18**, 1299.

THE D3 RECEPTOR

The D_3 receptor has a similar pharmacology to the D_2 receptor and the absence of selective ligands has prevented a full description of its distribution. It has also been called the D_{2B} receptor.

Distribution

mRNA encoding for D_3 receptor is present at one to two orders of magnitude lower than that for the D_2 receptor. It is expressed predominantly in the limbic area, including the olfactory tubercle, nucleus accumbens, islands of Calleja and hypothalamus, and is present in lower levels in caudate-putamen and cerebral cortex. It is also present in dopamine cell bodies in the substantia nigra.

The distribution of the D_3 receptor is consistent with a role in cognition and emotional functions, and it may therefore be the target of antipsychotic therapy involving dopamine antagonists.

Pharmacology

Agonists: Quinpirole is ~100-fold selective for D_3 relative to D_2 receptor; dopamine is ~20-fold selective.

Antagonists: AJ76 is ~fourfold selective relative to D_2.

Radioligands: See D_2 receptor.

Structures:

AJ76   *cis*-(+)-(1*S*,2*R*)-5methoxy-1-methyl-2-(*n*-propylamino)tetralin

Predominant effector pathways

Studies on the expressed receptor have not observed an effect on cAMP, inositol phosphate turnover, intracellular calcium, arachidonic acid metabolism or K^+ currents [1,2].

Amino acid sequence (rat) [1]

```
                                              TM 1
  1 MAPLSQISTH LNSTCGAENS TGVNRARPHA YYALSYCALI LAIIFGNGLV
                  TM 2
 51 CAAVLRERAL QTTTNYLVVS LAVADLLVAT LVMPWVVYLE VTGGVWNFSR
       TM 3                                               TM 4
101 ICCDVFVTLD VMMCTASILN LCAISIDRYT AVVMPVHYQH GTGQSSCRRV
                                              TM 5
151 ALMITAVWVL AFAVSCPLLF GFNTTGDPSI CSISNPDFVI YSSVVSFYVP

201 FGVTVLVYAR IYIVLRQRQR KRILTRQNSQ CISIRPGFPQ QSSCLRLHPI

251 RQFSIRARFL SDATGQMEHI EDKQYPQKCQ DPLLSHLQPP SPGQTHGGLK

301 RYYSICQDTA LRHPSLEGGA GMSPVERTRN SLSPTMAPKL SLEVRKLSNG
                              TM 6
351 RLSTSLRLGP LQPRGVPLRE KKATQMVVIV LGAFIVCWLP FFLTHVLNTH
              TM 7
401 CQACHVSPEL YRATTWLGYV NSALNPVIYT TFNVEFRKAF LKILSC
```

Amino acids	446
Molecular weight	49 515
Glycosylation	Asn12*, Asn19*, Asn97*, Asn173*
Disulfide bonds	Cys102/Cys103–Cys181*
Palmitoylation	None

The human receptor has 46 fewer amino acids in the third cytoplasmic loop than the rat receptor.

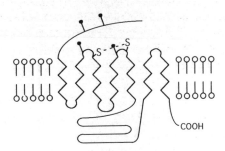

Database accession numbers

	PIR	SWISSPROT	EMBL/GENBANK	REFERENCES
Human				3
Rat		P19020	X53944	1
			M69189	
			M69190	
			M69191	
			M69192	
			M69193	
			M69194	

Gene

The coding region of the gene is interrupted by multiple introns which have a similar distribution to those in the D_2 gene.

Comment

Two nonfunctional splice variants of the rat D_3 receptor have been reported [4,5]. One of these lacks the sequence for the third transmembrane domain and has a shift in the reading frame which has introduced a stop codon such that it would encode for a protein consisting only of the first two transmembrane regions. The second variant lacks most of the second extracellular loop and ~33% of the fifth transmembrane domain; it does not appear to encode for a functional protein.

References

1 Sokoloff, P. *et al.* (1990) *Nature* **347**, 146–151.
2 Seabrook, G.R. *et al.* (1992) *FEBS Lett.* **312**, 123–126.
3 Giros, B. *et al.* (1990) *C.R. Acad. Sci. Paris*, **311**, 501–508.
4 Giros, B. *et al.* (1991) *Biochem. Biophys. Res. Commun.* **176**, 1584–1592.
5 Snyder, L.A. *et al.* (1991) *Biochem. Biophys. Res. Commun.* **180**, 1031–1035.

THE D_4 RECEPTOR

The D_4 receptor has a similar pharmacology to the D_2 receptor; it has also been called the D_{2C} receptor. The absence of selective ligands has prevented a full description of its distribution. The receptor displays polymorphic variation in the human population (see Comment).

Distribution

mRNA for the D_4 receptor is present at one to two orders of magnitude lower than that for the D_2 receptor. The highest levels of expression in brain are found in the medulla, amygdala, midbrain and frontal cortex; lower levels are present in striatum and olfactory tubercle. mRNA has also been detected in peripheral tissues, and the protein appears to be expressed preferentially in cardiovascular system in rat.

The high levels of the D_4 receptor in the limbic system and much lower levels in the basal ganglia may explain why the atypical antipsychotic, clozapine, which is selective to the D_4 receptor, has a low incidence of extrapyramidal side-effects.

Pharmacology

Agonists: The D_4 receptor has similar or lower affinities for agonists selective to D_2-receptors.

Antagonists: Clozapine has a selectivity of ~tenfold.

Radioligands: [^{3}H]Clozapine has weak selectivity, but also binds to some 5-HT receptor subtypes.

Structures:

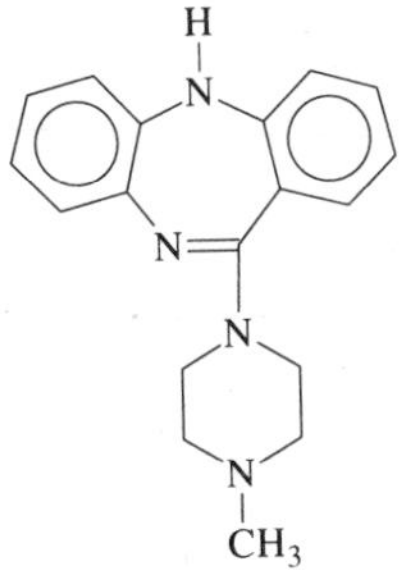

Clozapine

Predominant effector pathways
Inhibition of adenylyl cyclase through a pertussis-toxin-sensitive G-protein most likely of the G_i/G_o class [1].

Amino acid sequence [2]

```
                                                 TM 1
  1 MGNRSTADAD GLLAGRGPAA GASAGASAGL AGQGAAALVG GVLLIGAVLA
                               TM 2
 51 GNSLVCVSVA TERALQTPTN SFIVSLAAAD LLLALLVLPL FVYSEVQGGA
           TM 3
101 WLLSPRLCDA LMAMDVMLCT ASIFNLCAIS VDRFVAVAVP LRYNRQGGSR
      TM 4                                          TM 5
151 RQLLLIGATW LLSAAVAAPVL CGLNDVRGR DPAVCRLEDR DYVVYSSVCS

201 FFLPCPLMLL LYWATFRGLQ RWEVARRAKL HGRAPRRPSG PGPPSPTPPA

251 PRLPQDPCGP DCAPPAPGLP PDPCGSNCAP PDAVRAAALP PQTPPQTRRR
                          TM 6                          TM 7
301 RRAKITGRER KAMRVLPVVV GAFLLCWTPF FVVHITQALC PACSVPPRLV

351 SAVTWLGYVN SALNPVIYTV FNAEFRNVFR KALRACC
```

Amino acids	387
Molecular weight	40 893
Chromosome	11
Glycosylation	Asn3*
Disulfide bonds	Cys108–Cys185*
Palmitoylation	None

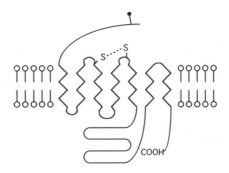

Database accession numbers

	PIR	SWISSPROT	EMBL/GENBANK	REFERENCES
Human		P21917	X58497	2
Rat			M84009	3

Gene

The coding region has introns in equivalent positions to those in the D_2 gene.

Comment

Different polymorphic forms of the D_4 receptor are present in human [4]. At least five distinct forms containing 1-, 2-, 4-, 5- or 7-fold repeats of a 48 base pair sequence in the third cytoplasmic domain have been identified (this region is absent in rat). These forms have similar agonist and antagonist pharmacology but exhibit different sensitivities to sodium ions. The possibility that genetic disposition to schizophrenia is related to the expression of the different length variants has been suggested.

References

1 Cohen, A.I. *et al.* (1992) *Proc. Natl Acad. Sci. USA* **89**, 12093–12097.
2 Van Tol, H.H.M. *et al.* (1991) *Nature* **350**, 610–614.
3 O'Malley, K.L. *et al.* (1992) *New Biologist* **4**, 1–9.
4 Van Tol, H.H.M. *et al.* (1992) *Nature* **358**, 149–152.

THE D_5 RECEPTOR

The D_5 receptor has a similar pharmacology to the D_1 receptor but is expressed in much lower levels in brain. It has also been named the D_{1B} or $D_{1\beta}$ receptor.

Distribution

The absence of ligands which are selective to D_1 or D_5 receptors prevents a full description of their distribution. D_5 mRNA is expressed at a much lower level in basal ganglia relative to D_1 mRNA; it is also present in frontal cortex, hippocampus and hypothalamus but does not appear to be expressed in the periphery.

The D_5 receptor may mediate arousal behaviour and perception of pain.

Pharmacology

Agonists: Dopamine is ~tenfold selective to D5 relative to D1.

Antagonists: "D1-selective antagonists" have similar affinities at the D5 receptor; (+)-butaclamol is a notable exception in that it has ~tenfold lower affinity at the D5 receptor.

Predominant effector pathways
Activation of adenylyl cyclase through G_s.

Amino acid sequence [1-3]

```
                                                      TM 1
  1 MLPPGSNGTA YPGQFALYQQ LAQGNAVGGS AGAPPLGPSQ VVTAQLLTLL
                                  TM 2
 51 IIWTLLGNVL VCAAIVRSRH LRANMTNVFI VSLAVSDLFV ALLVMPWKAV
                 TM 3
101 AEVAGYWPFG AFCDVWVAFD IMCSTASILN LCVISVDRYW AISRPFRYKR
            TM 4
151 KMTQRMALVM VGLAWTLSIL ISFIPVQLNW HRDQAASWGG LDLPNNLANW
                           TM 5
201 TPWEEDFWEP DVNAENCDSS LNRTYAISSS LISFYIPVAI MIVTYTRIYR
                                                      TM 6
251 IAQVQIRRIS SLERAAEHAQ SCRSSAACAP DTSLRASIKK ETKVLKTLSV
                                            TM 7
301 IMGVFVCCWL PFFILNCMVP FCSGHPEGPP AGFPCVSETT FDVFVWFGWA

351 NSSLNPVIYA FNADFQKVFA QLLGCSHFCS RTPVETVNIS NELISYNQDI

401 VFHKEIAAAY IHMMPNAVTP GNREVDNDEE EGPFDRMFQI YQTSPDGDPV

451 AESVWELDCE GEISLDKITP FTPNGFH
```

Amino acids	477
Molecular weight	52 976
Chromosome	4
Glycosylation	Asn7*, Asn199*
Disulfide bonds	Cys113–Cys217*
Palmitoylation	Cys355*, Cys359*

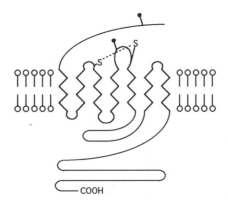

Database accession numbers

	PIR	SWISSPROT	EMBL/GENBANK	REFERENCES
Human		P21918	X58454	1–3
			M67439	
			M75867	
Rat	A41271	P25115	M69118	4

Gene (human)
The gene is intronless.

Comment
Two pseudogenes (EMBL/Genbank: M67440 and M67441) have also been described in human which are more than 90% homologous to the D_5 receptor genomic sequence [2,3]. They contain a number of insertions and deletions which introduce in frame stop codons, thereby preventing expression of a functional receptor; a mRNA transcript has been identified which is capable of producing a protein of 154 amino acids [5].

References
[1] Sunahara, R.K. et al. (1991) Nature 350, 614–619.
[2] Grandy, D.K. et al. (1991) Proc. Natl Acad. Sci. USA 88, 9175–9179.
[3] Weinshank, R.L. et al. (1991) J. Biol. Chem. 266, 22427–22435.
[4] Tiberi, M. et al. (1991) Proc. Natl Acad. Sci. USA 88, 7491–7495.
[5] Nguyen, T. et al. (1991) Biochem. Biophys. Res. Commun. 181, 16–21.

Endothelin

The endothelins (ET) and sarafotoxins (SRTX) are found in mammalian tissue and snake venom, respectively [1-14]. They represent two subclasses of a family of peptides, each of which has 21 amino acids and four cysteines linked by two disulfide bonds (the last six amino acids in the C-terminus of the endothelins are conserved).

Endothelins are believed to have an important physiological role in the regulation of the cardiovascular system. They are the most potent vasoconstrictors identified, stimulate cardiac contraction, regulate release of vasoactive substances (e.g. renin, atrial natriuretic peptide, EDRF) and stimulate mitogenesis in blood vessels in primary culture. In addition, they stimulate contraction in nearly all other smooth muscles that have been studied, including uterus, bronchus, vas deferens, stomach, and stimulate secretion in several tissues including kidney, liver and adrenals. Endothelin receptors are also found in brain, e.g. cerebral cortex, cerebellum and glial cells.

Endothelins have been implicated in several pathophysiological conditions associated with stress, including hypertension, myocardial infarction, subarachnoid haemorrhage and renal failure.

Endothelins and sarafotoxins

ET-1   Cys–Ser–Cys–Ser–Ser–Leu–Met–Asp–Lys–Glu–Cys–Val–Tyr–Phe–Cys–
            His–Leu–Asp–Ile–Ile–Trp

ET-2   Cys–Ser–Cys–Ser–Ser–Trp–Leu–Asp–Lys–Glu–Cys–Val–Tyr–Phe–Cys–
            His–Leu–Asp–Ile–Ile–Trp

ET-3   Cys–Thr–Cys–Phe–Thr–Tyr–Lys–Asp–Lys–Glu–Cys–Val–Tyr–Tyr–Cys–
            His–Leu–Asp–Ile–Ile–Trp

SRTX-a Cys–Ser–Cys–Lys–Asp–Met–Thr–Asp–Lys–Glu–Cys–Leu–Asn–Phe–Cys–
            His–Gln–Asp–Val–Ile–Trp

The four cysteine residues in each peptide are connected by disulfide bridges between positions 1–15 and 3–11.

Vasoactive intestinal contractor (VIC) may be the mouse and rat homologue of human ET-2. Endothelins share very close homology with the SRTX family from snake venom which includes SRTX-a, -b, -c and -d.

Distribution and synthesis

Endothelins are synthesized by proteolysis of large (~200 amino acids), isopeptide-specific proteins, preproendothelins, which are cleaved to "big endothelins" (~40 amino acids) before being processed to the mature peptide. Three structurally distinct preproendothelins encoded by three separate genes have been identified in human. Restriction mapping indicates that each gene is located on a separate chromosome.

ET-1, ET-2 and ET-3 are present in lung, kidney, adrenal gland, brain and other tissues. Todate, only ET-1 has been shown to be present in endothelial cells.

Reviews
[1] Doherty, A.M. (1992) Endothelin: a new challenge. *J. Med. Chem.* **35**, 1493–1508.

2 Furchgott, R.F. and Vanhoutte, P. (1989) Endothelium-derived relaxing factor and contracting factors. *FASEB. J.* **3**, 2007–2018.

3 Huggins, J. *et al.* (1993) Endothelin receptors. *Pharmacol. Ther.* in press.

4 Kloog, Y. and Sokolovsky, M. (1989) Similarities in mode and site of action of sarafotoxins and endothelins. *Trends Pharmacol. Sci.* **10**, 212–215.

5 Jones, C.R. *et al.* (1991) Endothelin receptor heterogeneity, structure activity, autoradiographic and functional studies. *J. Recept. Res.* **11**, 299–310.

6 Opgenorth, J.J. *et al.* (1992) Endothelin-converting enzymes. *FASEB J.* **6**, 2653–2659.

7 Randall, M.D. (1991) Vascular activities of the endothelins. *Pharmacol. Ther.* **50**, 73–93.

8 Sakurai, T. *et al.* (1992) Molecular characterisation of endothelin receptors. *Trends Pharmacol. Sci.* **13**, 103–108.

9 Simonson, M.S. and Dunn, M.J. (1990) Cellular signalling by peptides of the endothelin gene family. *FASEB. J.* **4**, 2989–3000.

10 Sokolovsky, M. (1991) Endothelins and sarafotoxins: physiological regulation, receptor subtypes and transmembrane signaling. *Trends Biochem. Sci.* **16**, 261–264.

11 Sokolovsky, M. (1992) Endothelins and sarafotoxins: physiological regulation, receptor subtypes and transmembrane signaling. *Pharmacol. Ther.* **54**, 129–151.

12 Sokolovsky, M. (1992) Structure-function relationships of endothelins, sarafotoxins, and their receptors. *J. Neurochem.* **59**, 809–818.

13 Yanagisawa, M. and Masaki, T. (1989) Endothelin, a novel endothelium-derived peptide: pharmacological activities, regulation and possible roles in cardiovascular control. *Biochem. Pharmacol.* **38**, 1877–1883.

14 Yanagisawa, M. and Masaki, T. (1989) Molecular biology and pharmacology of the endothelins. *Trends Pharmacol. Sci.* **10**, 374–378

THE ET$_A$ ENDOTHELIN RECEPTOR

ET-1 and ET-2 are approximately equipotent and are active in subnanomolar concentrations; ET-3 is ~1000 times less potent.

Distribution

The ET$_A$ receptor is believed to be the predominant type of endothelin receptor mediating contraction in blood vessels, bronchus, uterus and heart; it also inhibits aldosterone secretion. It has been identified in glial cells in the CNS.

Pharmacology

Agonists: The receptor can be identified by the rank order of potency of mammalian endothelins: ET-1 = ET-2 > ET-3.

Antagonists: The peptides BE 18257B [1] (pA$_2$ 5.9) and BQ-123 [2] (pA$_2$ 7.4), and the nonpeptide FR 139317 (pA$_2$ 8.2) [3] have selectivities of more than 100-fold.

Radioligands: [^{125}I]ET-1 (K_d 0.1-5 nM) and [^{125}I]SRTX S6b (K_d 0.1-5 nM) are nonselective.

Structures:

BE 18257B *cyc*(DGlu–Ala–*allo*–DIle–Leu–DTrp)
BQ-123 *cyc*(DTrp–DAsp–Pro–DVal–Leu)

FR 139317

Predominant effector pathways

Activation of the phosphoinositide pathway through a pertussis-toxin-insensitive G-protein most likely of the G_q/G_{11} class.

Amino acid sequence [4–6]

```
  1 METLCLRASF WLALVGC

                       VIS DNPERYSTNL SNHVDDFTTF RGTELSFLVT
                                      TM 1
 51 THQPTNLVLP SNGSMHNYCP QQTKITSAFK YINTVISCTI FIVGMVGNAT
               TM 2
101 LLRIIYQNKC MRNGPNALIA SLALGDLIYV VIDLPINVFK LLAGRWPFDH
               TM 3
151 NDFGVFLCKL FPFLQKSSVG ITVLNLCALS VDRYRAVASW SRVQGIGIPL
               TM 4
201 VTAIEIVSIW ILSFILAIPE AIGFVMVPFE YRGEQHKTCM LNATSKFMEF
                  TM 5
251 YQDVKDWWLF GFYFCMPLVC TAIFYTLMTC EMLNRRNGSL RIALSEHLKQ
            TM 6                                        TM 7
301 RREVAKTVFC LVVIFALCWF PLHLSRILKK TVYNEMDKNR CELLSFLLLM

351 DYIGINLATM NSCINPIALY FVSKKFKNCF QSCLCCCCYQ SKSLMTSVPM

401 NGTSIQWKNH DQNNHNTDRS SHKDSMN
```

The first 17 amino acids are thought to form a signal sequence.

Amino acids	427
Molecular weight (polypeptide)	48 722
Glycosylation	Asn29*, Asn62*
Disulfide bonds	Cys158–Cys239*

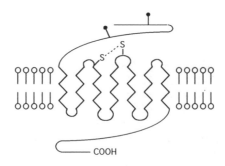

Database accession numbers

	PIR	SWISSPROT	EMBL/GENBANK	REFERENCES
Human	JS0607	P25101	D90348	4-6
Bovine	S13424	P21450	X57765	7
Rat	A40440	P26684	M60786	8,9

Gene
The organization of the gene is unknown.

References
1 Ihara, M. *et al.* (1991) *Biochem. Biophys. Res. Commun.* **178**, 132–137.
2 Nakamichi, K. *et al.* (1992) *Biochem. Biophys. Res. Commun.* **182**, 144–150.
3 Aramori, I. *et al.* (1993) *Mol. Pharmacol.* **43**, 127–131
4 Adachi, M. *et al.* (1991) *Biochem. Biophys. Res. Commun.* **180**, 1265–1272.
5 Hosoda, K. *et al.* (1991) *FEBS Lett.* **267**, 23–26.
6 Cyr, C. *et al.* (1991) *Biochem. Biophys. Res. Commun.* **181**, 184-190
7 Arai, H. *et al.* (1991) *Nature* **348**, 730–732.
8 Lin, H.Y. *et al.* (1991) *Proc. Natl Acad. Sci. USA* **88**, 3185–3189.
9 Hori, S. *et al.* (1992) *Endocrinology* **130**, 1885–1895.

THE ET$_B$ ENDOTHELIN RECEPTOR

All three mammalian endothelins have similar potencies and the receptor has therefore been described as ET isopeptide-nonselective.

Distribution

In the periphery, the ET$_B$ receptor is believed to have a major role in endothelium-dependent vasodilation and a more minor role in vasoconstriction; it is enriched in lung parenchyma. In CNS, the ET$_B$ receptor has been described in cerebral cortex, hippocampus, cerebellum and astrocytes.

Pharmacology

Agonists: [Ala^{1,3,11,15}]ET-1 *1* and SRTX-c *2* have a selectivity greater than 1000-fold and act in picomolar concentrations.

Antagonists: [Cys[11],Cys[15]]ET-11[11-21] (also known as IRL 1038) has been reported to have a selectivity of ~100-fold [3] and a K_d of ~10 nM.

Radioligands: [^{125}I]ET-1 (K_d 0.1–5 nM) and [^{125}I]SRTX S6b (K_d 0.1-5 nM) are nonselective; [^{125}I]BQ 3020 is selective.

Predominant effector pathways
Activation of the phosphoinositide pathway through a pertussis-toxin-insensitive G-protein. Some actions, however, are pertussis-sensitive.

Amino acid sequence [4-6]

```
  1 MQPPPSLCGR ALVALVLACG LSRIWG


                            EERG FPPDRATPLL QTAEIMTPPT

 51 KTLWPKGSNA SLARSLAPAE VPKGDRTAGS PPRTISPPPC QGPIEIKETF
        TM 1                                   TM 2
101 KYINTVVSCL VFVLGIIGNS TLLRIIYKNK CMRNGPNILI ASLALGDLLH
                               TM 3
151 IVIDIPINVY KLLAEDWPFG AEMCKLVPFI QKASVGITVL SLCALSIDRY
                          TM 4
201 RAVASWSRIK GIGVPKWTAV EIVLIWVVSV VLAVPEAIGF DIITMDYKGS
                          TM 5
251 YLRICLLHPV QKTAFMQFYK TAKDWWLFSF YFCLPLAITA FFYTLMTCEM
                          TM 6
301 LRKKSGMQIA LNDHLKQRRE VAKTVFCLVL VFALCWLPLH LSRILKLTLY
                     TM 7
351 NQNDPNRCEL LSFLLVLDYI GINMASLNSC INPIALYLVS KRFKNCFKSC

401 LCCWCQSFEE KQSLEEKQSC LKFKANDHGY DNFRSSNKYS SS
```

Amino acids	442
Molecular weight	49 643
Glycosylation	Asn59*
Disulfide bonds	Cys174–Cys255*

The first 26 amino acids are thought to be a signal sequence. There is a relatively long proline rich N-terminus which has little homology with that of the ET$_A$ receptor.

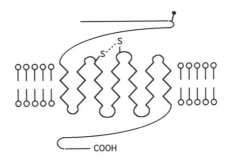

Database accession numbers

	PIR	SWISSPROT	EMBL/GENBANK	REFERENCES
Bovine			D90456	7,8
Human	JQ1042	P24530	M74921	4–6
			D90402	
Pig				9
Rat		P21451	X57764	10,11

Gene (bovine) [8]

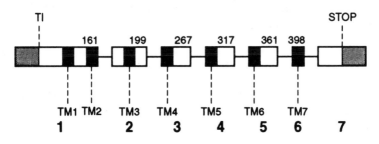

References

1 Saeki, T. *et al.* (1992) *Biochem. Biophys. Res. Commun.* **179**, 286–292.

2 Williams, D.L. *et al.* (1991) *Biochem. Biophys. Res. Commun.* **175**, 556–561.

3 Urade, Y. *et al.* (1992) *FEBS Lett.* **311**, 12–16.

4 Nakamuta, M. *et al.* (1991) *Biochem. Biophys. Res. Commun.* **177**, 34–39.

5 Ogawa, Y. *et al.* (1991) *Biochem. Biophys. Res. Commun.* **178**, 248–255.

6 Sakamoto, A. *et al.* (1991) *Biochem. Biophys. Res. Commun.* **178**, 656–663.

7 Saito, Y. *et al.* (1991) *J. Biol. Chem.* **266**, 23433–23437.

8 Mizuno, T. *et al.* (1992) *Biochem. J.* **287**, 305–309.

9 Elshourbagy, N.A. *et al.* (1992) *Mol. Pharmacol.* **41**, 465–473.

10 Sakurai, T. *et al.* (1990) *Nature* **348**, 732–735.

11 Hori, S. *et al.* (1992) *Endocrinology* **130**, 1885–1895.

Formyl-methionyl peptides

INTRODUCTION

The accumulation of phagocytic cells at the site of injury or infection is regulated by substances which stimulate chemotaxis, granule secretion, superoxide generation and upregulation of cell surface adhesion molecules, e.g. MAC-1, CR1, in cells of the immune system. The list of chemoattractant substances includes *N*-formyl-methionyl peptides of bacterial origin, the actions of which are typified by f-Met–Leu–Phe (fMLP). Other examples are complement component C5a, interleukin 8, leukotriene B4 and platelet activating factor.

Many of these substances participate in pathophysiological conditions, e.g. anaphylactoid and septic shock.

fMLP RECEPTORS

Three fMLP receptors have been identified in cloning studies. fMLP selectively activates the FPR1 receptor relative to the FPR2 receptor; its action on the FPR3 receptor is not known.

THE FPR1 RECEPTOR
Distribution

The FPR1 receptor is found in differentiated myeloid cells, i.e. neutrophils and monocytes, and in related cell lines, e.g. U937 and HL60.

Pharmacology

Agonists: fMLP and several other *N*-formyl-methionyl peptides induce activation in nanomolar concentrations.

Antagonists: None have been described.

Radioligands: [^{3}H]fMLP or [^{125}I]fMLPK-Pep12 [1] bind to high- (K_d 1.5 nM) and low- (K_d 10 nM) affinity sites.

Effector pathway
Activation of phosphoinositide metabolism and inhibition of adrenylyl cyclase through a pertussis-toxin-sensitive G-protein most likely of the G_i/G_o class; activation of phospholipase C may be mediated by βγ-subunits.

Amino acid sequence [1]

```
                               TM 1
   1 METNSSLPTN ISGGTPAVSA GYLFLDIITY LVFAVTFVLG VLGNGLVIWV
                   TM 2
  51 AGFRMTHTVT TISYLNLAVA DFCFTSTLPF FMVRKAMGGH WPFGWFLCKF
     TM 3                                         TM 4
 101 LFTIVDINLF GSVFLIALIA LDRCVCVLHP VWTQNHRTVS LAKKVIIGPW
```

```
151 VMALLLTLPV IIRVTTVPGK TGTVACTFNF SPWTNDPKER INVAVAMLTV
        TM 5                                           TM 6
201 RGIIRFIIGF SAPMSIVAVS YGLIATKIHK QGLIKSSRPL RVLSFVAAAF
                                        TM 7
251 FLCWSPYQVV ALIATVRIRE LLQGMYKEIG IAVDVTSALA FFNSCLNPML

301 YVFMGQDFRE RLIHALPASL ERALTEDSTQ TSDTATNSTL PSAEVALQAK
```

Amino acids	350
Molecular weight	
(polypeptide)	38 401
(SDS PAGE)	55–70 kDa
Chromosome	19
Glycosylation	Asn4*, Asn10*, Asn178*
Disulfide bonds	Cys98–Cys176*
Palmitoylation	None

There are two human fMLP receptor isoforms which differ in two amino acids (Val101 and Glu346 are replaced by Leu and Ala, respectively) and in their 5' and 3' untranslated regions. They probably represent allelic variations [2].

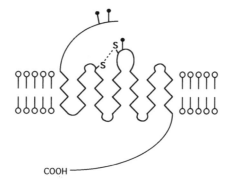

Database accession numbers

	PIR	SWISSPROT	EMBL/GENBANK	REFERENCES
Human	A35495	P21462	M60626	1,3
	A36309		M60627	
			M33537	
			M33538	

Gene
The encoding region resides on a single exon.

Comment
The cDNA reported to encode a rabbit fMLP receptor is now recognized to encode an IL-8 receptor [4].

References

[1] Boulay, F. *et al.* (1990) *Biochem. Biophys. Res. Commun.* **168**, 1103–1109.
[2] Boulay, F. *et al.* (1990) *Biochemistry* **29**, 11123–11133.
[3] Murphy, P.M. and McDermott, D. (1991) *J. Biol. Chem.* **266**, 12560–12567.
[4] Thomas, K.M. *et al.* (1990) *J. Biol. Chem.* **265**, 20061–20064; correction *ibid.* **267**, 13780.

THE FPR₂ RECEPTOR

A cDNA sequence encoding a protein with approximately 69% homology to the FPR₁ receptor has been identified in neutrophils. It has been reported to bind fMLP at greater than 1000 times higher concentrations than at the FPR₁ receptor [1], but may bind other *N*-formyl-methionyl peptides with higher affinity. It is also known as FPRL₁ [2] and FPRH₂ [3].

Distribution

mRNA is present only in differentiated myeloid cells, i.e. neutrophils and monocytes.

Amino acid sequence [1–3]

```
                                TM 1
  1 METNFSTPLN EYEEVSYESA GYTVLRILPL VVLGVTFVLG VLGNGLVIWV
            TM 2
 51 AGFRMTRTVT TICYLNLALA DFSFTATLPF LIVSMAMGEK WPFGWFLCKL
       TM 3                                            TM 4
101 IHIVVDINLF GSVFLIGFIA LDRCICVLHP VWAQNHRTVS LAMKVIVGPW

151 ILALVLTLPV FLFLTTVTIP NGDTYCTFNF ASWGGTPEER LKVAITMLTA
            TM 5                                     TM 6
201 RGIIRFVIGF SLPMSIVAIC YGLIAAKIHK KGMIKSSRPL RVLTAVVASF
                                          TM 7
251 FICWFPFQLV ALLGTVWLKE MLFYGKYKII DILVNPTSSL AFFNSCLNPM

301 LYVFVGQDFR ERLIHSLPTS LERALSEDSA PTNDTAANSA SPPAETELQA

351 M
```

Amino acids	351
Molecular weight	38 902
Chromosome	19
Glycosylation	Asn4*
Disulfide bond	Cys98–Cys176*
Palmitoylation	None

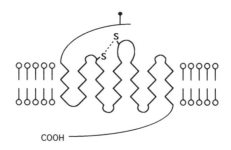

Database accession numbers

	PIR	SWISSPROT	EMBL/GENBANK	REFERENCES
Human		P25090	M84562	1–3
			M88107	
			M76672	

Gene

The organization of the gene is not known.

References

1 Ye, R.D. *et al.* (1992) *Biochem. Biophys. Res. Commun.* **184**, 582–589.
2 Murphy, P.M. *et al.* (1992) *J. Biol. Chem.* **267**, 7637–7643.
3 Lu, B. *et al.* (1992) *Genomics* **13**, 437–440.

THE FPR3 RECEPTOR

A human gene encoding a protein with approximately 56% homology to the FPR1 receptor has been identified. Its ligand specificity is unknown. It is also known as FPRH1 [1].

Amino acid sequence [7,8]

```
                                    TM 1
  1 METNFSIPLN ETEEVLPEPA GHTVLWIFSL LVHGVTFVFG VLGNGLVIWV
            TM 2
 51 AGFRMTRTVN TICYLNLALA DFSFSAILPF RMVSVAMREK WPFASFLCKL
    TM 4                                      TM 4
101 VHVMIDINLF VSVYLITIIA LDRCICVLHP AWAQNHRTMS LAKRVMTGLW
                                                     TM 5
151 IFTIVLTLPN FIFWTTISTT NGDTYCIFNF AFWGDTAVER LNVFITMAKV
                                                     TM 6
201 FLILHFIIGF TVPMSIITVC YGIIAAKIHR NHMIKSSRPL RVFAAVVASF
                                      TM 7
251 FICWFPYELI GILMAVWLKE MLLNGKYKII LVLINPTSSL AFFNSCLNPI

301 LYVFMGRNFQ ERLIRSLPTS LERALTEVPD SAQTSNTHTT SASPPEETEL

351 QAM
```

Amino acids	353
Molecular weight	40 015
Chromosome	19
Glycosylation	Asn4*, Asn10*
Disulfide bond	Cys98–Cys176*
Palmitoylation	None

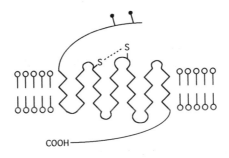

Database accession numbers

	PIR	SWISSPROT	EMBL/GENBANK	REFERENCE
Human		P25089	M76673	7

Gene
The organization of the gene is not known.

References
[1] Lu, B. et al. (1992) Genomics **13**, 437–440.

GABA

GABA is the major inhibitory neurotransmitter in the CNS and mediates the majority of its actions through the GABA$_A$ receptor which contains an intrinsic Cl-channel [1-3]. This receptor is described in The Ion Channel FactsBook [4]. Benzodiazepines potentiate the action of GABA on the GABA$_A$ receptor.

GABA$_B$ receptors are G-protein coupled and mediate the long-lasting synaptic hyperpolarization induced by GABA in many regions of the CNS, e.g. CA1 hippocampal pyrimidal cells. Identification of the physiological and pathophysiological role of GABA$_B$ receptors is hampered by the lack of highly selective, potent antagonists.

GABA$_B$ agonists may represent novel analgesics.

GABA

Distribution

GABA is found in CNS but not in peripheral tissues except in trace amounts. It is formed from glutamate by the action of glutamic acid decarboxylase.

Book
1 Bowery, N. *et al.* (eds) (1990) GABA$_B$ Receptors in Mammalian Function. Wiley, New York.

Reviews
2 Bowery, N. (1989) GABA$_B$ receptors and their significance in mammalian pharmacology. *Trends Pharmacol. Sci.* **10**, 401–407.
3 Bowery, N. (1993) GABA$_B$ receptor pharmacology. *Annu. Rev. Pharmacol. Toxicol.* **33**, 109–148.
4 Conley, E. (1994) The Ion Channel FactsBook, Academic Press, London, in press.

THE GABA$_B$ RECEPTOR

The receptor has not been cloned.

Distribution

GABA$_B$ receptors are widespread throughout the CNS with a distinct distribution to that of GABA$_A$ receptors. Many are presynaptic and inhibit neurotransmitter release, e.g. spinal cord, cerebral cortex, cerebellum. In the periphery, GABA$_B$ receptors inhibit neurotransmitter release in cardiac and smooth muscles, e.g. vas deferens, intestine, trachea, and mediate hyperpolarization in sympathetic ganglia.

Pharmacology

Agonists: 3-Aminopropylphosphinic acid and its methyl analogue are 10–100 times more potent than L-baclofen and act at nanomolar concentrations.

Antagonists: Saclofen (pA2 5.3) and 2-hydroxysaclofen (pA2 5.0) are selective antagonists; CGP 35348 (pA2 5.0) and CGP 36742 (pA2 5.1) are examples of selective antagonists that pass the blood–brain barrier.

Radioligands: [³H]L-Baclofen (K_d 100 nM) and [³H]3-aminopropylphosphinic acid (K_d 1 nM) are selective.

Structures:

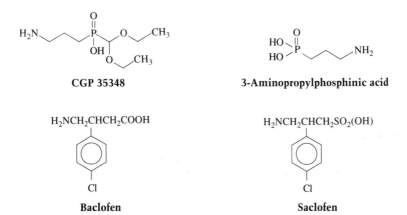

CGP 35348

3-Aminopropylphosphinic acid

H₂NCH₂CHCH₂COOH

Baclofen

H₂NCH₂CHCH₂SO₂(OH)

Saclofen

Predominant effector pathways

Activation of K⁺ channels, and inhibition of adenylyl cyclase and Ca²⁺ channels through a pertussis-toxin-sensitive G-protein, most likely of the G_i/G_o class.

Galanin

Galanin is a neurotransmitter in the peripheral and central nervous systems with a wide spectrum of activity [1-3]. Full activity resides in the N-terminal portion of the molecule, amino acids 1–16. It does not have strong sequence homology with other neuropeptides.

In the periphery, galanin inhibits insulin release induced by glucose and may be the sympathetic mediator of this effect during stress. In the CNS, it inhibits firing of locus coeruleus cells, is synergistic with opiates in inducing analgesia at the level of the spinal cord and stimulates feeding behaviour and release of growth hormone. The ability of galanin to inhibit acetylcholine release in hippocampus has led to the suggestion that galanin antagonists may be of use in the treatment of Alzheimer's disease. Galanin agonists may be novel analgesics.

Structure

Galanin   Gly–Trp–Thr–Leu–Asn–Ser–Ala–Gly–Tyr–Leu–Leu–Gly–Pro–His–
Ala–Val–Gly–Asn–His–Arg–Ser–Phe–Ser–Asp–Lys–Asn–Gly–Leu–
Thr–Ser

Human galanin has 30 amino acids and is not C-terminally amidated whereas bovine, chick, porcine, rat and sheep galanins are 29 amino acids in length and are C-terminally amidated. Species differences in amino acid sequence also exist.

Distribution and synthesis

It is synthesized in the form of a larger preprohormone that contains a signal sequence and an ~ 60 amino acid C-terminal flanking peptide (galanin message-associated peptide), the function of which is unknown. Galanin-like immunoreactivity is distributed widely in the central and the peripheral nervous systems, often coexisting with other neurotransmitters, e.g. 5-HT. In the CNS, it is found in locus coeruleus, spinal cord, hypothalamus, ventral hippocampus and striatum; in the periphery, it is present in the pituitary and in sensory neurons.

Reviews

1 Bartfai, T. et al. (1992) Galanin and galanin antagonists: molecular and biochemical perspectives. Trends Pharmacol. Sci. 13, 312–317.
2 Crawley, J. (1990) Coexistence of neuropeptides and "classical" neurotransmitters. Functional interactions between galanin and acetylcholine. Ann. N. Y. Acad. Sci. 579, 233–245.
3 Merchenthaler, I. et al. (1993) Anatomy and physiology of central galanin-containing pathways. Prog. Neurobiol. 40, 711–769.

THE GALANIN RECEPTOR

The receptor has not been cloned.

Distribution

The distribution of galanin receptors in the CNS corresponds to that of galanin, with high levels in locus coeruleus, spinal cord, hypothalamus, ventral hippocampus and striatum. In the periphery, galanin receptors are found in smooth muscle and in pancreas. They are also expressed in some hypothalamic tumours and in rat insulinoma Rin m5F cells.

Pharmacology

Agonists: N-terminal fragments, e.g. galanin$_{1-16}$, induce activation in low nanomolar concentrations.

Antagonists: A series of chimeric peptides based on the N-terminal portion of galanin and the C-terminal portion of other peptide neurotransmitters are claimed to act as antagonists. Examples include galantide (galanin$_{1-12}$–Pro–Glu–Glu–Phe–Phe–Gly–Leu–Met–NH$_2$) which has affinity of approximately 0.1 nM [1].

Radioligands: [^{125}I]Galanin (porcine).

Predominant effector pathways
Inhibition of adenylyl cyclase and Ca^{2+} channels, and activation of K$^+$ channels through a pertussis-toxin-sensitive G-protein most likely of the G$_i$/G$_o$ class.

Comment
There is evidence that galanin receptors in smooth muscle require both the N-terminal and C-terminal portions of the molecule for activation, suggesting that they may represent a subtype of galanin receptor [2].

References
1 Bartfai, T. *et al.* (1991) *Proc. Natl Acad. Sci. USA* **88**, 10961–10965.
2 Rossowski, W.J. *et al.* (1990) *Peptides* **11**, 333–338.

Glucagon

Glucagon has an essential role in the control of blood glucose levels. It stimulates glycogenolysis and gluconeogenesis in liver, leading to production of glucose for release into the bloodstream and causes lipolysis in liver and fat cells. Thus, its major actions are opposite to those of insulin.

Glucagon has a major role in the pathogenesis of diabetes. It has been occasionally used to increase rate and force of contraction in acute cardiac failure.

Structure

Glucagon   His–Ser–Gln–Gly–Thr–Phe–Thr–Ser–Asp–Tyr–Ser–Lys–Tyr–Leu–Asp– Ser–Arg–Arg–Ala–Gln–Asp–Phe–Val–Gln–Trp–Leu–Met–Asn–Thr

Its structure is conserved in all mammalian species. Glucagon shares limited sequence homology with members of the VIP family, e.g. 15 of the amino acids in glucagon are present in secretin.

Distribution and synthesis

Glucagon is synthesized in A cells in the endocrine pancreas and is released into the bloodstream in response to a fall in the plasma concentration of glucose, an increase in certain amino acids, in particular arginine, or in response to low concentrations of fatty acids. An increase in sympathetic activity also stimulates the release of glucagon.

THE GLUCAGON RECEPTOR
Distribution

Glucagon receptors are expressed in high levels in liver, and are also found in adipose tissue and in heart.

Pharmacology

Agonists: Glucagon induces activation in low nanomolar concentrations.

Antagonists: There are no antagonists.

Radioligands: [^{125}I]Glucagon (K_d 40 nM) binds to a single site.

Predominant effector pathways
Activation of adenylyl cyclase through G_s. It has also been reported to stimulate phosphoinositide metabolism.

Amino acid sequence (rat) [1]

```
  1 MLLTQLHCPY LLLLLVVLSC LPKAPSAQVM DFLFEKWKLY SDQCHHNLSL

 51 LPPPTELVCN RTFDKYSCWP DTPPNTTANI SCPWYLPWYH KVQHRLVFKR
                                                     TM 1
101 CGPDGQWVRG PRGQSWRDAS QCQMDDDEIE VQKGVAKMYS SYQVMYTVGY
```

```
                                   TM 2
151 SLSLGALLLA LVILLGLRKL HCTRNYIHGN LFASFVLKAG SVLVIDWLLK
                                   TM 3
201 TRYSQKIGDD LSVSVCLSDG AVAGCRVATV IMQYGIIANY CWLLVEGVYL
              TM 4
251 YSLLSITTFS EKSFFSLYLC IGWGSPLLFV IPWVVVKCLF ENVQCWTSND
            TM 5
301 NMGFWWILRI PVLLAILINF FIFARIIHLL VAKLRAHQMH YADYKFRLAR
    TM 6                                   TM 7
351 STLTLIPLLG VHEVVFAFVT DEHAQGTLRS TKLFFDLFFS SFQGLLVAVL

401 YCFLNKEVQA ELLRRWRRWQ EGKALQEERM ASSHGSHMAP AGTCHGDPCE

451 KLQLMSAGSS SGTGCEPSAK TSLASSLPRL ADSPT
```

Amino acids	485
Molecular weight	54 962
Glycosylation	Asn47*, Asn60*, Asn75*, Asn79*
Disulfide bonds	Cys216/Cys225–Cys287/Cys295*

The receptor is related to those for calcitonin, parathyroid hormone, secretin and VIP.

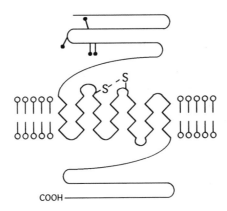

Database accession numbers

	PIR	SWISSPROT	EMBL/GENBANK	REFERENCE
Rat			L04796	1
			M96674	

Gene
The organization of the gene is not known.

Reference
1 Jelinek, L.J. et al. (1993) Science 259, 1614–1616.

Glugacon-like peptide 1 (GLP-1)

GLP-1 potentiates glucose-induced insulin secretion and suppresses acid secretion in the stomach [1]. Although derived from the same precursor as glucagon, it has a distinct structure and is inactive at the glucagon receptor.

Non-insulin-dependent diabetes mellitus is associated with a reduced stimulatory effect of GLP-1 on glucose-induced insulin secretion in rat.

Structure

GLP-1   His–Asp–Glu–Phe–Glu–Arg–His–Ala–Glu–Gly–Thr–Phe–Thr–Ser–Asp–
        Val–Ser–Ser–Tyr–Leu–Glu–Gly–Gln–Ala–Ala–Lys–Glu–Phe–Ile–Ala–
        Trp–Leu–Val–Lys–Gly–Arg–Gly

Shorter forms of GLP-1$_{(1-37)}$, GLP-1$_{(7-36)}$ and GLP-1$_{(7-37)}$, are also produced.

Distribution and synthesis

It is derived from preproglucagon and is made and secreted by intestinal L cells.

Review
[1] Goke, R. *et al.* (1991) Glucagon-like peptide-1 (17–36) amide is a new incretin/enterogastrone candidate. *Eur. J. Invest.* **21**, 135–144.

THE GLP-1 RECEPTOR
Distribution

mRNA is present in pancreas, stomach and lung but is absent in other peripheral tissues and in brain. It is also found in rat insulinoma cells. This is in agreement with the distribution of binding sites for [^{125}I]GLP-1.

Pharmacology

Agonists: GLP-1 induces receptor activation in picomolar concentrations. Glucagon has been reported to bind to the receptor in micromolar concentrations.

Antagonists: There are no antagonists.

Radioligands: [^{125}I]GLP-1 binds to high- (K_d 0.1–0.5 nM) and low (K_d > 10 nM) affinity sites.

Predominant effector pathways
Formation of cAMP by activation of G_s.

Amino acid sequence (rat) [1]

```
  1 MAVTPSLLRL ALLLLGAVGR AGPRPQGATV SLSETVQKWR EYRHQCQRFL

 51 TEAPLLATGL FCNRTFDDYA CWPDGPPGSF VNVSCPWYLP WASSVLQGHV
                                                      TM 1
101 YRFCTAEGIW LHKDNSSLPW RDLSECEESK QGERNSPEEQ LLSLYIIYTV
```

```
                                TM 2
151 GYALSFSALV IASAILVSFR HLHCTRNYIH LNLFASFILR ALSVFIKDAA
                                TM 3
201 LKWMYSTAAQ QHQWDGLLSY QDSLGCRLVF LLMQYCVAAN YYWLLVEGVY
                 TM 4
251 LYTLLAFSVF SEQRIFKLYL SIGWGVPLLF VIPWGIVKYL YEDEGCWTRN
         TM 5
301 SNMNYWLIIR LPILFAIGVN FLVFIRVICI VIAKLKANLM CKTDIKCRLA
       TM 6                                    TM 7
351 KSTLTLIPLL GTHEVIFAFV MDEHARGTLR FVKLFTELSF TSFQGFMVAV

401 LYCFVNNEVQ MEFRKSWERW RLERLNIQRD SSMKPLKCPT SSVSSGATVG

451 SSVYAATCQN SCS
```

Amino acids	463
Glycosylation	Asn63*, Asn82*, Asn 115*
Disulfide bonds	Cys226–296*

The receptor is related to those for calcitonin, parathyroid hormone, secretin and VIP.

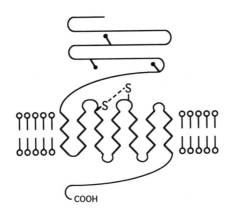

Database accession numbers

	PIR	SWISSPROT	EMBL/GENBANK	REFERENCE
Rat			M97797	1

Gene
The organization of the gene is not known.

Reference
1 Thorens, B. (1992) Proc. Natl Acad. Sci. USA 89, 8641–8645.

Glutamate

INTRODUCTION

Glutamate is the major excitatory neurotransmitter in the CNS and is believed to have important roles in neuronal plasticity, cognition, memory, learning and some neurological disorders such as epilepsy, stroke and neurodegeneration [1-9]. It mediates its actions through two distinct classes of receptor termed ionotropic and metabotropic. Ionotropic receptors mediate "fast" excitatory actions of glutamate and at least three subtypes exist all of which contain an intrinsic cation channel; they are described in The Ion Channels FactsBook [1]. Five metabotropic receptors have been identified and many of their actions can be described as modulatory; for example, they increase the membrane excitability of neurons by inhibiting Ca^{2+}-dependent K^+ conductances, inhibit and potentiate excitatory transmission supported by ionotropic glutamate receptors and inhibit the afterhyperpolarization that follows a burst of actions potentials in dentate gyrus and CA1 neurons in hippocampus. They are involved in long-term potentiation. A full understanding of their physiological roles, however, is hampered by the lack of selective agonists and antagonists.

$$\text{HOOCCCH}_2\text{CH}_2-\overset{\overset{\displaystyle NH_2}{|}}{\underset{\underset{\displaystyle H}{\blacktriangle}}{C}}\text{···COOH}$$

Glutamate

Book
[1] Conley, E. (1994) The Ion Channel FactsBook, Academic Press, London, in press.

Reviews
[2] Baskys, A. (1992) Metabotropic receptors and 'slow' excitatory actions of glutamate agonists in the hippocampus. *Trends Neurosci.* **15**, 92–96.
[3] Collingridge, G. and Lester, R. (1989) Excitatory amino acid receptors in the vertebrate nervous system. *Pharmacol. Rev.* **40**, 145–210.
[4] Collingridge, G. and Singer, W. (1990) Excitatory amino acid receptors and synaptic plasticity. *Trends Pharmacol. Sci.* **11**, 290–296.
[5] Meldrum, B. and Garthwaite, J. (1990) Excitatory amino acid neurotoxicity and neurodegenerative disease. *Trends Pharmacol. Sci.* **11**, 379–387.
[6] Monoghan, D.T. *et al.* (1990) The excitatory amino acid receptors: their classes, pharmacology, and distinct properties in the function of the central nervous system. *Annu. Rev. Pharmacol. Toxicol.* **29**, 365–402.
[7] Nakanishi, S. *et al.* (1992) Molecular diversity of glutamate receptors and implications for brain function. *Science* **258**, 597–603.
[8] Schoepp, D.D. *et al.* (1990) Pharmacological and functional characteristics of metabotropic excitatory amino acid receptors. *Trends Pharmacol. Sci.* **11**, 508–515.
[9] Schoepp, D.D. and Conn, P.J. (1993) Metabotropic glutamate receptors in brain function and pathology. *Trends Pharmacol. Sci.* **14**, 13–20.

GLUTAMATE RECEPTORS

Ionotropic receptors

These are described in further detail in the Ion Channels FactsBook.

AMPA receptor

This receptor is identified by high affinity for the selective agonist AMPA (D,L-α-amino-3-hydroxy-5-methyl-4-isoxalone propionic acid) and for the selective antagonist NBQX (6-nitro-7-sulfamobenzo(f)quinoxaline-2,3-dione). It contains an intrinsic cation channel enabling movement of Na^+ and K^+ ions.

Kainate receptor

This receptor is identified by high affinity for the selective agonists kainate and domoate; certain quinoxalinediones antagonists, e.g. CNQX and DNQX, show limited selectivity relative to AMPA receptors. It contains an intrinsic cation channel enabling movement of Na^+ and K^+ ions.

NMDA receptor

This receptor is identified by high affinity for the selective agonist NMDA (N-methyl-D-aspartate); several selective antagonists have been described that act either at the agonist binding site, e.g. D-AP5 (D-amino-5-phosphonopentanoate) or at the channel, e.g. MK801 [(+)-5-methyl-10,11-dihydro-5H-dibenzo(a,d) cyclohepten-5,10-imine maleate]. It contains an intrinsic cation channel enabling movement of Ca^{2+}, Na^+ and K^+ ions.

Metabotropic receptors

Five subtypes have been identified, all of which are seven transmembrane proteins; they are described below.

mGluR$_1$, mGluR$_2$, mGluR$_3$, mGluR$_4$ AND mGluR$_5$ METABOTROPIC RECEPTORS

Five subtypes of metabotropic receptor have been identified in cloning studies [1–4], together with three alternatively spliced variants of mGluR$_1$ (mGluR$_{1\alpha}$; mGluR$_{1\beta}$ and mGluR$_{1C}$). There is also evidence for a further receptor, mGluR$_6$, although this has not yet been shown to be functional [5]. They are characterized by a unique structural architecture consisting of large N-terminal and C-terminal regions.

The five receptor types differ in their agonist pharmacology and signal transduction pathways; however, the lack of selective, potent agonists and antagonists has prevented characterization of their individual physiological functions and distributions. Nearly all of the comparative pharmacology and study of signalling pathways has been performed on receptors expressed in cell lines. The receptors can be subdivided into three groups using this information and on the basis of their sequence: mGluR$_{1/5}$; mGluR$_{2/3}$ and mGluR$_4$.

Distribution

mRNA for mGluR$_1$ is widespread in brain and is abundant in neuronal cells in hippocampal dentate gyrus and CA2-3 regions, cerebellum Purkinje cells, olfactory bulb and thalamic nuclei [1].

mRNA for mGluR$_2$ is also widespread in brain with a unique distribution; it is found in high levels in neurons in olfactory bulb, cerebral cortex, cerebellum Golgi cells and dentate gyrus granule cells [3].

mRNA for mGluR$_5$ is also widespread in brain with a unique distribution; it is found in high levels in striatum, cerebral cortex, hippocampus and olfactory bulb [4].

The distribution of mRNA for mGluR$_3$ and mGluR$_4$ is not known.

Pharmacology

Agonists: Many agonists at the metabotropic receptor have affinity for certain subtypes of ionotropic receptor; for example, quisqualate and ibotenate interact with AMPA and NMDA receptors. ACPD is the only selective agonist to metabotropic receptors that has been described.

mGluR$_1$ and mGluR$_5$ share a similar profile of agonist selectivity: quisqualate > ibotenate ~ glutamate > ACPD.

mGluR$_2$ and mGluR$_3$ share a similar profile of agonist selectivity: glutamate ~ APCD > ibotenate > quisqualate [4]. (2S,1'S,2'S)CCG has a selectivity of ~ten fold to mGluR$_2$ relative to mGluR$_1$ and mGluR$_4$ (mGluR$_3$ and mGluR$_5$ have not been studied) [5].

APCD, ibotenate and quisqualate have very weak/negligible activity at mGluR$_4$ receptors [6]; L-AP4 (L-amino-4-phosphobutanoate) appears to be weakly selective [5].

Antagonists: There are no selective antagonists. The ionotropic receptor antagonists AP3 (2-amino-4-phosphopropionate) acts as an antagonist at metabotropic receptors in low millimolar concentrations; other antagonists at ionotropic receptors are inactive at metabotropic receptors.

Radioligands: There are no selective radioligands. [^{3}H]Glutamate can be used to label mGluRs in the presence of saturating concentrations of ionotropic agents [7].

Structures:

Aminocyclopentanedicarboxylic acid (1S,3R-ACPD)

(2S,1'S,2'S)2-(Carboxycyclopropyl)glycine (CCG-I)

L-AP4

Ibotenate

Quisqualate

Predominant effector pathways

The mGluR receptor subtypes mediate their effects through G-proteins.

mGluR$_1$ (all three forms) and mGluR$_5$ receptors activate the phosphoinositide pathway, most likely through a G-protein of the Gq/G$_{11}$ class; pertussis toxin partially inhibits the response to mGluR$_1$ but not to mGluR$_5$.

mGluR$_2$, mGluR$_3$ and mGluR$_4$ receptors inhibit adenylyl cyclase through a pertussis-toxin-sensitive G-protein, most likely of the G$_i$/G$_o$ class; mGluR$_2$ has also been reported to cause a weak stimulation of phosphoinositide metabolism.

mGluR$_{1\alpha}$ amino acid sequence (rat) [1–2]

```
  1 MVRLLLIFFP MIFLEMSILP

               RMPDRKVLLA GASSQRSVAR MDGDVIIGAL

 51 FSVHHQPPAE KVPERKCGEI REQYGIQRVE AMFHTLDKIN ADPVLLPNIT

101 LGSEIRDSCW HSSVALEQSI EFIRDSLISI RDEKDGLNRC LPDGQTLPPG

151 RTKKPIAGVI GPGSSSVAIQ VQNLLQLFDI PQIAYSATSI DLSDKTLYKY

201 FLRVVPSDTL QARAMLDIVK RYNWTYVSAV HTEGNYGESG MDAFKELAAQ

251 EGLCIAHSDK IYSNAGEKSF DRLLRKLRER LPKARVVVCF CEGMTVRGLL

301 SAMRRLGVVG EFSLIGSDGW ADRDEVIEGY EVEANGGITI KLQSPEVRSF

351 DDYFLKLRLD TNTRNPWFPE FWQHRFQCRL PGHLLENPNF KKVCTGNESL

401 EENYVQDSKM GFVINAIYAM AHGLQNMHHA LCPGHVGLCD AMKPIDGRKL

451 LDFLIKSSFV GVSGEEVWFD EKGDAPGRYD IMNLQYTEAN RYDYVHVGTW

501 HEGVLNIDDY KIQMNKSGMV RSVCSEPCLK GQIKVIRKGE VSCCWICTAC
                                                 TM 1
551 KENEFVQDEF TCRACDLGWW PNAELTGCEP IPVRYLEWSD IESIIAIAFS
                                          TM 2
601 CLGILVTLFV TLIFVLYRDT PVVKSSSREL CYIILAGIFL GYVCPFTLIA
       TM 3
651 KPTTTSCYLQ RLLVGLSSAM CYSALVTKTN RIARILAGSK KKICTRKPRF
       TM 4
701 MSAWAQVIIA SILISVQLTL VVTLIIMEPP MPILSYPSIK EVYLICNTSN
```

```
                    TM 5                                    TM 6
     751  LGVVAPVGYN GLLIMSCTYY AFKTRNVPAN FNEAKYIAFT MYTTCIIWLA
                    TM 7
     801  FVPIYFGSNY KIITTCFAVS LSVTVALGCM FTPKMYIIIA KPERNVRSAF

     851  TTSDVVRMHV GDGKLPCRSN TFLNIFRRKK PGAGNANSNG KSVSWSEPGG

     901  RQAPKGQHVW QRLSVHVKTN ETACNQTAVI KPLTKSYQGS GKSLTFSDAS

     951  TKTLYNVEEE DNTPSAHFSP PSSPSMVVHR RGPPVATTPP LPPHLTAEET

    1001  PLFLADSVIP KGLPPPLPQQ QPQQPPPQQP PQQPKSLMDQ LQGVVTNFGS

    1051  GIPDFHAVLA GPGTPGNSLR SLYPPPPPPQ HLQMLPLHLS TFQEESISPP

    1101  GEDIDDDSER FKLLQEFVYE REGNTEEDEL EEEEDLPTAS KLTPEDSPAL

    1151  TPPSPFRDSV ASGSSVPSSP VSESVLCTPP NVTYASVILR DYKQSSSTL
```

The first 20 amino acids are thought to be a signal sequence.

mGluR$_{1\alpha}$ is cleaved at Arg906 to yield a polypeptide, mGluR$_{1\beta}$ [3] (906 amino acids, molecular weight 101 630). A third alternative splice variant exists, mGluR$_{1C}$ [8].

Amino acids	1199
Molecular weight	133 229
Glycosylation	Asn98*, Asn223*, Asn397*, Asn515*, Asn747*
Disulfide bonds	Cys257–Cys746

mGluR$_{1\alpha}$

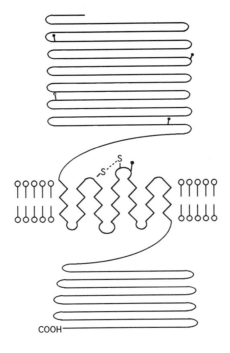

mGluR₂ amino acid sequence (rat) [3]

```
  1 MESLLGFLAL LLLWGAVA

                      EG PAKKVLTLEG DLVLGGLFPV HQKGGPAEEC

 51 GPVNEHRGIQ RLEAMLFALD RINRDPHLLP GVRLGAHILD SCSKDTHALE

101 QALDFVRASL SRGADGSRHI CPDGSYATHS DAPTAVTGVI GGSYSDVSIQ

151 VANLLRLFQI PQISYASTSA KLSDKSRYDY FARTVPPDFF QAKAMAEILR

201 FFNWTYVSTV ASEGDYGETG IEAFELEARA RNICVATSEK VGRAMSRAAF

251 EGVVRALLQK PSARVAVLFT RSEDARELLA ATQRLNASFT WVASDGWGAL

301 ESVVAGSERA AEGAITIELA SYPISDFASY FQSLDPWNNS RNPWFREFWE

351 ERFHCSFRQR DCAAHSLRAV PFEQESKIMF VVNAVYAMAH ALHNMHRALC

401 PNTTHLCDAM RPVNGRRLYK DFVLNVKFDA PFRPADTDDE VRFDRFGDGI

451 GRYNIFTYLR AGSGRYRYQK VGYWAEGLTL DTSFIPWASP SAGPLPASRC

501 SEPCLQNEVK SVQPGEVCCW LCIPCQPYEY RLDEFTCADC GLGYWPNASL
                            TM 1
551 TGCFELPQEY IRWGDAWAVG PVTIACLGAL ATLFVLGVFV RHNATPVVKA
       TM 2                                      TM 3
601 SGRELCYILL GGVFLCYCMT FVFIAKPSTA VCTLRRLGLG TAFSVCYSAL
                            TM 4
651 LTKTNRIARI FGGAREGAQR PRFISPASQV AICLALISGQ LLIVAAWLVV
                            TM 5
701 EAPGTGKETA PERREVVTLR CNHRDASMLG SLAYNVLLIA LCTLYAFKTR
            TM 6                                    TM 7
751 KCPENFNEAK FIGFTMYTTC IIWLAFLPIF YVTSSDYRVQ TTTMCVSVSL

801 SGSVVLGCLF APKLHIILFQ PQKNVVSHRA PTSRFGSAAP RASANLGQGS

851 GSQFVPTVCN GREVVDSTTS SL
```

The first 18 amino acids are thought to be a signal sequence.

Amino acids	872
Molecular weight	95 770
Glycosylation	Asn203*, Asn286*, Asn338*, Asn402*, Asn547*
Disulphide bonds	Cys632–Cys721*

mGluR₂

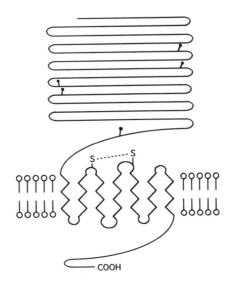

mGluR₃ amino acid sequence (rat) [3]

```
  1 MKMLTRLQIL MLALFSKGFL LS

                           LGDHNFMR REIKIEGDLV LGGLFPINEK

 51 GTGTEECGRI NEDRGIQRLE AMLFAIDEIN KDNYLLPGVK LGVHILDTCS

101 RDTYALEQSL EFVRASLTKV DEAEYMCPDG SYAIQENIPL LIAGVIGGSY

151 SSVSIQVANL LRLFQIPQIS YASTSAKLSD KSRYDYFART VPPDFYQAKA

201 MAEILRFFNW TYVSTVASEG DYGETGIEAF EQEARLRNIC IATAEKVGRS

251 NIRKSYDSVI RELLQKPNAR VVVLFMRSDD SRELIAAANR VNASFTWVAS

301 DGWGAQESIV KGSEHVAYGA ITLELASHPV RQFDRYFQSL NPYNNHRNPW

351 FRDFWEQKFQ CSLQNKRNHR QVCDKHLAID SSNYEQESKI MFVVNAVYAM

401 AHALHKMQRT LCPNTTKLCD AMKILDGKKL YKEYLLKINF TAPFNPNKGA

451 DSIVKFDTFG DGMGRYNVFN LQQTGGKYSY LKVGHWAETL SLDVDSIHWS

501 RNSVPTSQCS DPCAPNEMKN MQPGDVCCWI CIPCEPYEYL VDEFTCMDCG
                                  TM 1
551 PGQWPTADLS GCYNLPEDYI KWEDAWAIGP VTIACLGFLC TCIVITVFIK
               TM 2                                   TM 3
601 HNNTPLVKAS GRELCYILLF GVSLSYCMTF FFIAKPSPVI CALRRLGLGT
```

```
                                              TM 4
651 SFAICYSALL TKTNCIARIF DGVKNGAQRP KFISPSSQVF ICLGLILVQI
                                     TM 5
701 VMVSVWLILE TPGTRRYTLP EKRETVILKC NVKDSSMLIS LTYDVVLVIL
                 TM 6
751 CTVYAFKTRK CPENFNEAKF IGFTMYTTCI IWLAFLPIFY VTSSDYRVQT
         TM 7
801 TTMCISVSLS GFVVLGCLFA PKVHIVLFQP QKNVVTHRLH LNRFSVSGTA

851 TTYSQSSAST YVPTVCNGRE VLDSTTSSL
```

The first 22 amino acids are thought to be a signal sequence.

Amino acids   879
Molecular weight   98 960
Glycosylation   Asn209*, Asn292*, Asn414*, Asn439*
Disulfide bond   Cys641–Cys730*

mGluR₃

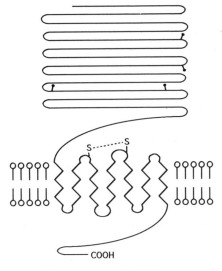

mGluR₄ amino acid sequence (rat) [3]

```
  1 MSGKGGWAWW WARLPLCLLL SLYAPWVPSS LG

                                        KPKGHPHM NSIRIDGDIT

 51 LGGLFPVHGR GSEGKACGEL KKEKGIHRLE AMLFALDRIN NDPDLLPNIT

101 LGARILDTCS RDTHALEQSL TFVQALIEKD GTEVRCGSGG PPIITKPERV

151 VGVIGASGSS VSIMVANILR LFKIPQISYA STAPDLSDNS RYDFFSRVVP

201 SDTYQAQAMV DIVRALKWNY VSTLASEGSY GESGVEAFIQ KSRENGGVCI
```

```
251 AQSVKIPREP KTGEFDKIIK RLLETSNARG IIIFANEDDI RRVLEAARRA

301 NQTGHFFWMG SDSWGSKSAP VLRLEEVAEG AVTILPKRMS VRGFDRYFSS

351 RTLDNNRRNI WFAEFWEDNF HCKLSRHALK KGSHIKKCTN RERIGQDSAY

401 EQEGKVQFVI DAVYAMGHAL HAMHRDLCPG RVGLCPRMDP VDGTQLLKYI

451 RNVNFSGIAG NPVTFNENGD APGRYDIYQY QLRNGSAEYK VIGSWTDHLH

501 LRIERMQWPG SGQQLPRSIC SLPCQPGERK KTVKGMACCW HCEPCTGYQY
                                                    TM 1
551 QVDRYTCKTC PYDMRPTENR TSCQPIPIVK LEWDSPWAVL PLFLAVVGIA
                                           TM 2
601 ATLFVVVTFV RYNDTPIVKA SGRELSYVLL AGIFLCYATT FLMIAEPDLG
       TM 3
651 TCSLRRIFLG LGMSISYAAL LTKTNRIYRI FEQGKRSVSA PRFISPASQL
         TM 4
701 AITFILISLQ LLGICVWFVV DPSHSVVDFQ DQRTLDPRFA RGVLKCDISD
        TM 5                                        TM 6
751 LSLICLLGYS MLLMVTCTVY AIKTRGVPET FNEAKPIGFT MYTTCIVWLA
                        TM 7
801 FIPIFFGTSQ SADKLYIQTT TLTVSVSLSA SVSLGMLYMP KVYIILFHPE

851 QNVPKRKRSL KAVVTAATMS NKFTQKGNFR PNGEAKSELC ENLETPALAT

901 KQTYVTYTNH AI
```

The first 32 amino acids are thought to be a signal sequence.

Amino acids	912
Molecular weight	101 810
Glycosylation	Asn98*, Asn301*, Asn454*, Asn484*
Disulfide bonds	Cys652–Cys746*

mGluR4

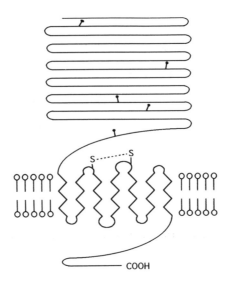

COOH

mGluR5 amino acid sequence (rat) [4]

```
   1 MVLLLILSVL LLKEDVRGSA

                         QSSERRVVAH MPGDIIIGAL FSVHHQPTVD

  51 KVHERKCGAV REQYGIQRVE AMLHTLERIN SDPTLLPNIT LGCEIRDSCW

 101 HSAVALEQSI EFIRDSLISS EEEEGLVRCV DGSSSFRSKK PIVGVIGPGS

 151 SSVAIQVQNL LQLFNIPQIA YSATSMDLSD KTLFKYFMRV VPSDAQQARA

 201 MVDIVKRYNW TYVSAVHTEG NYGESGMEAF KDMSAKEGIC IAHSYKIYSN

 251 AGEQSFDKLL KKLRSHLPKA RVVACFCEGM TVRGLLMAMR RLGLAGEFLL

 301 LGSDGWADRY DVTDGYQREA VGGITIKLQS PDVKWFDDYY LKLRPETNLR

 351 NPWFQEFWQH RFQCRLEGFA QENSKYNKTC NSSLTLRTHH VQDSKMGFVI

 401 NAIYSMAYGL HNMQMSLCPG YAGLCDAMKP IDGRKLLDSL MKTNFTGVSG

 451 DMILFDENGD SPGRYEIMNF KEMGKDYFDY INVGSWDNGE LKMDDDEVWS

 501 KKNNIIRSVC SEPCEKGQIK VIRKGEVSCC WTCTPCKENE YVFDEYTCKA
                                     TM 1
 551 CQLGSWPTDD LTGCDLIPVQ YLRWGDPEPI AAVVFACLGL LATLFVTVIF
                TM 2                                   TM 3
 601 IIYRDTPVVK SSSRELCYII LAGICLGYLC TFCLIAKPKQ IYCYLQRIGI
                                                       TM 4
 651 GLSPAMSYSA LVTKTNRIAR ILAGSKKKIC TKKPRFMSAC AQLVIAFILI
                                                 TM 5
 701 CIQLGIIVAL FIMEPPDIMH DYPSIREVYL ICNTTNLGVV TPLGYNGLLI
                TM 6
 751 LSCTFYAFKT RNVPANFNEA KYIAFTMYTT CIIWLAFVPI YFGSNYKIIT
        TM 7
 801 MCFSVSLSAT VALGCMFVPK VYIILAKPER NVRSAFTTST VVRMHVGDGK

 851 SSSAASRSSS LVNLWKRRGS SGETLSSNGK SVTWAQNEKS TRGQHLWQRL

 901 SVHINKKENP NQTAVIKPFP KSTENRGPGA AAGGGSGPGV AGAGNAGCTA

 951 TGGPEPPDAG PKALYDVAEA EESFPAAARP RSPSPISTLS HLAGSAGRTD

1001 DDAPSLHSET AARSSSSQGS LMEQISSVVT RFTANISELN SMMLSTAATP

1051 GPPGTPICSS YLIPKEIQLP TTMTTFAEIQ PLPAIEVTGG AQGATGVSPA

1101 QETPTGAESA PGKPDLEELV ALTPPSPFRD SVDSGSTTPN SPVSESALCI

1151 PSSPKYDTLI IRDYTQSSSS L
```

The first 20 amino acids are thought to be a signal sequence.

Amino acids   1171
Molecular weight   128 289
Glycosylation   Asn88*, Asn209*, Asn377*, Asn381*, Asn444*, Asn733*
Disulfide bonds   Cys643–Cys732*

mGluR5

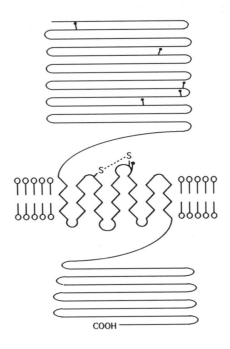

Database accession numbers

	PIR	SWISSPROT	EMBL/GENBANK	REFERENCES
Rat mGluR1	S15362	P23385	X57569	1,2
			M61099	
Rat mGluR2			M92075	3
Rat mGluR3			M92076	3
Rat mGluR4			M92077	3
			M90518	
Rat mGluR5			D10891	4

Gene
The organization of the genes for each receptor is not known.

Comment
The previously named L-AP4 receptor, identified by high affinity for the selective agonist L-AP4 (L-amino-4-phosphobutanoate), is thought to be a subtype of a metabotropic receptor, most likely mGluR4 [9].

References

1 Masu, M. *et al.* (1991) *Nature* **349**, 760–765.
2 Houamed, K.M. *et al.* (1991) *Science* **252**, 1318–1321.
3 Tanabe, Y. *et al.* (1992) *Neuron* **8**, 169–179.
4 Abe, T. *et al.* (1992) *J. Biol. Chem.* **267**, 13361–13368.
5 Nakanishi, S. (1992) *Science* **258**, 597–603.
6 Hayashi, Y. *et al.* (1992) *Br. J. Pharmacol.* **107**, 539–543.
7 Schoepp, D. and True, R.A. (1992) *Neurosci. Lett.* **145**, 100–104.
8 Pin, J.P. *et al.* (1992) *Proc. Natl Acad. Sci USA.* **89**, 10331–10335
9 Trombley, P.Q. and Westbrook, G.L. (1992) *J. Neurosci.* **12**, 2043–2050.

Glycoprotein hormones

The gonadotrophins [luteinizing hormone (LH), human choriogonadotrophin (hCG), follicle-stimulating hormone (FSH)] and thyroid-stimulating hormone (TSH) are 28–38 kDa heterodimeric glycoproteins composed of a common α-subunit and distinct β-subunits [1-4]. The carbohydrate moiety of the hormones has an essential role in receptor recognition. Their receptors share close sequence homology (~ 40%) and are characterized by large extracellular domains believed to be involved in hormone binding.

FSH

FSH is released from the anterior pituitary under the influence of gonadotrophin-releasing hormone and oestrogens, and from the placenta during pregnancy. In the female it acts on the ovaries promoting development of follicles and is the major hormone regulating secretion of oestrogens. In the male it is responsible for the integrity of the seminiferous tubules and acts on Sertoli cells to support gametogenesis.

FSH is used clinically to treat infertility in females and for some types of failure of spermatogenesis in males.

LH/hCG

The β-subunits of LH (also known as lupotrophin) and hCG are closely related in sequence and elicit their biological actions through the same receptor.

LH is released from the anterior pituitary under the influence of gonadotrophin-releasing hormone and progesterones. hCG is released by the placenta in pregnancy. In the female LH stimulates ovulation and is the major hormone involved in the regulation of progesterone secretion by the corpus luteum. In the male it stimulates Leydig cells to secrete androgens, particularly testosterone.

TSH

TSH (also known as thyrotrophin) is synthesized in the thyrotroph cells of the anterior pituitary under the influence of TRH and thyroid hormones. It acts on the thyroid gland, stimulating iodine uptake, synthesis and release of thyroid hormones, hypertrophy and hyperplasia.

Graves' disease is caused by stimulatory anti-TSH receptor antibodies.

Reviews
[1] Pierce, J.G. and Parsons, T.F. (1981) Glycoprotein hormones: structure and function. *Annu. Rev. Biochem.* **50**, 466–495.
[2] Rees Smith, B. *et al.* (1988) Autoantibodies to the thyrotropin receptor. *Endocrinol. Rev.* **9**, 106–121.
[3] Reichert, L.E. *et al.* (1991) Structure-function relationships of the glycoprotein hormones and their receptors. *Trends Pharmacol. Sci.* **12**, 199–203.
[4] Ryan, R.J. *et al.* (1989) The glycoprotein hormones: structure and function. *FASEB J.* **2**, 2661–2669.

THE FSH RECEPTOR

TSH and LH/hCG are inactive.

Distribution

The FSH receptor is found in ovary and testis.

Pharmacology

Agonists: FSH acts in nanomolar concentrations.

Antagonists: The deglycosylated hormone has been reported to act as an antagonist.

Radioligands: [125I]FSH (K_d 1nM).

Predominant effector pathways
Activation of adenylyl cyclase through G_s.

Amino acid sequence [1]

```
  1 MALLLVSLLA FLSLGSG

            CHH RICHCSNRVF LCQESKVTEI PSDLPRNAIE

 51 LRFVLTKLRV IQKGAFSGFG DLEKIEISQN DVLEVIEADV FSNLPKLHEI

101 RIEKANNLLY ITPEAFQNLP NLQYLLISNT GIKHLPDVHK IHSLQKVLLD

151 IQDNINIHTI ERNSFVGLSF ESVILWLNKN GIQEIHNCAF NGTQLDAVNL

201 SDNNNLEELP NDVFHGASGP VILDISRTRI HSLPSYGLEN LKKLRARSTY

251 NLKKLPTLEK LVALMEASLT YPSHCCAFAN WRRQISELHP ICNKSILRQE

301 VDYMTQARGQ RSSLAEDNES SYSRGFDMTY TEFDYDLCNE VVDVTCSPKP
                        TM 1                          TM 2
351 DAFNPCEDIM GYNILRVLIW FISILAITGN IIVLVILTTS QYKLTVPRFL
                                                        TM 3
401 MCNLAFADLC IGIYLLLIAS VDIHTKSQYH NYAIDWQTGA GCDAAGFFTV
                                                   TM 4
451 FASELSVYTL TAITLERWHT ITHAMQLDCK VQLRHAASVM VMGWIFAFAA
                        TM 5
501 ALFPIFGISS YMKVSICLPM DIDSPLSQLY VMSLLVLNVL AFVVICGCYI
                                   TM 6
551 HIYLTVRNPN IVSSSDTRI  AKRMAMLIFT DFLCMAPISF FAISASLKVP
              TM 7
601 LITVSKAKIL LVLFHPINSC ANPFLYAIFT KNFRRDFFIL LSKCGCYEMQ

651 AQIYRTETSS TVHNTHPRNG HCSSAPRVTS GSTYILVPLS HLAQN
```

The first 17 amino acids may be a signal sequence.

Amino acids	695
Molecular weight	
polypeptide)	78 152
(protein)	~240 kDa
Glycosylation	Asn191*, Asn199*, Asn293*, Asn318*
Disulfide bonds	Cys442–Cys517*
Palmitoylation	Cys644*, Cys646*
Phosphorylation (PKA)	None

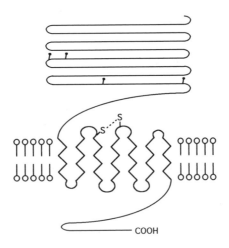

Database accession numbers

	PIR	SWISSPROT	EMBL/GENBANK	REFERENCES
Human	JN0122	P23945	M65085	1
			M95489	
Rat	A34548	P20395		2

Gene
The organization of the gene is not known.

References
1 Minegish, T. et al. (1991) Biochem. Biophys. Res. Commun. **175**, 1125–1130.
2 Sprengel, R. et al. (1990) Mol. Endocrinol. **4**, 525–530.

THE LH/HCG RECEPTOR

FSH and TSH are inactive on the LH/hCG receptor.

Pharmacology

Agonists: LH and hCG induce activation in subnanomolar concentrations.

Antagonists: None have been described.

Radioligands: [^{125}I]hCG (K_d 200 pM).

Distribution

The receptor is found in organs involved in reproductive physiology including testicular Leydig cells, ovarian theca, granulosa, luteal and interstitial cells.

Predominant effector pathways

Activation of adenylyl cyclase through G_s and stimulation of the phosphoinositide pathway through a pertussis-toxin-insensitive G-protein [1].

Amino acid sequence [1-3]

```
  1 MKQRFSALQL LKLLLLLQPP LPRALREALC PEPCNCVPDG ALRCPGPTAG

 51 LTRLSLAYLP VKVIPSQAFR GLNEVIKIEI SQIDSLERIE ANAFDNLLNL

101 SEILIQNTKN LRYIEPGAFI NLPGLKYLSI CNTGIRKFPD VTKVFSSESN

151 FILEICDNLH ITTIPGNAFQ GMNNESVTLK LYGNGFEEVQ SHAFNGTTLT

201 SLELKENVHL EKMHNGAFRG ATGPKTLDIS STKLQALPSY GLESIQRLIA

251 TSSYSLKKLP SRETFVNLLE ATLTYPSHCC AFRNLPTKEQ NFSHSISENF

301 SKQCESTVRK VSNKTLYSSM LAESELSGWD YEYGFCLPKT PRCAPEPDAF
                       TM 1                              TM 2
351 NPCEDIMGYD FLRVLIWLIN ILAIMGNMTV LFVLLTSRYK LTVPRFLMCN
                                                       TM 3
401 LSFADFCMGL YLLLIASVDS QTKGQYYNHA IDWQTGSGCS TAGFFTVFAS
                                                TM 4
451 ELSVYTLTVI TLERWHTITY AIHLDQKLRL RHAILIMLGG WLFSSLIAML
                                 TM 5
501 PLVGVSNYMK VSICFPMDVE TTLSQVYILT ILILNVVAFF IICACYIKIY
                   TM 6
551 FAVRNPELMA TNKDTKIAKK MAILIFTDFT CMAPISFFAI SAAFKVPLIT
           TM 7
601 VTNSKVLLVL FYPINSCANP FLYAIFTKTF QRDFFLLLSK FGCCKRRAEL

651 YRRKDFSAYT SNCKNGFTGS NKPSQSTLKL STLHCQGTAL LDKTRYTEC
```

Amino acids	699
Molecular weight	78 516
Glycosylation	Asn99*, Asn174*, Asn195*, Asn291*, Asn299*, Asn313*
Disulfide bonds	Cys424–Cys499
Palmitoylation	Cys643*, Cys644*
Phosphorylation (PKA)	None

The 341 amino acid extracellular domain displays a number of imperfectly repeated sequences of 25 amino acids which have been termed leucine-rich repeats and are believed to be involved in agonist binding [4].

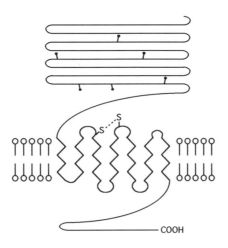

Database accession numbers

	PIR	SWISSPROT	EMBL/GENBANK	REFERENCES
Human	A23728	P22888	M63108	1–3
	A36243			
	B36243			
	B36120			
Mouse			M81318	1
Porcine	A41344	P16582	M29525	5
			M29526	
			M29527	
			M29528	
Rat	A32460	P16235	M26199	6,7

Gene (rat) [7]

The arrangement of the gene suggests that it has been assembled by addition of ten exons to an intronless gene. The porcine and human genes have similar exon–intron structures.

mRNA corresponding to three short forms of the LH receptor are derived by alternative splicing [7]. Some of these may encode for soluble forms of the receptor.

Comment

LH/hCG receptor mRNA is also present in thyroid cells in an incompletely spliced form [1]; it has not been possible to express the message, indicating that it may not encode for a functional protein [8].

References

[1] Frazier, A.L. et al. (1990) Mol. Endocrinology **4**, 1264–1273.
[2] Gudermann, T. et al. (1992) J. Biol. Chem. **267**, 4479–4488.
[3] Minegish, T. et al. (1990) Biochem. Biophys. Res. Commun. **172**, 1049–1054.
[4] Braun, T. et al. (1991) EMBO J. **7**, 1885–1890.
[5] Loosfelt, H. et al. (1990) Science **245**, 525–528.
[6] McFarland, K.C. et al. (1989) Science **245**, 494–499.
[7] Koo, Y.B. et al. (1991) Endocrinology **128**, 2279–2308.
[8] Oikawa, M. et al. (1991) Mol. Endocrinol. **5**, 759–768.

THE TSH RECEPTOR

FSH is inactive on the TSH receptor; hCG has been reported to stimulate thyroid cells but it is unclear whether this is through a TSH receptor or the LH/hCG receptor [1].

Distribution

It is found in thyroid follicular cells and related cell lines, e.g. FRTL5 cells.

Pharmacology

Agonists: TSH induces receptor activation in low nanomolar concentrations.

Antagonists: None have been described.

Radioligands: [^{125}I]BovineTSH (K_d 1 nM).

Predominant effector pathways
Activation of adenylyl cyclase through G$_s$.

Amino acid sequence 1–4

```
  1 MRPADLLQLV LLLDLPRDLG GMGCSSPPCE CHQEEDFRVT CKDIQRIPSL

 51 PPSTQTLKLI ETHLRTIPSH AFSNLPNISR IYVSIDVTLQ QLESHSFYNL

101 SKVTHIEIRN TRNLTYIDPD ALKELPLLKF LGIFNTGLKM FPDLTKVYST

151 DIFFILEITD NPYMTSIPVN AFQGLCNETL TLKLYNNGFT SVQGYAFNGT

201 KLDAVYLNKN KYLTVIDKDA FGGVYSGPSL LDVSQTSVTA LPSKGLEHLK

251 ELIARNTWTL KKLPLSLSFL HLTRADLSYP SHCCAFKNQK KIRGILESLM

301 CNESSMQSLR QRKSVNALNS PLHQEYEENL GDSIVGYKEK SKFQDTHNNA

351 HYYVFFEEQE DEIIGFGQEL KNPQEETLQA FDSHYDYTIC GDSEDMVCTP
                    TM 1
401 KSDEFNPCED IMGYKFLRIV VWFVSLLALL GNVFVLLILL TSHYKLNVPR
     TM 2                                             TM 3
451 FLMCNLAFAD FCMGMYLLLI ASVDLYTHSE YYNHAIDWQT GPGCNTAGFF
                                                    TM 4
501 TVFASELSVY TLTVITLERW YAITFAMRLD RKIRLRHACA IMVGGWVCCF
                                              TM 5
551 LLALLPLVGI SSYAKVSICL PMDTETPLAL AYIVFVLTLN IVAFVIVCCC
                                   TM 6
601 HVKIYITVRN PQYNPGDKDT KIAKRMAVLI FTDFICMAPI SFYALSAILN
            TM 7
651 KPLITVSNSK ILLVLFYPLN SCANPFLYAI FTKAFQRDVF ILLSKFGICK

701 RQAQAYRGQR VPPKNSTDIQ VQKVTHDMRQ GLHNMEDVYE LIENSHLTPK

751 KQGQISEEYM QTVL
```

Amino acids	764
Molecular weight	86 803
Glycosylation	Asn77*, Asn99*, Asn113*, Asn177*, Asn198*, Asn302*
Disulfide bonds	Cys494–Cys569*
Palmitoylation	Cys699*
Phosphorylation (PKA)	None

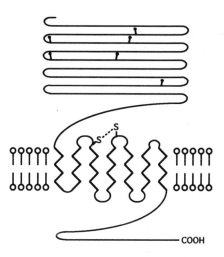

Database accession numbers

	PIR	SWISSPROT	EMBL/GENBANK	REFERENCES
Canine	S06933	P14763	X17147 M29957	5
Human	A34052 A33786 A33789	P16473	M31774 M32215	1–4
Rat	A35956	P21463	M34842	6

Gene
The genomic organization is not known.

References
1 Frazier, A.C. et al. (1990) Mol. Endocrinol. **4**, 1264–1276.
2 Nagayama, Y. et al. (1989) Biochem. Biophys. Res. Commun. **165**, 1184–1190.
3 Libert, F. et al. (1989) Biochem. Biophys. Res. Commun. **165**, 1250–1255.
4 Misrahi, M. et al. (1990) Biochem. Biophys. Res. Commun. **166**, 394–403.
5 Parmentier, M. et al. (1989) Science **246**, 1620–1622.
6 Akamizu, T. et al. (1990) Proc. Natl Acad. Sci. USA **87**, 5677–5681.

Gonadotrophin-releasing hormone

Gonadotrophin-releasing hormone (GnRH) is an important hypothalamic regulator of the reproductive process [1-3]. Its major site of action is in the regulation of the synthesis and release of gonadotrophins (FSH and LH) from the anterior pituitary.

GnRH agonists are of potential therapeutic use in the suppression of prostate cancer, precocious puberty and endometriosis.

Structure

GnRH   Pyr–Glu–His–Trp–Ser–Tyr–Gly–Leu–Arg–Pro–Gly–NH$_2$

Distribution and synthesis

It is synthesized by hypothalamic neurons and transported by the hypophyseal portal system to the anterior pituitary.

Review

1 Andreyko, J.L. *et al.* (1987) Therapeutic uses of gonadotrophin-releasing hormone analogues. *Obstet. Gynecol. Surv.* **42**, 1–21.
2 Conn, P.M. and Crowley, W.F. (1991) Gonadotrophin-releasing hormone and its analogues *N. Engl. J. Med.* **324**, 93–103.
3 Hodgen, G.D. (1989) General application of GnRH agonists in gynecology: past, present and future. *Obstet. Gynecol. Surv.* **44**, 293–296.

THE GnRH RECEPTOR

GnRH receptors are found in anterior pituitary, placenta, ovary, testis, prostate, breast and in certain tumours and immortalized cell lines, e.g. αT3 pituitary gonadotroph cells. They are also present in the nervous system and gonads of certain species.

Pharmacology

Agonists: GnRH induces activation in nanomolar concentrations.

Antagonists: None have been described.

Radioligands: [^{125}I]-[DAla^6]des-Gly^{10}-GnRH (K_d = 0.5 nM).

Predominant effector pathways
Activation of the phosphoinositide pathway through a pertussis-toxin-insensitive G-protein, most likely of the G_q/G_{11} class.

Amino acid sequence (mouse) [1]

```
                                        TM 1
    1 MANSASPEQN QNHCSAINNS IPLMQGNLPT LTLSGKIRVT VTFFLFLLSA
                              TM 2
   51 TFNASFLLKL QKWTQKKEKG KKLSRMKLLL KHLTLANLLE TLIVMPLDGM
         TM 3
  101 WNITVQWYAG ELLCKVLSYL KLFSMYAPAF MMVVISLDRS LAITRPLALK
```

```
      TM 4
151 SNSKVGQSMV GLAWILSSVF AGPQLYIFRM IHLADSSGQT KVFSQCVTHC
               \TM 5
201 SFSQWWHQAF YNFFTFSCLF IIPLFIMLIC NAKIIFTLTR VLHQDPHELQ
                  TM 6
251 LNQSKNNIPR ARLKTLKMTV AFATSFTVCW TPYYVLGIWY WFDPEMLNRL
      TM 7
301 SDPVNHFFFL FAFLNPCFDP LIYGYFSL
```

This is the only known G-protein coupled receptor which is predicted to lack a C-terminal cytoplasmic sequence.

Amino acids	328
Molecular weight	
(polypeptide)	37 730
(protein)	~60 kDa
Glycosylation	Asn18*, Asn102*
Disulfide bonds	Cys114–Cys195/Cys200*
Palmitoylation	None
Phosphorylation (PKC)	Thr64*, Ser74*, Ser153*, Thr237*, Thr265*

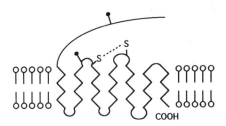

Database accession numbers

	PIR	SWISSPROT	EMBL/GENBANK	REFERENCE
Human			L03380	1
Mouse			L01119	2,3

Gene
The organization of the genome is not known.

References
1 Kaker, S.S. et al. (1992) Biochem. Biophys. Res. Commun. **189**, 289–295.
2 Reinhart, J. et al. (1992) J. Biol. Chem. **267**, 21281–21284.
3 Tsutsumi, M. et al. (1992) Mol. Endocrinol. **16**, 1163–1169.

Histamine

INTRODUCTION

Histamine is distributed within mast cells in all peripheral tissues and is an established mediator of inflammation and allergy [1-6]. It also regulates the release of gastric acid from parietal cells in the gastric mucosa.

It is distributed widely in brain, both in neuronal and non-neuronal cells. Although histamine perikarya are confined to the posterior hypothalamus, histaminergic neurons project to almost all brain regions including spinal cord and cerebral cortex. Histamine has been implicated in a variety of CNS functions including arousal, sexual behaviour and analgesia.

Histamine

Reviews

[1] Goot, H. van der *et al.* (1991) Structural requirements of histamine H_2-agonists and H_2-antagonists. *Handbook Exp. Pharmacol.* **97**, 573–748.

[2] Haaksma, E.E.J. *et al.* (1990) Histamine receptors: subclasses and specific ligands. *Pharmacol. Ther.* **77**, 73–104.

[3] Hill, S.J. (1987) Histamine receptors in the central nervous system: biochemical studies. *Prog. Med. Chem.* **24**, 29–84.

[4] Hill, S.J. (1990) Distribution properties and functional characteristics of three classes of histamine receptor. *Pharmacol. Rev.* **42**, 45–83.

[5] Leurs, R. *et al.* (1991) Histamine agonists and antagonists: recent developments. *Adv. Drug. Res.* **20**, 218–304.

[6] Schwartz, J.C. *et al.* (1991) Histaminergic transmission in the mammalian brain. *Physiol. Rev.* **71**, 1–51.

HISTAMINE RECEPTORS

H_1 and H_2 receptors

These are seven transmembrane spanning proteins and are described below.

The H_3 receptor

The H_3 receptor has not been cloned and its signal transduction mechanism is not known, although radioligand binding studies suggest a possible link to a G-protein [1]. R-α-Methylhistamine and thioperamide (pA2 8.4) are examples of a selective agonist and antagonist, respectively. The receptor is found within the CNS and periphery and is involved in the presynaptic inhibition of neurotransmitter release.

Reference
[1] Kilpatrick, G.J. and Michel, A.D. (1991) *Agents Actions Suppl.* **33,** 69–75.

THE H_1 RECEPTOR

Histamine is active in low micromolar concentrations. The H_1 receptor mediates the increase in vascular permeability induced by histamine at sites of inflammation. H_1 antagonists are used clinically in the treatment of allergic and anaphylactic reactions, and various inflammatory conditions, e.g. hay fever, itching. They are also used in the treatment of motion sickness.

Distribution

The H_1 receptor is distributed widely in the periphery, notably in smooth muscle where it stimulates contraction, e.g. intestine, trachea, bladder and blood vessels, and also in adrenal medulla, vascular endothelium and heart. The H_1 receptor is found throughout the CNS, including cerebral cortex, spinal cord and cerebellum.

Pharmacology

Agonists: No highly selective agonist has been reported; 2-thiazolylethylamine and 2-methylhistamine have a selectivity of less than one order of magnitude.

Antagonists: There are a large number of antagonists of which mepyramine (pA$_2$ 9.1) and triprolidine (pA$_2$ 9.9) are the most selective (> 1000-fold). Examples of antagonists which do not cross the blood–brain barrier are terfenadine and temelastine. A number of H_1 antagonists have additional actions, e.g. promethazine blocks muscarinic receptors; amitriptyline is a tricyclic antidepressant.

Radioligands: The antagonists [^{3}H]mepyramine (K_d 0.8 nM) and [^{125}I]iodobolpyramine [1] (K_d 0.1 nM) are highly selective. [^{3}H](+)-4-Methyldiphenhydramine (K_d 0.8 nM) is a quaternary ligand.

Structures

2-Thiazolylethylamine

2-Methylhistamine

Mepyramine

Triprolidine

Terfenadine

Iodobolpyramine

Predominant effector pathways

Activation of the phosphoinositide pathway through a pertussis-toxin-insensitive G-protein most likely of the G_q/G_{11} class.

Amino acid sequence (bovine) [1]

```
                                              TM 1
  1 MTCPNSSCVF EDKMCQGNKT APANDAQLTP LVVVLSTISL VTVGLNLLVL
                TM 2
 51 YAVRSERKLH TVGNLYIVSL SVADLIVGVV VMPMNILYLL MSRWSLGRPL
      TM 3                                              TM 4
101 CLFWLSMDYV ASTASIFSVF ILCIDRYRSV QQPLKYLRYR TKTRASITIL
                                                    TM 5
151 AAWFLSFLWI IPILGWRHFQ PKTPEPREDK CETDFYNVTW FKVMTAIINF

201 YLPTLLMLWF YAKIYKAVRQ HCQHRELING SFPSFSDMKM KPENLQVGAK

251 KPGKESPWEV LKRKPKDTGG GPVLKPPSQE PKEVTSPGVF SQEKEEKDGE

301 LGKFYCFPLD TVQAQPEAEG SGRGYATINQ SQNQLEMGEQ GLSMPGAKEA
```

```
351 LEDQILGDSQ SFSRTDSDTP AEPAPAKGKS RSESSTGLEY IKFTWKRLRS
                TM 6
401 HSRQYVSGLH MNRERKAAKQ LGFIMAAFII CWIPYFIFFM VIAFCESCCN
        TM 7
451 QHVHMFTIWL GYINSTLNPL IYPLCNENFK KTFKKILHIR S
```

Amino acids	491
Molecular weight	55 954
Glycosylation	Asn5*, Asn18*, Asn187*
Disulfide bonds	Cys101–Cys181*
Palmitoylation	None

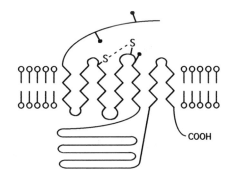

Database accession numbers

	PIR	SWISSPROT	EMBL/GENBANK	REFERENCE
Bovine			D90430	1

Reference

[1] Yamashita, M. et al. (1991) Proc. Natl Acad. Sci. USA **88**, 11515–11519.

THE H₂ RECEPTOR

Histamine is active in high nanomolar/low micromolar concentrations.

Distribution

The H_2 receptor is found in high levels in stomach and heart. It has a limited distribution in smooth muscle, e.g. uterus, and cells of the immune system.
    H_2 antagonists are used clinically in the treatment of peptic ulceration.

Pharmacology

Agonists: Dimaprit is highly selective (> 10 000-fold); other, less selective agonists have been described, e.g. 4-methylhistamine. Impromidine also has high affinity but is an antagonist at the H_3 receptor.

Antagonists: A large number of H$_2$ antagonists are available, with ranitidine (pA$_2$ 7.2) and tiotidine (pA$_2$ 7.2) among the most selective (> 1000-fold).

Radioligands: The selective antagonist, [^{125}I]iodoaminopotentidine [1] (K_d 0.3 nM), has high affinity. [^{3}H]Tiotidine (K_d 15 nM) is also selective but has lower affinity and this may account for its inability to detect H$_2$ receptors in peripheral tissues.

Structures:

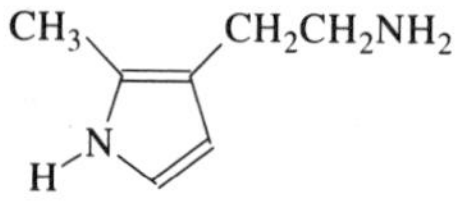

4-Methylhistamine

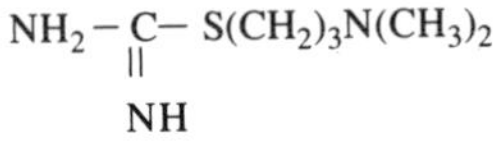

Dimaprit

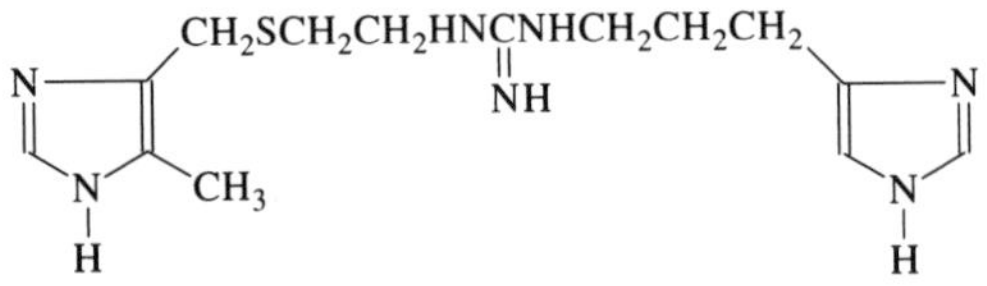

Impromidine

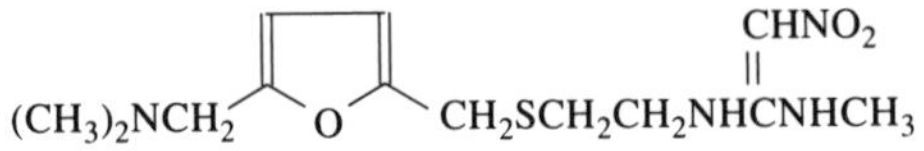

Ranitidine

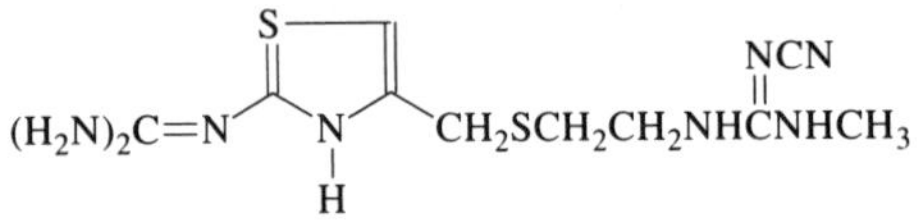

Tiotidine

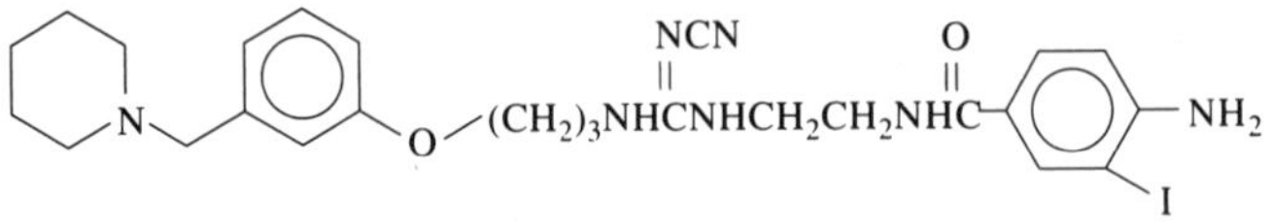

Iodoaminopotentidine

Predominant effector pathways

Formation of cAMP by activation of G_s. Studies with the cloned rat H_2 receptor have also suggested a negative coupling to phospholipase A_2 which is independent of cAMP [2].

Amino acid sequence [3]

```
                           TM 1
  1 MAPNGTASSF CLDSTACKIT ITVVLAVLIL ITVAGNVVVC LAVGLNRRLR
     TM 2                                          TM 3
 51 NLTNCFIVSL AITDLLLGLL VLPFSAIYQL SCKWSFGKVF CNIYTSLDVM
                                     TM 4
101 LCTASILNLF MISLDRYCAV MDPLRYPVLV TPVRVAISLV LIWVISITLS
                                     TM 5
151 FLSIHLGWNS RNETSKGNHT TSKCKVQVNE VYGLVDGLVT FYLPLLIMCI
                                     TM 6
201 TYYRIFKVAR DQAKRINHIS SWKAATIREH KATVTLAAVM GAFIICWFPY
                TM 7
251 FTAFVYRGLR GDDAINEVLE AIVLWLGYAN SALNPILYAA LNRDFRTGYQ

301 QLFCCRLANR NSHKTSLRSN ASQLSRTQSR EPRQQEEKPL KLQVWSGTEV

351 TAPQGATDR
```

Amino acids	359
Molecular weight	40 098
Glycosylation	Asn4*
Disulfide bonds	Cys91–Cys174*
Palmitoylation	Cys304*, Cys305*

The negatively charged Asp186 in TM 5 may be involved in the interaction of the receptor with the positively charged imidazole ring of histamine [4].

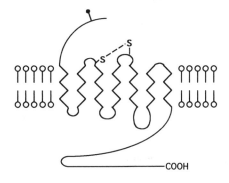

COOH

Database accession numbers

	PIR	SWISSPROT	EMBL/GENBANK	REFERENCES
Canine		P17124	M32701	4
Human	JH0449	P25021	M64799	3
Rat	JQ1278	P25102	S57565	5

Gene (human)
The gene is intronless.

References
[1] Gantz, I. *et al.* (1991) *Biochem. Biophys. Res. Commun.* **178**, 1386–1392.
[2] Traiffort, E. *et al.* (1992) *Proc. Natl Acad. Sci. USA* **89**, 2649–2653.
[3] Ruat, M. *et al.* (1990) *Proc. Natl Acad. Sci. USA* **87**, 1658–1672.
[4] Gantz, I. *et al.* (1991) *Proc. Natl Acad. Sci. USA* **88**, 429–433.
[5] Ruat, M. *et al.* (1991) *Biochem. Biophys. Res. Commun.* **179**, 1470–1478.

5-Hydroxytryptamine   5-HT

INTRODUCTION

5-HT (also known as serotonin) has a widespread distribution in plants and animals [1-15]. It is an important neurotransmitter and local hormone in the CNS and intestine, and is implicated in a vast array of physiological and pathophysiological pathways.

In the periphery, 5-HT contracts a number of smooth muscles including large blood vessels, intestine and uterus; it also induces endothelium-dependent vasodilation through the formation of NO. 5-HT stimulates sensory nerve endings and is one of the components of "nettle sting". It is a mediator of peristalsis and may be involved in platelet aggregation and haemostasis. It may also have a role as an inflammatory mediator and involvement in microvascular control.

In the CNS, 5-HT is thought to be involved in a wide range of functions including control of appetite, mood, anxiety, hallucinations, sleep, vomiting and pain perception. 5-HT receptor ligands have found clinical use in the treatment of depression, migraine and post-operative vomiting, and there is a strong potential for their use in other conditions.

$$HO-\text{[indole ring]}-CH_2CH_2NH_2$$

5-HT

Distribution and synthesis

About 90% of 5-HT in the body is present in chromaffin cells in the wall of the intestine with a small amount in the myenteric plexus. 5-HT is also found in blood where it is present in high concentration in platelets. In the CNS, 5-HT neurons originate primarily in the raphe nuclei of the brainstem and project to most areas of the brain.

It is synthesized from the amino acid tryptophan, via 5-hydroxytryptophan.

Books
1 Fozard, J.R. (ed.) (1989) The Peripheral Actions of 5-Hydroxytryptamine, Oxford University Press, Oxford.
2 Fozard, J.R. and Saxena, P.R. (eds) (1991) Serotonin: Molecular Biology, Receptors and Functional Effects, Birkhäuser Verlag, Basel.
3 Harmon, S. (ed.) (1992) Central and Peripheral 5-HT$_3$ Receptors. Academic Press, London.
4 Sanders-Bush, E. (Ed; 1989) The Serotonin Receptors, Humana Press, New Jersey.
5 Peroutka, S.J. (ed.) (1991) Serotonin Receptor Subtypes: Basic and Clinical Aspects, Wiley-Liss, New York.

Reviews
6 Bockaert, J. *et al.* (1992) The 5-HT$_4$ receptor: a place in the sun *Trends Pharmacol. Sci.* **13**, 141–145.

6 Kalkman, H.O. and Fozard, J.R. (1991) 5HT receptor types and their role in disease. *Curr. Opinion Neurol. Neurosurgery* **4**, 560–565.

7 Frazer, A.S. *et al.* (1990) Subtypes of receptors for serotonin. *Annu. Rev. Pharmacol. Toxicol.* **30**, 307–348.

8 Hartig. P. (1989) Molecular biology of 5-HT receptors. *Trends Pharmacol. Sci.* **10**, 64–69.

9 Hartig. P. *et al.* (1992) A subfamily of 5-HT$_{1D}$ receptor genes. *Trends Pharmacol. Sci.* **13**, 152–159.

10 Hen, R. (1992) Of mice and flies: commonalities among 5-HT receptors. *Trends Pharmacol. Sci.* **13**, 160–165.

11 Hoyer, D. and Schoeffter, P. (1991) 5-HT receptors: subtypes and second messengers. *J. Recept. Res.* **11**, 197–214.

12 Julius, D. (1991) Molecular biology of serotonin receptors. *Annu. Rev. Neurosci.* **14**, 335–360.

13 Peroutka, S.J. (1988) 5-hydroxytryptamine receptor subtypes. *Annu. Rev. Neurosci.* **11**, 45–60.

14 Peroutka, S.J. (1993) 5-hydroxytryptamine receptors. *J. Neurochem.* **60**, 408–416.

15 Waeber, C.P. *et al.* (1990) The serotonin 5-HT$_{1D}$ receptor: a progress review. *Neurochem. Res.* **15**, 567–582.

5-HT RECEPTORS

5-HT$_{1A}$, 5-HT$_{1B}$, 5-HT$_{1D}$, 5-HT$_{1E}$ and 5-HT$_{1F}$ receptors

5-HT$_{1\text{-LIKE}}$ receptors were originally defined by their nanomolar affinity for 5-HT, susceptibility to antagonism by methiothepin and/or methysergide, resistance to antagonism by 5-HT$_2$ and 5-HT$_3$ antagonists and high affinity for the agonist 5-carboxamidotryptamine [1].

Five subtypes of 5-HT$_{1\text{-LIKE}}$ receptors are now identified, although they do not all satisfy the above criteria, while the 5-HT$_{1C}$ receptor has been reclassified as 5-HT$_{2C}$. They are all seven transmembrane proteins and are linked to the inhibition of adenylyl cyclase. They share a high degree of sequence homology (~50%) and have overlapping pharmacological specificities. In view of the recent identification of some of these receptors in cloning studies, the selectivity of several of the ligands shown overleaf is uncertain.

5-HT$_{2A}$, 5-HT$_{2B}$ and 5-HT$_{2C}$ receptors

The 5-HT$_2$ receptor was originally defined by its ability to display micromolar affinity for 5-HT, to be labelled with [^{3}H]spiperone and by its susceptibility to 5-HT$_2$ antagonists. It is now established that at least three members of the family exist (including the reclassified 5-HT$_{1C}$ receptor), which share ~70% homology in their transmembrane regions (overall homology ~50 %) and stimulate the phosphoinositide pathway. The agonist α-methyl-5-HT and antagonist LY 53857 are selective to the 5-HT$_2$ class of receptor.

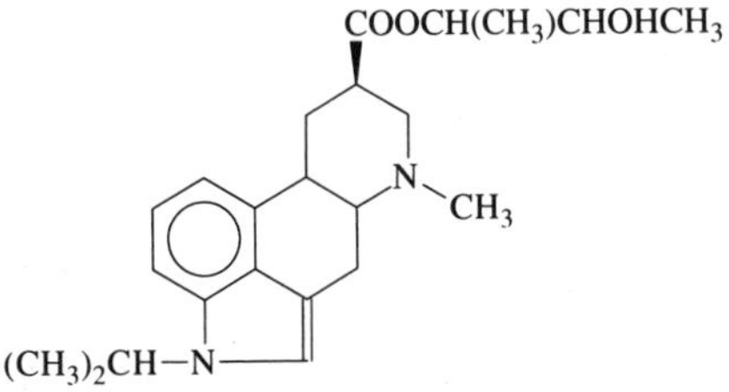

COOCH(CH$_3$)CHOHCH$_3$

LY 53857

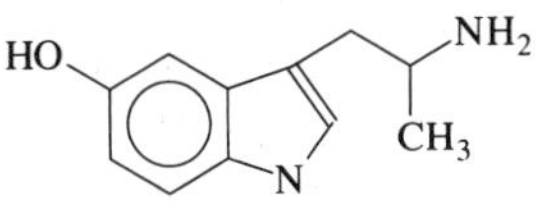

α-Methyl-5-HT

5-HT$_3$ receptor

The 5-HT$_3$ receptor (originally M receptor) contains an intrinsic cation channel; it is not G-protein coupled. It is localized exclusively to neurons in the central and peripheral nervous systems. A large number of selective agonists (e.g. 2-methyl-5-HT) and antagonists (e.g. tropisetron and ondansetron) have been described. It is described in further detail in the Ion Channels FactsBook [1].

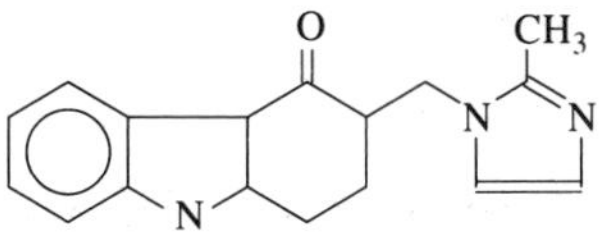

Ondansetron

5-HT$_4$ receptor

This receptor was recently identified by the high potency of certain benzamide agonists, e.g. cisapride, and the ability of the 5-HT$_3$ receptor antagonist, tropisetron (in contrast to other 5-HT$_3$ antagonists), to act as a low-affinity antagonist. Ligands with increased selectivity are beginning to appear. It is described below.

5-HT$_5$ receptor

This receptor was identified recently following isolation of its cDNA [2]. It has little sequence homology with other 5-HT receptor subtypes and is therefore believed to represent a new class of receptor.

Drosophila 5-HT receptors

A putative 5-HT receptor has been identified in *Drosophila*, 5-HT$_{dro1}$, which has closest resemblence to the 5-HT$_{1A}$ receptor but is linked to activation of adenylyl cyclase [3]. Two additional *Drosophila* 5-HT receptors, 5-HT$_{dro2B}$ and 5-HT$_{dro2B}$, inhibit adenylyl cyclase and activate phospholipase C, respectively [4].

References
1 Conley, E. (1994) The Ion Channel FactsBook, Academic Press, London.
2 Bradley, P.R. *et al.* (1986) *Neuropharmacology* **25**, 563–576.
3 Witz, P. *et al.* (1990) *Proc. Natl Acad. Sci. USA* **87**, 8940–8944.
4 Saudou, F. *et al.* (1992) *EMBO J.* **11**, 7–17.

THE 5-HT₁A RECEPTOR

5-HT induces activation in nanomolar concentrations.

Distribution

The 5-HT1A receptor is found presynaptically and postsynapically in neurons in the CNS, e.g. dorsal raphe, hippocampus, medulla and cerebral cortex, and in the periphery, e.g. ileum. It is also abundant in foetal lymphatic tissue, particularly lymph nodes, spleen and thymus.

Clinically, 5-HT1A receptor ligands represent potential anxiolytic and anti-hypertensive agents.

Pharmacology

Agonists: A large number of potent and selective agonists for the 5-HT1A receptor have been described. 8-OH-DPAT has a selectivity of more than 100-fold, while ipsapirone (partial agonist) and buspirone also have high selectivity.

Antagonists: No selective antagonist has been described.

Radioligands: [3H]8-OH-DPAT (K_d 0.1 and 10 nM) binds to high- and low-affinity sites; [3H]ipsapirone (K_d 20 nM) is also selective.

Structures:

8-OH-DPAT

Ipsapirone

Predominant effector pathways
Inhibition of adenylyl cyclase and activation of K+ channels through a pertussis-toxin-sensitive G-protein most likely of the G_i/G_o class.

Amino acid sequence [1]

```
                                              TM 1
  1 MDVLSPGQGN NTTSPPAPFE TGGNTTGISD VTVSYQVITS LLLGTLIFCA
                      TM 2
 51 VLGNACVVAA IALERSLQNV ANYLIGSLAV TDLMVSVLVL PMAALYQVLN
          TM 3
101 KWTLGQVTCD LFIALDVLCC TSSILHLCAI ALDRYWAITD PIDYVNKRTP
       TM 4                                      TM 5
151 RPRALISLTW LIGFLISIPP ILGWRTPEDR SDPDACTISK DHGYTIYSTF

201 GAFYIPLLLM LVLYGRIFRA ARFRIRKTVK KVEKTGADTR HGASPAPQPK

251 KSVNGESGSR NWRLGVESKA GGALCANGAV RQGDDGAALE VIEVHRVGNS
                                                        TM 6
301 KEHLPLPSEA GPTPCAPASF ERKNERNAEA KRKMALARER KTVKTLGIIM
                              TM 7
351 GTFILCWLPF FIVALVLPFC ESSCHMPTLL GAIINWLGYS NSLLNPVIYA

401 YFNKDFQNAF KKIIKCNFCR
```

Amino acids	421
Molecular weight	46 029
Chromosome	5
Glycosylation	Asn10*, Asn11*, Asn24*
Disulfide bonds	Cys109–Cys186*

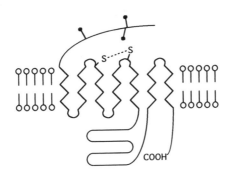

Database accession numbers

	PIR	SWISSPROT	EMBL/GENBANK	REFERENCES
Human		P08908	X13556	1, 2
Rat	A35181	P19327	J05276	3
	JH0315			

Gene (human)
The gene is intronless.

References

1 Kobilka, B.K. *et al.* (1987) *Nature* **329**, 75–79.
2 Fargin, A. *et al.* (1988) *Nature* **335**, 358–360.
3 Albert, P. *et al.* (1990) *J. Biol. Chem.* **265**, 5825–5832.

THE 5-HT$_{1B}$ RECEPTOR

5-HT induces activation in high nanomolar concentrations. This receptor has also been named the 5-HT$_{1D\beta}$[1] and 5-HT$_{S12}$[2] receptor.

The 5-HT$_{1B}$ receptor was originally identified in rat where it has a distinct pharmacological profile. In human, however, it shares an almost identical pharmacology to the 5-HT$_{1D}$ receptor (see 5-HT$_{1D}$ receptor) and the lack of selective ligands has prevented identification of their individual physiological roles.

Distribution

In the CNS it is found in striatum, medulla, hippocampus, frontal cortex and amygdala. In the periphery it is found in vascular smooth muscle.

The 5-HT$_{1B}$ / 5-HT$_{1D}$ receptor may be the therapeutic substrate of the anti-migraine drug, sumatriptan; these sites are also implicated in feeding behaviour, anxiety, depression, cardiac function and movement.

Pharmacology

Agonists: Sumatriptan induces activation of human 5-HT$_{1B}$ and 5-HT$_{1D}$ receptors in low nanomolar concentrations, but has low affinity for rat 5-HT$_{1B}$ receptors. CP 93129 is selective to rat 5-HT$_{1B}$ receptors.

Antagonists: No selective antagonist has been described.

Radioligands: In rat, but not in human, the receptor can be labelled with [^{3}H]CP 96 501 [1] or [^{125}I]cyanopindolol in the presence of a β-adrenoceptor antagonist.

Structures

Cyanopindolol   **CP 96501**

Predominant effector pathways

Inhibition of adenylyl cyclase and activation of K$^+$ channels through a pertussis-toxin-sensitive G-protein most likely of the G$_i$/G$_o$ class.

Amino acid sequence [2-8]

```
                                                            TM 1
  1 MEEPGAQCAP PPPAGSETWV PQANLSSAPS QNCSAKDYIY QDSISLPWKV
                                        TM 2
 51 LLVMLLALIT LATTLSNAFV IATVYRTRKL HTPANYLIAS LAVTDLLVSI
               TM 3
101 LVMPISTMYT VTGRWTLGQV VCDFWLSSDI TCCTASILHL CVIALDRYWA
               TM 4
151 ITDAVEYSAK RTPKRAAVMI ALVWVFSISI SLPPFFWRQA KAEEEVSECV
          TM 5
201 VNTDHILYTV YSTVGAFYFP TLLLIALYGR IYVEARSRIL KQTPNRTGKR

251 LTRAQLITDS PGSTSSVTSI NSRVPDVPSE SGSPVYVNQV KVRVSDALLE
                        TM 6                          TM 7
301 KKKLMAARER KATKTLGIIL GAFIVCWLPF FIISLVMPIC KDACWFHLAI

351 FDFFTWLGYL NSLINPIIYT MSNEDFKQAF HKLIRFKCTS
```

Amino acids	390
Molecular weight	43 539
Chromosome	6
Glycosylation sites	Asn24*, Asn32*
Disulfide bonds	None

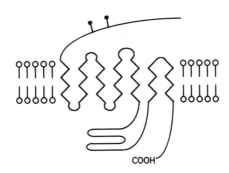

Database accession numbers

	PIR	SWISSPROT	EMBL/GENBANK	REFERENCES
Human	JN0268	P28222	M81590	2-8
			M83180	
			D10995	
			M84986	
			M75128	
			M89478	
Mouse	S22191	P28334	M85151	9,10
			Z11597	
Rat	S18637	P28564	M89954	10,11
			X62944	

Gene (human)
The gene is intronless.

References
[1] Koe, B.K. et al. (1992) J. Neurochem. **58**, 1268–1276.
[2] Weinshank, R.L. et al. (1992) Proc. Natl Acad. Sci. USA **89**, 3630–3634.
[3] Levy, F.O. et al. (1992) J. Biol. Chem. **267**, 7553–7562.
[4] Jin, H. et al. (1992) J. Biol. Chem. **267**, 5735–5738.
[5] Demchyshyn, L. et al. (1992) Proc. Natl Acad. Sci. USA **89**, 5522–5526.
[6] Hamblin, M.W. and Metcalf, M.A. (1992) Biochem. Biophys. Res. Commun. **184**, 752–759.
[7] Mochizuki, D. et al. (1992) Biochem. Biophys. Res. Commun. **185**, 517–523.
[8] Veldman, S.A. and Bienkowski, M.J. (1992) Mol. Pharmacol. **42**, 439–444.
[9] Maroteaux, L. et al. (1992) Proc. Natl Acad. Sci. USA **89**, 3020–3024.
[10] Voigt, M.M. et al. (1991) EMBO J. **10**, 4017–4023.
[11] Zgombick, J.M. et al. (1991) Mol. Pharmacol. **40**, 1036–1042.

THE 5-HT$_{1D}$ RECEPTOR

5-HT induces activation in high nanomolar concentrations. It has also been named the 5-HT$_{1D\alpha}$ receptor [1].

Human 5-HT$_{1B}$ and 5-HT$_{1D}$ receptors share an almost identical pharmacology and no selective ligands have been described (see Comment). This has prevented identification of their physiological role and distribution in human.

Distribution

5-HT$_{1D}$ receptors are found in neurons in the CNS, with the highest levels in substantia nigra, basal ganglia and subiculum. In the periphery, they are found in various vascular smooth muscles, e.g. coronary artery, saphenous vein.

The 5-HT$_{1B}$ / 5-HT$_{1D}$ receptor may be the therapeutic substrate of the anti-migraine drug, sumatriptan; they are also implicated in feeding behaviour, anxiety, depression, cardiac function and movement.

Pharmacology

Agonists: Sumatriptan induces activation of human 5-HT$_{1B}$ and 5-HT$_{1D}$ receptors in low nanomolar concentrations; it also has reasonable affinity for the newly identified 5-HT$_{1F}$ receptor.

Antagonists: No selective antagonist has been described.

Radioligands: 5-HT-O-carboxymethylglycyl[^{125}I]iodotyrosinamide (K_d 2–5 nM) is claimed to be a selective ligand for 5-HT$_{1B}$/5-HT$_{1D}$ receptors [1].

Structures:

Sumatriptan

5-HT-*O*-carboxymethylglycyltyrosinamide

Predominant effector pathways

Inhibition of adenylyl cyclase and activation of K⁺ channels through a pertussis-toxin-sensitive G-protein, most likely of the G_i/G_o class.

Amino acid sequence [2,3]

```
                                                TM 1
  1 MSPLNQSAEG LPQEASNRSL NATETSEAWD PRTLQALKIS LAVVLSVITL
                         TM 2
 51 ATVLSNAFVL TTILLTRKLH TPANYLIGSL ATTDLLVSIL VMPISIAYTI
            TM 3
101 THTWNFGQIL CDIWLSSDIT CCTASILHLC VIALDRYWAI TDALEYSKRR
       TM 4                                         TM 5
151 TAGHAATMIA IVWAISICIS IPPLFWRQAK AQEEMSDCLV NTSQISYTIY

201 STCGAFYIPS VLLIILYGRI YRAARNRILN PPSLYGKRFT TAHLITGSAG

251 SSLCSLNSSL HEGHSHSAGS PLFFNHVKIK LADSALERKR ISAARERKAT
       TM 6                                 TM 7
301 KILGIILGAF IICWLPFFVV SLVLPICRDS CWIHPALFDF FTWLGYLNSL

351 INPIIYTVFN EEFRQAFQKI VPFRKAS
```

Amino acids	377
Molecular weight	41 906
Chromosome	1
Glycosylation sites	Asn5*, Asn17*, Asn21*
Disulfide bonds	Cys111–Cys188*
Palmitoylation	None

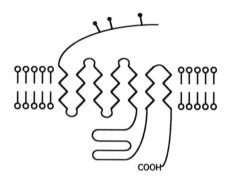

Database accession numbers

	PIR	SWISSPROT	EMBL/GENBANK	REFERENCES
Canine	B30341	P11614	X14049	4
	S12822			
Human		P28221	M81589	2,3
			M89955	
Rat		P28565	M89953	unpublished

The canine gene was originally identified as an orphan receptor and named *RDC4* [4]; it was later identified as the 5-HT$_{1D}$ receptor [5,6].

Gene (human)
The gene is intronless [2,3].

Comment
The 5-HT$_{1B}$ receptor subtype was identified in rodent tissue but an equivalent receptor appeared to be absent in man. Subsequently, cloning studies revealed a receptor in human with ~93% sequence homology to the rat 5-HT$_{1B}$ receptor but with a pharmacology similar to the human 5-HT$_{1D}$ receptor. Mutation of Thr355 in the seventh transmembrane region of the human 5-HT$_{1B}$ receptor to asparagine located in the equivalent position in the rat receptor yields a protein with an almost identical pharmacology to the rat 5-HT$_{1B}$ receptor [7,8]. Thus a single amino acid underlies the major pharmacological differences between rat and human 5-HT$_{1B}$ receptors.

References
1 Boulenguez, P. *et al.* (1991) *J. Neurochem.* **58**, 951–959.
2 Weinshank, R.L. *et al.* (1992) *Proc. Natl Acad. Sci.* USA **89**, 3630–3634.
3 Hamblin, M.W. and Metcalf, M.A. (1991) *Mol. Pharmacol.* **40**, 143–148.
4 Libert, F. *et al.* (1989) *Science* **244**, 569–572.
5 Zgombick, J.M. *et al.* (1991) *Mol. Pharmacol.* **40**, 1036–1042.
6 Maenhaut, C. *et al.* (1991) *Biochem. Biophys. Res. Commun.* **180**, 1460–1468.
7 Oksenberg, D. *et al.* (1992) *Nature* **360**, 161–163.
8 Parker, E.M. *et al.* (1993) *J. Neurochem.* **60**, 380–383.

THE 5-HT₁E RECEPTOR

The existence of this receptor was suggested from binding studies [1] and was confirmed by cloning of the protein [2,3]. 5-HT induces activation in nanomolar concentrations. It has also been named 5-HTS31 [3].

Distribution

The 5-HT₁E receptor was originally described in cerebral cortex, but the lack of selective ligands has prevented a full description of its distribution.

Pharmacology

Agonists: No selective agonist exists; it can be distingushed from other 5-HT₁-LIKE receptors by low affinity (micromolar) for 5-carboxamidotryptamine [1].

Antagonists: No selective antagonist has been described.

Radioligands: It can be labelled with [³H]5-HT [1] in the presence of selective inhibitors of other receptor subtypes.

Predominant effector pathways
It is linked to inhibition of adenylyl cyclase although the effect of pertussis toxin on this response has not been investigated.

Amino acid sequence [2,3]

```
                              TM 1
  1 MNITNCTTEA SMAIRPKTIT EKMLICMTLV VITTLTTLLN LAVIMAIGTT
       TM 2                                            TM 3
 51 KKLHQPANYL ICSLAVTDLL VAVLVMPLSI IYIVMDRWKL GYFLCEVWLS
                                         TM 4
101 VDMTCCTCSI LHLCVIALDR YWAITNAIEY ARKRTAKRAA LMILTVWTIS
                                    TM 5
151 IFISMPPLFW RSHRRLSPPP SQCTIQHDHV IYTIYSTLGA FYIPLTLILI

201 LYYRIYHAAK SLYQKRGSSR HLSNRSTDSQ NSFASCKLTQ TFCVSDFSTS
                                                    TM 6
251 DPTTEFEKFH ASIRIPPFDN DLDHPGERQQ ISSTRERKAA RILGLILGAF
                         TM 7
301 ILSWLPFFIK ELIVGLSIYT VSSEVADFLT WLGYVNSLIN PLLYTSFNED

351 FKLAFKKLIR CREHT
```

Amino acids	365
Molecular weight	41 682
Glycosylation	Asn2*, Asn5*
Disulfide bonds	None

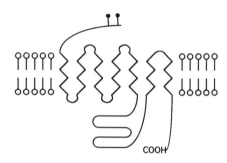

Database accession numbers

	PIR	SWISSPROT	EMBL/GENBANK	REFERENCES
Human	S20579	P28566	M91467	2,3

Gene

The organization of the gene is not known.

References

1 Leonhardt, S. et al. (1989) J. Neurochem. **53**, 465–471.
2 McAllister, G. et al. (1992) Proc. Natl Acad. Sci. USA **89**, 5517–5521.
3 Levy, F.O. et al. (1992) FEBS Lett. **296**, 201–206.

THE 5-HT$_{1F}$ RECEPTOR

This receptor was identified recently following isolation of its cDNA [1]. It is also known as 5-HT$_{1E\beta}$ and 5-HT$_6$. It appears to be a member of the 5-HT$_{1-LIKE}$ family of receptors, with 5-HT$_2$ ligands having low affinity.

Distribution

It has a limited distribution in peripheral tissues and has been detected only in uterus and mesentary. In the CNS, mRNA is present in cerebral cortex, hippocampus, raphe and spinal cord. The elucidation of its physiological role will require development of selective ligands.

Pharmacology

Agonists: there are no selective agonists; the 5-HT$_{1D}$ agonist sumatriptan has reasonable affinity.

Antagonists: There are no selective antagonists.

Radioligands: [^{125}I]LSD (K_d 1 nM) is a nonselective radioligand with high affinity.

Predominant effector pathways

Inhibition of adenylyl cyclase through an uncharacterized G-protein, most likely of the G$_i$/G$_o$ class.

Amino acid sequence [1]

```
                              TM 1
  1 MDFLNSSDQN LTSEELLNRM PSKILVSLTL SGLALMTTTI NSLVIAAIIV
            TM 2                                       TM 3
 51 TRKLHHPANY LICSLAVTDF LVAVLVMPFS IVYIVRESWI MGQVVCDIWL
                                                    TM 4
101 SVDITCCTCS ILHLSAIALD RYRAITDAVE YARKRTPKHA GIMITIVWII
                              TM 5
151 SVFISMPPLF WRHQGTSRDD ECIIKHDHIV STIYSTFGAF YIPLALILIL

201 YYKIYRAAKT LYHKRQASRI AKEEVNGQVL LESGEKSTKS VSTSYVLEKS
                                                    TM 6
251 LSDPSTDFDK IHSTVRSLRS EFKHEKSWRR QKISGTRERK AATTLGLILG
                              TM 7
301 AFVICWLPFF VKELVVNVCD KCKISEEMSN FLAWLGYLNS LINPLIYTIF

351 NEDFKKAFQK LVRCRC
```

Amino acids	366
Molecular weight	41 977
Glycosylation	Asn5*, Asn10*
Disulfide bonds	Cys95–Cys172*

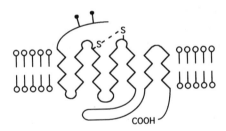

Database accession numbers

	PIR	SWISSPROT	EMBL/GENBANK	REFERENCES
Human			L04962	1
Mouse			Z114224	2

Gene
There are no introns in the coding domain.

References
[1] Adham, N. et al. (1993) Proc. Natl Acad. Sci. USA 90, 408–412.
[2] Amlaiky, N. et al. (1992) J. Biol. Chem. 267, 19761–19764.

THE 5-HT2A RECEPTOR

5-HT induces activation in low micromolar concentrations. It was previously named 5HT2 or D subtype of 5-HT receptor.

Distribution

It is less widely expressed in the CNS relative to the 5-HT2C receptor with the highest levels in the cerebral cortex; it is also present in olfactory tubercle, hypothalamus, hippocampus and spinal cord. In the periphery, it is found in a number of vascular and nonvascular smooth muscles and in platelets.

Pharmacology

Agonists: There are no selective agonists which distinguish 5-HT2 receptor subtypes.

Antagonists: Ritanserin (pA2 9.5) and ketanserin (pA2 9.0) have a selectivity of less than 100-fold.

Radioligands: The dopamine receptor antagonist, [³H]spiperone (K_d 0.5 nM), and [³H]ketanserin (K_d 2 nM) are selective.

Structures

Ketanserin

Ritanserin

Spiperone

Predominant effector pathways

Activation of phosphoinositide hydrolysis through a pertussis-toxin-insensitive G-protein most likely of the G_q/G_{11} class.

Amino acid sequence [1]

```
  1 MDILCEENTS LSSTTNSLMQ LNDDTRLYSN DFNSGEANTS DAFNWTVDSE
                                  TM 1
 51 NRTNLSCEGC LSPSCLSLLH LQEKNWSALL TAVVIILTIA GNILVIMAVS
              TM 2         NWSALL    TAVVIILTIA            TM 3
101 LEKKLQNATN YFLMSLAIAD MLLGFLVMPV SMLTILYGYR WPLPSKLCAV
              YFLMSLAIAD MLLGFLVMPV SM                    CAV
                                                        TM 4
151 WIYLDVLFST ASIMHLCAIS LDRYVAIQNP IHHSRFNSRT KAFLKIIAVW
    WIYLDVLFST ASIMHLCAIS                      KAFLKIIAVW
                                                  TM 5
201 TISVGISMPI PVFGLQDDSK VFKEGSCLLA DDNFVLIGSF VSFFIPLTIM
    TISVGISMPI PVF                      NFVLIGSF VSFFIPLTIM

251 VITYFLTIKS LQKEATLCVS DLGTRAKLAS FSFLPQSSLS SEKLFQRSIH
    VITYFLTIKS                                  TM 6
301 REPGSYTGRR TMQSISNEQK ACKVLGIVFF LFVVMWCPFF ITNIMAVICK
              TM 7              GIVFF LFVVMWCPFF ITNIMA
351 ESCNEDVIGA LLNVFVWIGY LSSAVNPLVY TLFNKTYRSA FSRYIQCQYK
              LLNVFVWIGY LSSAVNPLVY TLFN

401 ENKKPLQLIL VNTIPALAYK SSQLQMGQKK NSKQDAKTTD NDCSMVALGK

451 QHSEEASKDN SDGVNEKVSC V
```

Amino acids	471
Molecular weight	52 603
Glycosylation	Asn8*, Asn38*, Asn44*, Asn51*, Asn54*
Disulfide bonds	Cys148–Cys227*
Palmitoylation	Cys397*

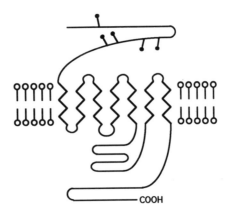

Database accession numbers

	PIR	SWISSPROT	EMBL/GENBANK	REFERENCES
Hamster	S11280	P18599	X53797	2
Human	JS0615	P28223	X57830	1
			M86841	
Rat	A34863	P14842	X13971	3,4
			M36966	
	S02011		M30705	

Gene (human)

The gene has two introns within the coding region which are also present in complementary positions in the 5-HT$_{2B}$ and 5-HT$_{2C}$ receptor [5].

References

1 Saltzman, A.G. et al. (1991) Biochem. Biophys. Res. Commun. **181**, 1469–1478.
2 Chambard, J.C. et al. (1990) Nucleic Acids Res. **18**, 5282.
3 Pritchett, D.B. et al. (1988) EMBO J. **7**, 4135–4140.
4 Julius, D. et al. (1990) Proc. Natl Acad. Sci. USA **87**, 928–932.
5 Foguet, M. et al. (1992) EMBO J. **11**, 3481–3487.

THE 5-HT$_{2B}$ RECEPTOR

5-HT induces activation in low micromolar concentrations. It was previously called the 5-HT$_{2F}$ receptor.

Distribution

The 5-HT$_{2B}$ receptor was cloned from rat stomach fundus, a tissue which had been previously observed to have a unique pharmacological profile to 5-HT and its analogues. mRNA is absent or present in very low levels in other peripheral tissues and in the CNS.

Pharmacology

Agonists: There are no selective agonists which distinguish 5-HT$_2$ receptor subtypes.

Antagonists: There are no selective antagonists.

Predominant effector pathways
Activation of the phosphoinositide pathway through an uncharacterized G-protein, most likely of the G$_q$/G$_{11}$ class.

Amino acid sequence (rat) [1]

```
  1 MASSYKMSEQ STISEHILQK TCDHLILTDR SGLKAESAAE EMKQTAENQG
            TM 1                                      TM 2
 51 NTVHWAALLI FAVIIPTIGG NILVILAVSL EKRLQYATNY FLMSLAVADL
                                TM 3
101 LVGLFVMPIA LLTIMFEATW PLPLALCPAW LFLDVLFSTA SIMHLCAISL
                                TM 4
151 DRYIAIKKPI QANQCNSRTT AFVKITVVWL ISIGIAIPVP IKGIEADVVN
                 TM 5
201 AHNITCELTK DRFGSFMLFG SLAAFFAPLT IMIVTYFLTI HALRKKAYLV

251 RNRPPQRLTR WTVSTVLQRE DSSFSSPEKM VMLDGSHKDK ILPNSTDETL
                                          TM 6
301 MRRMSSAGKK PAQTISNEQR ASKVLGIVFL FFLLMWCPFF ITNVTLALCD
             TM 7
351 SCNQTTLKTL LQIFVWVGYV SSGVNPLIYT LFNKTFREAF GRYITCNYQA

401 TKSVKVLRKC SSTLYFGNSM VENSKFFTKH GIRNGINPAM YQSPVRLRSS

451 TIQSSSIILL NTFLTENDGD KVEDQVSYI
```

Amino acids	479
Molecular weight	53 651
Glycosylation	Asn203*
Disulfide bonds	None
Palmitoylation	Cys396*
Phosphorylation (PKA)	Ser305*,Ser411*, Ser449*
Phosphorylation (PKC)	None

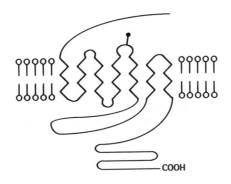

175

Database accession numbers

	PIR	SWISSPROT	EMBL/GENBANK	REFERENCE
Human			X66842	1

This protein is highly homologous to a partial amino acid sequence cloned from mouse stomach [2].

Gene

There are two introns in the coding domain in positions corresponding to those in the 5-HT$_{2A}$ receptor.

References

1 Kursar, J.D. *et al.* (1992) *Mol. Pharmacol.* **42**, 549–557.
2 Foguet, M.H. *et al.* (1992) *Neuroreport* **3**, 345–348.

THE 5-HT$_{2C}$ RECEPTOR

5-HT induces activation in high nanomolar concentrations. It was originally named 5-HT$_{1C}$ receptor.

Distribution

It is distributed widely both pre- and postsynaptically in neurons in the CNS, with the highest levels in the choroid plexus. It is also found in hippocampus, globus pallidus, substantia nigra and raphe, with lower levels in the cerebral cortex.

Pharmacology

Agonists: There are no selective agonists which distinguish 5-HT$_2$ receptor subtypes.

Antagonists: Mesulergine (pA$_2$ 9.1) has a selectivity of ~ tenfold.

Radioligands: [^{3}H]Mesulergine (K_d 3 nM) has low selectivity.

Structures:

Mesulergine

Predominant effector pathways

Activation of the phosphoinositide pathway through a pertussis-toxin-insensitive G-protein most likely of the G$_q$/G$_{11}$ class.

Amino acid sequence [1]

```
  1 MVNLRNAVHS FLVHLIGLLV WQCDISVSPV AAIVTDIFNT SDGGRFKFPD
       TM 1                                        TM 2
 51 GVQNWPALSI VIIIMTIGG NILVIMAVSM EKKLHNATNY FLMSLAIADM
                                    TM 3
101 LVGLLVMPLS LLAILYDYVW PLPRYLCPVW ISLDVLFSTA SIMHLCAISL
                                    TM 4
151 DRYVAIRNPI EHSRFNSRTK AIMKIAIVWA ISIGVSVPIP VIGLRDEEKV
             TM 5
201 FVNNTTCVLN DPNFVLIGSF VAFFIPLTIM VITYCLTIYV LRRQALMLLH

251 GHTEEPPGLS LDFLKCCKRN TAEEENSANP NQDQNARRRK KKERRPRGTM
             TM 6                                        TM 7
301 QAINNERKAS KVLGIVFFVF LIMWCPFFIT NILSVLCEKS CNQKLMEKLL

351 NVFVWIGYVC SGINPLVYTL FNKIYRRAFS NYLRCNYKVE KKPPVRQIPR

401 VAATALSGRE LNVNIYRHTN EPVIEKASDN EPGIEMQVEN LELPVNPSSV

451 VSERISSV
```

There is a unique hydrophobic sequence in the N-terminus which is absent in the other 5-HT_2 receptors

Amino acids	458
Molecular weight	51 821
Glycosylation	Asn39*, Asn204*
Disulfide bonds	Cys127–Cys207*
Palmitoylation	Cys385*

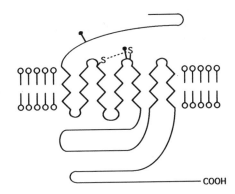

Database accession numbers

	PIR	SWISSPROT	EMBL/GENBANK	REFERENCES
Human	J50616	P28335	M81778	1
Mouse			M63685	2
Rat	A32605	P08908	M21410	3

Gene

There are two introns in the coding domain in positions corresponding to those in the 5-HT$_{2A}$ receptor and a third intron in the hydrophobic region of the N-terminus [4].

References

[1] Saltzman, A.G. et al. (1991) *Biochem. Biophys. Res. Commun.* **181**, 1469–1478.
[2] Yu, L. et al. (1991) *Mol. Brain Res.* **11**, 143–149.
[3] Julius, D. et al. (1988) *Science* **241**, 558–564.
[4] Foguet, M. et al. (1992) *NeuroReport* **3**, 345–348.

THE 5-HT$_4$ RECEPTOR

In the absence of high-affinity selective radioligands, the characterization of this receptor has been achieved primarily through functional assays. It has not been cloned.

Distribution

It is found in neurons in the CNS with the highest levels in colliculus and hippocampus. In the periphery it is found in myenteric neurons in ileum and in smooth muscle, e.g. rat oesophagus and heart muscle.

The established gastrointestinal effects of certain susbtituted benzamides, e.g metoclopramide, may result from activation of 5-HT$_4$ receptors.

Pharmacology

Agonists: A number of benzamides, e.g. zacopride, and benzimidazoles, e.g. BIMU8, are selective partial agonists to the 5-HT$_4$ receptor but are also potent antagonists at the 5-HT$_3$ receptor [1]. SC 49518 is a benzamide derivative with partial agonist activity and low affinity at 5-HT$_3$ receptors [2].

Antagonists: RS 23597-190 (pA$_2$ 7.8) [3] has a selectivity of ~100-fold relative to 5-HT$_3$ receptors.

Radioligands: No high-affinity selective ligands have been described.

Structures:

SC 49518

Predominant effector pathways

It is linked to the adenylyl cyclase pathway through G$_S$.

References

1 Bockhaert, J. *et al.* (1992) *Trends Pharmacol. Sci.* **13**, 141–145.
2 Gullikson, G.W. *et al.* (1992) *J. Pharmacol. Ther.* **264**, 240–248.
3 Eglen, R. *et al.* (1992) *Br. J. Pharmacol.* **107**, 439P.

THE 5-HT₅ RECEPTOR

Although the 5-HT$_5$ receptor has little sequence homology to the other 5-HT receptors, it has a similar pharmacology to the 5-HT$_{1D}$ receptor.

Distribution

In the CNS, its mRNA is found in cerebral cortex, hippocampus, habenula, olfactory bulb and granular layer of the cerebellum. Its distribution in the periphery has not been investigated.

Pharmacology

Agonists: There are no selective agonists.

Antagonists: There are no selective antagonists.

Radioligands: [^{125}I]LSD (K_d 0.34 pM) is a nonselective radioligand with high affinity.

Predominant effector pathways
Although high-affinity agonist binding is inhibited by GTP, indicating that the receptor is coupled to a G-protein, it does not appear to be linked to the adenylyl cyclase or phosphoinositide pathways [1].

Amino acid sequence (mouse) [1]

```
                                                      TM 1
  1 MDLPVNLTSF SLSTPSSLEP NRSLDTEVLR PSRPFLSAFR VLVLTLLGFL
                                     TM 2
 51 AAATFTWNLL VLATILKVRT FHRVPHNLVA SMAISDVLVA VLVMPLSLVH
                 TM 3
101 ELSGRRWQLG RRLCQLWIAC DVLCCTASIW NVTAIALDRY WSITRHLEYT
         TM 4                                          TM 5
151 LRTRKRVSNV MILLTWALST VISLAPLLFG WGETYSEPSE ECQVSREPSY

201 TVFSTVGAFY LPLWLVLFVY WKIYRAAKFR MGSRKTNSVS PVPEAVEVKN
                                                 TM 6
251 ATQHPQMVFT ARHATVTFQT EGDTWREQKE QRAALMVGIL IGVFVLCWFP
                                     TM 7
301 FFVTELISPL CSWDVPAIWK SIFLWLGYSN SFFNPLIYTA FNRSYSSAFK

351 VFFSKQQ
```

Amino acids	367
Molecular weight	40 804
Glycosylation	Asn6*, Asn21
Disulfide bonds	Cys114–Cys162*
Palmitoylation	None
Phosphorylation (PKA)	Ser158*
Phosphorylation (PKC)	Thr150*, Thr153*, Ser233*, Ser260*, Ser264*

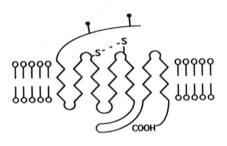

Database accession numbers

	PIR	SWISSPROT	EMBL/GENBANK	REFERENCE
Mouse			Z18278	1

Gene

Its genomic organization is not known.

Reference

1 Plussat, J.-L. *et al.* (1992) *EMBO J.* **11**, 4779–4786.

Leukotrienes

INTRODUCTION

Leukotrienes (LT) have important physiological roles in the cardiovascular, respiratory and immune systems [1-10]. They are divided into cysteinyl-containing leukotrienes (LTC4, LTD4, LTE4 and LTF4), previously known as the slow reacting substance of anaphylaxis (SRS-A), and other lipoxygenase metabolites (LTB4, 12-HETE and lipoxins). LTB4 and 12-HETE are important chemoattractants and are found in high levels in inflammatory conditions, e.g. septic shock, inflammatory bowel disease and allergic asthma. The cysteinyl-containing leukotrienes stimulate bronchoconstriction and induce vasodilation in most vessels, although vasoconstriction occurs in coronary arteries; they also increase vascular permeability and induce cardiac depression. They are found in high levels in bronchial tissue and lung where they may have a pathological role in allergic asthma and respiratory distress syndrome. There is evidence that leukotrienes act as neuromodulators in the CNS.

Structures

LTB4

LTC4

LTD4

Distribution and synthesis

Leukotrienes are not stored within cells but are synthesized *de novo* in response to an appropriate stimulus. They are derived from arachidonate which is esterified in the 2-position of membrane phospholipids and liberated by the action of PLA2 or the combined actions of phospholipase C and diacylglycerol lipase. Arachidonate is converted to leukotrienes by lipoxygenase enzymes which are present in lung, platelets and white blood cells. LTB4 is produced mainly in neutrophils and the cysteinyl-containing leukotrienes in eosinophils, basophils, macrophages and mast cells.

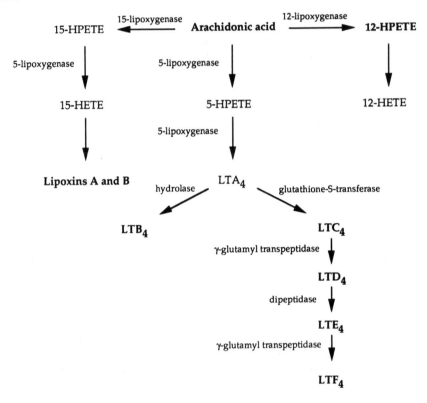

Pathway of synthesis of leukotrienes from arachidonate. Compounds with biological activity are shown in bold.

Books

1 Crooke, S.T. and Wong, A. (eds) (1991) Lipoxygenase and their Products, Academic Press, London.
2 Levi, R. and Krell, R.D. (eds) (1990) Biology of the leukotrienes. *Ann. N.Y. Acad. Sci.* **524**.
3 Rokach (ed.) (1989) Bioactive Molecular II: Leukotrienes and Lipoxygenases: Chemical, Biological and Clinical Aspects, Elsevier, Amsterdam.

Reviews

4 Aharony, D. (1990) Assessment of leukotriene D4 receptor antagonists. *Methods Enzymol.* **187**, 414–421.
5 Ford-Hutchinson, A. (1990) Leukotriene B4 in inflammation. *Crit. Rev. Immunol.* **10**, 1–12.

[6] Halushka, P.V. *et al.* (1989) Thromboxane, prostaglandin and leukotriene receptors. *Annu. Rev. Pharmacol. Toxicol.* **29**, 213–239.

[7] Lewis, R.A. *et al.* (1990) Leukotrienes and other products of the 5-lipoxygenase pathway *N. Engl. J. Med.* **323**, 646–655.

[8] Piper, P. (1989) Leukotrienes and the airways. *Eur. J. Anaesth.* **6**, 241–255.

[9] Piper, P. and Samhoun, P. (1987) Leukotrienes. *Br. Med. Bull.* **43**, 297–311.

[10] Snyder, D.W. and Fleish, J.H. (1989) Leukotriene receptor antagonists as potential therapeutic agents. *Annu. Rev. Pharmacol. Toxicol.* **29**, 123–143.

LEUKOTRIENE RECEPTORS

LTB$_4$ receptor

This receptor is named after the most potent of the endogenous leukotrienes and is described below.

LTC$_4$ receptor

In guinea-pig trachea, contractions-induced LTC$_4$ are not blocked by the LTD$_4$ receptor antagonist, FPL 55712 [1,2], providing evidence for an LTC$_4$ receptor. Characterization of this receptor is hampered by a lack of selective agonists and antagonists and by rapid conversion of LTC$_4$ to LTD$_4$. However, N-methyl LTC$_4$ is reported to be a metabolically stable LTC$_4$-mimetic[3]. The LTD$_4$ receptor antagonist LY 170680 inhibits LTC$_4$-induced contractions in guinea-pig trachea with a pA$_2$ of 7.0. Binding sites for [^{3}H]LTC$_4$ are present in a wide range of tissues but many of these represent enzymes involved in leukotriene metabolism [4].

LTD$_4$ receptor

This receptor is named after the most potent of the endogenous leukotrienes and is described below.

It has been suggested that a separate receptor may exist for LTE$_4$, but current evidence indicates that most of the actions of LTE$_4$ are mediated through the LTD$_4$ receptor.

References
[1] Snyder, D.W. and Krell, A.D. (1984) *J. Pharmacol. Exp. Ther.* **231**, 616–622.

[2] Weichman, B.M. and Tucker, S.S. (1985) *Prostaglandins* **29**, 503–513.

[3] Baker, S.R. *et al.* (1990) *Prostaglandins* **40**, 445–461.

[4] Sun, F.F. *et al.* (1987) *Fed. Proc.* **46**, 204–207.

THE LTB₄ RECEPTOR

LTB$_4$ has a slightly higher potency than 12-HETE; other leukotrienes are inactive.

Distribution

Activation of the LTB$_4$ receptor induces chemotaxis and adhesion of neutrophils to vascular endothelium.

Pharmacology

Agonists: LTB$_4$ and 12-HETE induce activation in high nanomolar concentrations.

Antagonists: LY 255283 (pA$_2$ 8.3), SC 41930 ($K_d \simeq 300$ nM) and U-75303 ($K_d \simeq 1$ μM) have selectivities of more than 100-fold; several LTB$_4$ receptor antagonists have partial agonist activity.

Radioligands: [³H]LTB$_4$ labels a single binding site in neutrophils.

Structures:

LY 255283

SC 41930

U-75302

Predominant effector pathways

Activation of the phosphoinositide pathway through a pertussis-toxin-sensitive G-protein.

Amino acid sequence

The receptor has not been cloned.

THE LTD$_4$ RECEPTOR

LTC$_4$ and LTD$_4$ are approximately equipotent and have a slightly higher potency than LTE$_4$; lipoxin A is also active but LTB$_4$ is inactive.

Distribution

Activation of the LTD$_4$ receptor induces contraction of gastrointestinal, pulmonary, reproductive and vascular smooth muscles, and stimulates mucus secretion in bronchial tissue.

Pharmacology

Agonists: LTD$_4$ induces activation in high nanomolar concentrations.

Antagonists: A wide range of structurally distinct antagonists have been described of which the most selective is ICI 198615 [1] (pA$_2$ 9.8; selectivity > 1000-fold); other examples include SKF 104353 [2] (pA$_2$ 8.3) and LY 170680 [3] (pA$_2$ 8.1–9.0). FPL 55712 (pA$_2$ 7.0) was the first LTD$_4$ receptor antagonist to be identified.

Radioligands: [^{3}H]LTD$_4$ and [^{3}H]ICI 198615 [4] (K_d 0.1 nM) are selective radioligands.

Structures:
LY   170680   5-(3-[2(*R*)(Carboxyethylthio)-1(*S*)-hydroxy-pentadeca3(*E*)5(*Z*)-dienyl] phenyl) 1H tetrazole

ICI 198615

SKF 104353

FPL 55712

Predominant effector pathways

Activation of the phosphoinositide pathway through a pertussis-toxin-insensitive G-protein most likely of the G_q/G_{11} class. Some of the actions of LTD_4 are inhibited by pertussis toxin.

Amino acid sequence

The receptor has not been cloned.

References
1 Snyder, D.W. *et al.* (1987) *J. Pharmacol. Exp. Ther.* **243**, 548–556.
2 Hay, D.W.P. *et al.* (1987) *J. Pharmacol. Exp. Ther.* **243**, 474–481.
3 Boot, J.R. *et al.* (1989) *Br. J. Pharmacol.* **98**, 259–267.

Melanocortins

Pro-opiomelanocortin is present in high levels in pituitary and is processed into three major peptide families: adrenocorticotrophin (ACTH), α-, β-and γ-melanocyte-stimulating hormones (MSH) and β-endorphin [1-3].

ACTH regulates the synthesis and release of glucocorticoids and, to a lesser extent, aldosterone in the adrenal cortex; it also has a trophic action on these cells. ACTH and β-endorphin are synthesized and released in response to corticotrophin-releasing factor at times of stress, i.e. heat, cold, infections, etc.; their release leads to increased metabolism and analgesia, respectively. A 24 amino acid peptide, tetracosactrin, based on the N-terminus of ACTH is used clinically to diagnose adrenal cortical insufficiency and, occasionally, to stimulate adrenal cortex function or as an alternative to glucocorticoids to treat inflammatory disorders.

MSH has a trophic action on melanocytes and, in fish and amphibia, regulates pigment production. Their action in man is poorly understood, although there is evidence for a role in temperature regulation.

Structures (human)

ACTH   Ser–Tyr–Ser–Met–Glu–His–Phe–Arg–Trp–Gly–Lys–Pro–Val–Gly–Lys–
        Lys–Arg–Arg–Pro–Val–Lys–Val–Tyr–Pro–Asn–Gly–Ala–Glu–Asp–Glu–
        Ser–Ala–Glu–Ala–Phe–Pro–Leu–Glu–Phe

α-MSH   Ac-Ser–Tyr–Ser–Met–Glu–His–Phe–Arg–Trp–Gly–Lys–Pro–Val–NH$_2$

β-MSH   Ala–Glu–Lys–Lys–Asp–Glu–Gly–Pro–Tyr–Arg–Met–Glu–His–Phe–Arg–
        Trp–Gly–Ser–Pro–Pro–Lys–Asp

γ-MSH   Tyr–Val–Met–Gly–His–Phe–Arg–Trp–Asp–Arg–Phe–Gly

Full activity of ACTH resides in the first 20 N-terminal amino acids. Residues 1-13 of ACTH are identical to those in α-MSH. There is species variation in the above sequences. For the structure of β-endorphin see Opioid peptides (p. 205).

Distribution and synthesis

Pro-opiomelanocortin is expressed primarily in the pituitary and in limited amounts in brain and peripheral tissues. It is processed into the three major peptide activities described above.

Reviews

[1] Fisher, L.A. (1989) Corticotrophin-releasing factor: endocrine and autonomic integration of responses to stress. *Trends Pharmacol. Sci.* **10**, 189–193.
[2] Munck, A. *et al.* (1984) Physiological functions of glucocorticoids in stress and their relation to pharmacological actions. *Endocrinol. Rev.* **5**, 25–44.
[3] Quinn, S.J. and Williams, G.H. (1988) Regulation of aldosterone secretion. *Annu. Rev. Physiol.* **50**, 409–426.

THE ACTH RECEPTOR

Distribution

The ACTH receptor is found in high levels in adrenal cortex. Binding sites are present in lower levels in the CNS.

Pharmacology

Agonists: ACTH induces receptor activation in nanomolar concentrations.

Antagonists: none have been described.

Radioligands: [^{125}I]ACTH.

Predominant effector pathway
Activation of adenylyl cyclase through G_s.

Amino acid sequence [1]

```
                             TM 1
    1 MKHIINSYEN INNTARNNSD CPRVVLPEEI FFTISIVGVL ENLIVLLAVF
          TM 2
   51 KNKNLQAPMY FFICSLAISD MLGSLYKILE NILIILRNMG YLKPRGSFET
        TM 3                                          TM 4
  101 TADDIIDSLF VLSLLGSIFS LSVIAADRYI TIFHALRYHS IVTMRRTVVV
                                          TM 5
  151 LTVIWTFCTG TGITMVIFSH HVPTVITFTS LFPLMLVFIL CLYVHMFLLA
                      TM 6
  201 RSHTRKISTL PRANMKGAIT LTILLGVFIF CWAPFVLHVL LMTFCPSNPY
          TM 7
  251 CACYMSLFQV NGMLIMCNAV IDPFIYAFRS PELRDAFKKM IFCSRYW
```

Amino acids   297
Molecular weight   33 926
Glycosylation   Asn12*, Asn17
Disulfide bonds   None
Palmitoylation   Cys293*

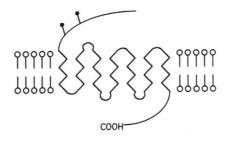

COOH

Database accession numbers

	PIR	*SWISSPROT*	*EMBL/GENBANK*	*REFERENCE*
Human		X65633		*1*

This is the smallest seven transmembrane protein that has been identified.

Gene
There are no introns in the coding sequence.

Reference
[1] Mountjoy, K.G. *et al.* (1992) *Science* **257**, 1248–1251.

THE MSH RECEPTOR

α- and β-MSH have similar affinities and induce activation in low nanomolar concentrations; ACTH has an ~ fivefold lower affinity.

Distribution

The MSH receptor is expressed in high levels in melanocytes, melanomas and their derived cell lines. Receptors are found in low levels in the CNS. MSH regulates temperature control in the septal region of the brain and releases prolactin from the pituitary.

Predominant effector pathway
Activation of adenylyl cyclase through G_s.

Amino acid sequence [1]

```
                                            TM 1
  1 MAVQGSQRRL LGSLNSTPTA IPQLGLAANQ TGARCLEVSI SDGLFLSLGL
                                    TM 2
 51 VSLVENALVV ATIAKNRNLH SPMYCFICCL ALSDLLVSGT NVLETAVILL
                      TM 3
101 LEAGALVARA AVLQQLDNVI DVITCSSMLS SLCFLGAIAV DRYISIFYAL
                  TM 4                                TM 5
151 RYHSIVTLPR APRAVAAIWV ASVVFSTLFI AYYDHVAVLL CLVVFFLAML
                                              TM 6
201 VLMAVLYVHM LARACQHAQG IARLHKRQRP VHQGFGLKGA VTLTILLGIF
                      TM 7
251 FLCWGPFFLH LTLIVLCPEH PTCGCIFKNF NLFLALIICN AIIDPLIYAF

301 HSQELRRTLK EVLTCSW
```

Amino acids	317
Molecular weight	34 660
Glycosylation	Asn15*, Asn29*
Disulfide bonds	None

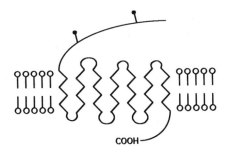

Database accession numbers

	PIR	SWISSPROT	EMBL/GENBANK	REFERENCES
Human			X65634	1
Mouse			X65635	1

Gene

There are no introns in the coding sequence. Southern blot analysis suggests the presence of as many as six members of the receptor family in the human genome.

References

1 Mountjoy, K.G. *et al.* (1992) *Science* **257**, 1248–1251.

Melatonin

Melatonin is secreted by the pineal gland during darkness. It regulates a variety of neuroendocrine functions and is believed to have an essential role in circadian rhythms [1-8]. It condenses pigment granules in skin melanophores of amphibians.

Drugs which modify the action of melatonin are of potential clincial importance in the modification of circadian cycles, e.g. for the treatment of "jet lag".

Structure

Melatonin

Distribution and synthesis

It is made from 5-hydroxytryptamine (5-HT) and [N-acetyl]5-HT in the pineal gland. Small amounts are synthesized in the retina and have a local action.

Reviews

1 Cardinali, D.P. (1981) Melatonin: A mammalian pineal hormone. *Endocrinol. Rev.* **2**, 327–346.
2 Cassone, V.M. (1990) Effects of melatonin on vertebrate circadian systems. *Trends Neurosci.* **13**, 457–464.
3 Dubocovich, M.L. (1988) Pharmacology and function of melatonin receptors. *FASEB J.* **2**, 2765–2773.
4 Krause, D.N. and Dubocovich, M.L. (1990) Regulatory sites in the melatonin system of mammals. *Trends Neurosci.* **13**, 464–470.
5 Krause, D.N. and Dubocovich, M.L. (1991) Melatonin receptors. *Annu. Rev. Pharmacol. Toxicol.* **31**, 549–568.
6 Stankov, B. and Reitner, R.J. (1990) Minireview: Melatonin receptors: current status, facts, and hypotheses. *Life Sci.* **46**, 971–982.
7 Sugden, D. (1989) Melatonin biosynthesis in the mammalian pineal gland. *Experentia* **45**, 922–932.
8 Wiechmann, A.F. (1986) Melatonin: Parallels in pineal gland and retina. *Exp. Eye Res.* **42**, 507–527.

THE MELATONIN RECEPTOR

Until recently, the study of the melatonin receptor was hampered by the lack of quantitative bioassays and specific, high-affinity ligands. The receptor has not been cloned.

Distribution

It is found in retina, the pars tuberalis of the pituitary and in discrete areas of the brain including hypothalamic nuclei and area postrema. In amphibia, it is also found in melanophores.

Pharmacology

Agonists: Melatonin induces activation in high picomolar concentrations; 5-HT is several orders of magnitude less potent.

Antagonists: Luzindole (pA2 7.7) is a competitive and specific antagonist [1].

Radioligands: 2-[125I]Iodomelatonin is a specific ligand which binds to high-affinity (K_d 45 pM) and low-affinity (K_d 400 pM) sites which represent two states of the same receptor [2]. It also binds to a further site (K_d 3 nM) although no functional correlate has been identified [3].

Structures:

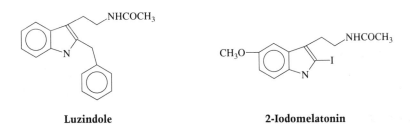

Luzindole   **2-Iodomelatonin**

Predominant effector pathways
Inhibition of adenylyl cyclase through a pertussis-toxin-sensitive G-protein most likely of the G_i/G_o class.

References
1 Dubocovich, M.L. (1988) *J. Pharmacol. Exp. Ther.* **246**, 902–910.
2 Dubocovich, M.L. and Takahashi, J.S. (1987) *Proc. Natl Acad. Sci. USA* **84**, 3916–3920.
3 Duncan, M.J. *et al.* (1988) *Endocrinology* **122**, 1825–1833.

Neuropeptide Y   NPY

INTRODUCTION

NPY is the best known member of a family of neuropeptides which share close structural homology and possess a common secondary structure sometimes referred to as the PP fold [1-6]. It is one of the most abundant peptides in mammalian brain, inducing a variety of behavioural effects, e.g. stimulation of food intake, anxiety, facilitation of learning and memory, and regulation of the cardiovascular and neuroendocrine systems. In the periphery, NPY stimulates vascular smooth muscle contraction and modulates hormone secretion.

NPY has been implicated in the pathophysiology of hypertension, congestive heart failure, affective disorders and appetite regulation.

Structures (human)

NPY
: Tyr–Pro–Ser–Lys–Pro–Asp–Asn–Pro–Gly–Glu–Asp–Ala–Pro–Ala– Glu–Asp–Leu–Ala–Arg–Tyr–Tyr–Ser–Ala–Leu–Arg–His–Tyr–Ile– Asn–Leu–Ile–Thr–Arg–Gln–Arg–Tyr–NH2

Pancreatic polypeptide (PP)
: Ala–Pro–Leu–Glu–Pro–Val–Tyr–Pro–Gly–Asp–Asn–Ala–Thr–Pro– Glu–Gln–Met–Ala–Gln–Tyr–Ala–Ala–Asp–Leu–Arg–Arg–Tyr–Ile– Asn–Met–Leu–Thr–Arg–Pro–Arg–Tyr–NH2

Peptide YY (PYY)
: Tyr–Pro–Ala–Lys–Pro–Glu–Ala–Pro–Gly–Glu–Asp–Ala–Ser–Pro– Glu–Glu–Leu–Ser–Arg–Tyr–Tyr–Ser–Ala–Leu–Arg–His–Tyr–Leu– Asn–Leu–Val–Thr–Arg–Gln–Arg–Tyr–NH2

Shared amino acids are underlined. Fragments have also been isolated, e.g. NPY_{12-36}, NPY_{1-30} or PYY_{3-36}, which may represent degradation products.

Distribution and synthesis

NPY is distributed widely within the CNS and is found in particularly high amounts in hypothalamus and cerebral cortex. It is often co-localized with catecholamines in the CNS and in peripheral nerves.

PYY is found mainly in endocrine cells of the intestine. It is also present within the brain stem.

PP is not found in brain.

Reviews

[1] Danger, J.M. et al. (1990) Neuropeptide Y: localization in the central nervous system and neuroendocrine functions. *Fundam. Clin. Pharmacol.* **4**, 307–340.
[2] Dumont, Y. et al. (1992) Neuropeptide Y and neuropeptide Y receptor subtypes in brain and peripheral tissues. *Prog. Neurobiol.* **38**, 125–169.
[3] Heilig, M. and Widerlov, E. (1990) Neuropeptide Y: an overview of central distribution, functional aspects and possible involvement in neuropsychiatric illnesses. *Acta Psychiatr. Scand.* **82**, 92–114.
[4] Michel, M.C. (1991) Receptors for neuropeptide Y: multiple subtypes and multiple second messengers. *Trends Pharmacol. Sci.* **12**, 389–394 .

5 Wahlestedt, C. *et al.* (1990) Neuropeptide Y receptor subtypes Y_1 and Y_2. *Ann. N.Y. Acad. Sci.* **611**, 7–26.

6 Wahlestedt, C. and Reis, D.J. (1993) Neuropeptide Y (NPY)-related peptides and their receptors – are the receptors potential therapeutic drug targets? *Annu. Rev. Pharmacol. Toxicol.* **33**, 309–352.

NPY RECEPTORS

Y_1 and Y_2 receptors

Both receptors are predicted to be G-protein coupled and are described below.

Y_3 receptor

The existence of a Y_3 receptor characterized by a low affinity for PYY has been proposed [1]; NPY_{18-36} may be an antagonist at the receptor. It is thought to be linked to inhibition of adenylyl cyclase through a pertussis-toxin-sensitive G-protein.

A cDNA sequence (LCR1) which may encode the Y_3 receptor has been described [2]; however, it has little homology with the Y_1 receptor and further work is required to substantiate this claim. It is described in the section on Orphan receptors (page 223).

Other receptors

A NPY receptor has been cloned in *Drosophila melanogaster* [3]. The potency order on receptor expressed in oocytes is PYY > NPY >>PP.

References
1 Michel, A.D. (1991) *Trends Pharmacol. Sci.* **12**, 389–394.
2 Rimland, J. *et al.* (1991) *Mol. Pharmacol.* **40**, 869–875.
3 Li, Y.-J. *et al.* (1992) *J. Biol. Chem.* **267**, 9–12.

THE Y_1 RECEPTOR

NPY and PYY are approximately equipotent and induce activation in low nanomolar concentrations; C-terminal fragments of NPY, e.g. NPY_{13-36}, have much lower affinity [1].

Distribution

The receptor is present in smooth muscle, e.g. intestine and blood vessels, and has also been described in cell lines, e.g. MC-IXC human neuroblastoma and Hel cells. The lack of highly selective agonists or antagonists hampers characterization of NPY receptors in CNS, although the majority are thought to be of the Y_2 subtype. Y_1 receptors are believed to have a predominantly postsynaptic location.

Pharmacology

Agonists: [Leu^{31},Pro^{34}]NPY [2] and [Pro^{34}]NPY have a selectivity of more than 100-fold and act in low nanomolar concentrations. The receptor is most easily identified by the following rank order of potencies: NPY ~ PYY ≥ [Leu^{31},Pro^{34}]NPY >> C-terminal fragments; C-terminal fragments may be partial agonists.

Antagonists: No high-affinity, selective antagonist is available. A number of compounds have antagonist action but suffer from low affinity (> 100 μM), a lack of specificity or partial agonist activity.

Radioligands: [^{3}H] or [^{125}I] analogues of NPY (K_d 0.1 – 5 nM) or PYY (K_d 0.1 – 5 nM) are nonselective. Low-affinity binding sites for NPY (K_d ~200 nM) have also been reported.

Predominant effector pathways

Inhibition of adenylyl cyclase through a pertussis-toxin-sensitive G-protein, most likely of the G_o/G_i class. There is also evidence that NPY can stimulate an increase in intracellular Ca^{2+} independent of the phosphoinositide pathway.

Amino acid sequence [3,4]

```
                                            TM 1
  1 MNSTLFSQVE NHSVHSNFSE KNAQLLAFEN DDCHLPLAMI FTLALAYGAV
                             TM 2
 51 IILGVSGNLA LIIIILKQKE MRNVTNILIV NLSFSDLLVA IMCLPFTFVY
                      TM 3
101 TLMDHWVFGE AMCKLNPFVQ CVSITVSIFS LVLIAVERHQ LIINPRGWRP
         TM 4
151 NNRHAYVGIA VIWVLAVASS LPFLIYQVMT DEPFQNVTLD AYKDKYVCFD
            TM 5
201 QFPSDSHRLS YTTLLLVLQY FGPLCFIFIC YFKIYIRLKR RNNMMDKMRD
             TM 6
251 NKYRSSETKR INIMLLSIVV AFAVCWLPLT IFNTVFDWNH QIIATCNHNL
         TM 7
301 LFLLCHLTAM ISTCVNPIFY GFLNKNFQRD LQFFFNFCDF RSRDDDYETI

351 AMSTMHTDVS KTSLKQASPV AFKKINNNDD NEKI
```

Amino acids	384
Molecular weight	44 392
Glycosylation	Asn2*, Asn11*, Asn17*, Asn186*
Disulfide bonds	Cys113–Cys198*
Palmitoylation	Cys338*

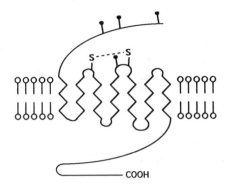

Database accession numbers

	PIR	SWISSPROT	EMBL/GENBANK	REFERENCES
Human		P25929	M88461	3,4
			M84755	
Rat	S12863	P21555	Z11504	5,6

The Y_1 receptor was originally identified as an orphan receptor, FC5 [5,6].

Gene
The organization of the gene is unknown.

References
1 Wahlestedt, C. et al. (1986) Regul. Pept. 13, 307–318.
2 Fuhlendorff, J. et al. (1990) Proc. Natl Acad. Sci. USA 87, 182–186.
3 Larhammar, D. et al. (1992) J. Biol. Chem. 267, 10935–10938.
4 Herzog, H. et al. (1992) Proc. Natl Acad. Sci. USA 89, 5794–5798.
5 Eva, C. et al. (1990) FEBS Lett. 271, 81–84.
6 Krause, J. et al. (1991) Mol. Pharmacol. 41, 817–821.

THE Y_2 RECEPTOR

NPY and PYY are approximately equipotent and induce activation in low nanomolar concentrations. The receptor has not been cloned.

Distribution

Binding sites for $[^{125}I]$NPY or $[^{125}I]$PYY are widespread throughout the CNS. High densities of Y_2 receptors are present in rat hippocampus and are also found in high levels in superficial layers of cortex, certain thalamic nuclei, lateral septum, anterior olfactory nuclei; lower levels are found in striatum. Y_2 receptors are present in high levels in smooth muscle, e.g. rat vas deferens and intestine, kidney proximal tubules and in cell lines, e.g. SM-MSN and CHP-234 neuroblastoma cells. Y_2 receptors are believed to have a predominantly presynaptic location.

Pharmacology

Agonists: C-terminal fragments of NPY or PYY, e.g. NPY_{13-36} and NPY_{18-36}, show weak selectivity of between 10- and 100-fold [1]. The receptor can be identified by the rank order of potencies: $NPY \simeq PYY \geq NPY_{13-36} >> [Leu^{31},Pro^{34}]NPY$.

Antagonists: No high-affinity, selective antagonist is available. A number of compounds have antagonist action but suffer from low affinity (> 100 µM), a lack of specificity or partial agonist activity.

Radioligands: [^{3}H] or [^{125}I] analogues of NPY (K_d 0.1 – 5 nM) or PYY (K_d 0.1 – 5 nM) are nonselective.

Predominant effector pathways
Inhibition of adenylyl cyclase and voltage-dependent Ca^{2+} channels through a pertussis-toxin-sensitive G-protein, most likely of the G_i/G_o class.

References
1 Wahlestedt, C. *et al.* (1986) *Regul. Pept.* **13**, 307–318.

Neurotensin

Neurotensin is the best known member of a family of peptide transmitters which share close homology in their last six C-terminal amino acids [1,2]. Full biological activity resides in this region with the N-terminal sequence having a modulatory role.

In the periphery, neurotensin stimulates smooth muscle contraction, e.g. intestine, and secretion, e.g. exocrine pancreas. In the CNS, neurotensin induces a variety of effects including antinoception, hypothermia and increased locomotor activity; it is involved in the regulation of nigrostriatal and mesolimbic dopamine pathways.

Structure (human)

Neuromedin N	H–Lys–Ile–Pro–Tyr–Ile–Leu–OH
Neurotensin	Pyr–Leu–Tyr–Glu–Asn–Lys–Pro–Arg–Arg–Pro–Tyr–Ile–Leu–OH

There are small differences in structure between species. Additional members of the family exist in lower organisms, notably amphibia, e.g. xenopsin.

Distribution

High levels of neurotensin are found in endocrine N cells throughout the mucosa of the small intestine. It is also distributed throughout the CNS, being found in high levels in hypothalamus, amygdala and nucleus accumbens.

Gene

Neurotensin and neuromedin N are derived from a 170 amino acid precursor.

Reviews

1 Kasckow, S. and Nemeroff, C. (1991) The neurobiology of neurotensin: focus on neurotensin-dopamine interactions. *Regul. Pept.* **36**, 153–164.
2 Stowe, Z.N. and Nemeroff, C. (1991) The electrophysiological actions of neurotensin in the central nervous system. *Life Sci.* **49**, 987–1002.

THE NEUROTENSIN RECEPTOR

The neurotensin receptor is distributed widely on neuronal and non-neuronal cells in peripheral tissues and in CNS. High levels of the receptor are found in ileum and hypothalamus. The receptor is also found in cell lines, e.g. HT29 cells.

Pharmacology

Agonists: Neurotensin and its analogues induce activation in nanomolar concentrations.

Antagonists: SR 48692 (pA$_2$ 8.1) is a potent, nonpeptide antagonist [1].

Radioligands: The receptor can be radiolabelled with [^{3}H]neurotensin (K_d 0.3 – 5 nM), [^{125}I-Tyr^3]neurotensin (K_d 0.3 – 5 nM) or the antagonist [^{3}H]SR 48692.

Structures

SR 48692

SR 48692

Predominant effector pathways

Activation of the phosphoinositide pathway through a pertussis-toxin-insensitive G-protein most likely of the G_q/G_{11} class and inhibition of adenylyl cyclase.

Amino acid sequence [2]

```
  1 MRLNSSAPGT PGTPAADPFQ RAQAGLEEAL LAPGFGNASG NASERVLAAP
                TM 1                                      TM 2
 51 SSELDVNTDI YSKVLVTAVY LALFVVGTVG NTVTAFTLAR KKSLQSLQST
                                                      TM 3
101 VHYHLGSLAL SDLLTLLLAM PVELYNFIWV HHPWAFGDAG CRGYYFLRDA
                                                  TM 4
151 CTYATALNVA SLSVERYLAI CHPFKAKTLM SRSRTKKFIS AIWLASALLT
                                                  TM 5
201 VPMLFTMGEQ NRSADGQHAG GLVCTPTIHT ATVKVVIQVN TFMSFIFPMV

251 VISVLNTIIA NKLTVMVRQA AEQGQVCTVG GEHSTFSMAI EPGRVQALRH
        TM 6                                      TM 7
301 GVRVLRAVVI AFVVCWLPYH VRRLMFCYIS DEQWTPFLYD FYHYFYMVTN

351 ALFYVSSTIN PILYNLVSAN FRHIFLATLA CLCPVWRRRR KRPAFSRKAD

401 SVSSNHTLSS NATRETLY
```

Amino acids	418
Molecular weight	46 288
Glycosylation	Asn4*, Asn26*, Asn31*, Asn211*
Disulfide bonds	Cys140 – Cys224*
Palmitoylation	Cys370*, Cys372*

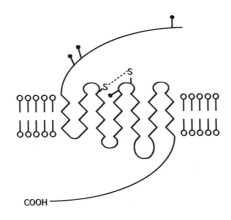

Database accession numbers

	PIR	SWISSPROT	EMBL/GENBANK	REFERENCES
Human			S54181	2
Rat	JH0164	P20789		3

Comment

A low-affinity binding site for neurotensin has been described which is inhibited by the antihistamine levocabastine but not by GTP [4]. There is no supporting evidence to suggest that it is a functional receptor.

Xenopsin and [Trp11]neurotensin, which both possess a Trp residue in position 11, have the same affinity as neurotensin for receptors in mouse and rat but are 10–20-fold less potent in other species, including man [5].

References

[1] Gully, D. et al. (1993) Proc. Natl Acad. Sci. USA 90, 65–69.
[2] Vita, N. et al. (1993) FEBS Lett. 317, 139–142.
[3] Tanaka, K. et al. (1990) Neuron 4, 847–854 .
[4] Kitabgi, P. et al. (1987) Eur. J. Pharmacol. 140, 285–293.
[5] Quirion, R. et al. (1980) Br. J. Pharmacol. 69, 689–692.

Odorant receptors

The olfactory system is responsible for transmitting odorous signals from the environment to the olfactory bulb and from there to cortical regions of the brain. Odour detection is thought to occur when an odorous ligand binds to a receptor on the surface of an olfactory sensory neuron leading to generation of action potentials and the transmission of signals to the olfactory bulb [1].

The discovery of a multigene family expressed in the olfactory epithelium encoding seven transmembrane domain proteins has suggested that the diversity and specificity of odour detection results from the binding of odorous ligands to distinct receptors [2].

Distribution

Odorant receptors have been detected only in olfactory epithelium [2,3]. However, a homologous family of more than 20 proteins is expressed in testis, where they are thought to encode receptors responsible for sperm cell chemotaxis during fertilization [4]. There are data to support the expression of the same receptor in testis and olfactory tissue.

Pharmacology

The existence of a large number of olfactory receptor genes suggests that individual receptors may be highly specific for structurally related odorant stimuli, a mechanism that may underlie the ability to recognize a wide variety of structurally diverse stimuli. Functional studies on individually expressed proteins have shown that each protein appears to be activated by a unique spectrum of odorous compounds.

Effector pathway
Both rapid formation of cAMP and activation of the phosphoinositide pathway have been reported in olfactory cilia and in cell lines expressing odorant receptors [1]. These two responses appear to be mediated by distinct receptors [5].

Example amino acid sequence (rat: M64376) [3]

```
                                     TM 1
    1 MDSSNRTRVS EFLLLGFVEN KDLQPLIYGL FLSMYLVTVI GNISIIVAII
          TM 2
   51 SDPCLHTPMY FFLSNLSFVD ICFISTTVPK MLVNIQTQNN VITYAGCITQ
          TM 3                                      TM 4
  101 IYFFLLFVEL DNFLLTIMAY DRYVAICHPM HYTVIMNYKL CGFLVLVSWI
                                                         TM 5
  151 VSVLHALFQS LMMLALPFCT HLEIPHYFCE PNQVIQLTCS DAFLNDLVIY
                                                  TM 6
  201 FTLVLLATVP LAGIFYSYFK IVSSICAISS VHGKYKAFST CASHLSVVSL
                              TM 7
  251 FYCTGLGVYL SSAANNSSQA SATASVMYTV VTPMVNPFIY SLRNKDVKSV

  301 LKKTLCEEVI RSPPSLLHFF LVLCHLPCFI FCY
```

Amino acids	333
Molecular weight	37 540
Glycosylation	Asn5*
Disulfide bonds	Cys97–Cys189*
Palmitoylation	Cys306*

Accession No. M64376

Database accession numbers

	PIR	SWISSPROT	EMBL/ GENBANK	AMINO ACIDS	MOLECULAR WEIGHT	REFERENCES
Mouse	-	P23275	M84005			6
Rat	A23701	P23265	M64376	333	37 540	3
		P23266	M64377	313	34 703	3
	C23701	P23267	M64378	311	34 168	3
	D23701	P23268	M64381	317	35 655	3
	E23701	P23269	M64385	310	35 247	3
	F23701	P23270	M64386	327	36 265	3
	G23701	P23271	M64387	312	35 315	3
	H23701	P23272	M64388	314	35 455	3
	I23701	P23273	M64391	312	35 718	3
	A37286	P23274	M64392	314	35 563	3

Comment

Thirty-six distinct cDNA clones were isolated by Buck and Axel from rat olfactory cDNA libraries of which at least 18 have unique sequences [3]. To date, ten of these have been sequenced and their accession numbers are shown above. Five additional members of this family have recently been sequenced [5]. It is predicted that as many as 500–1000 distinct genes encoding for members of the odorant family exist in rat, canine and human genomes [2,4].

In contrast to other receptor subfamilies, the family of odorant receptors shows a great deal of diversity within the transmembrane domains but not outside of these regions. Since the transmembrane domains are believed, in many cases, to be the site of agonist binding, this may account for the ability of the odorant receptor family to recognize a wide variety of structurally diverse stimuli. There are subfamilies of receptors within the odorant family which share close sequence homology. In addition, odorant receptors share a number of sequence motifs which are not seen in other seven transmembrane proteins.

Gene

Coding regions of the genes are not interrupted by introns; introns may be present in the 5' untranslated region [3,4].

Reviews
[1] Ronnett, G.V. and Snyder, S.H. (1992) Molecular messengers of olfaction. *Trends Neurosci.* **15**, 508–513.

References
[2] Buck, L. (1992) *Curr. Opinion Neurobiol.* **2**, 282–288.
[3] Buck, L. and Axel, R. (1991) *Cell* **65**, 175–187.
[4] Parmentier, M. *et al.* (1992) *Nature* **355**, 453–455.
[5] Raming, K. *et al.* (1993) *Nature* **361**, 353–356.
[6] Nef, P. *et al.* (1992) *Proc. Natl Acad. Sci USA.* **89**, 8948–8952.

Opioid peptides

INTRODUCTION

Opioid peptides are important neurotransmitters within the peripheral and central nervous systems [1-5]. In brain, they have important roles in the regulation of sensory function (including pain), neuroendocrine activity, the central control of respiration and mood. In the periphery, they are involved in the regulation of gut motility.

Nonpeptide agonists at opioid receptors include codeine, morphine and related substances. Many of these are used clinically in the treatment of pain and constipation, and as drugs of abuse.

Structures (human)

β-Endorphin	Tyr–Gly–Gly–Phe–Met–Thr–Ser–Glu–Lys–Ser–Gln–Thr–Pro–Leu–Val–Thr–Leu–Phe–Lys–Asn–Ala–Ile–Ile–Lys–Asn–Ala–Tyr–Lys–Lys–Gly–Glu
Dynorphin A	Tyr–Gly–Gly–Phe–Leu–Arg–Arg–Ile–Arg–Pro–Lys–Leu–Lys–Trp–Asp–Asn–Gln
Leu-enkephalin	Tyr–Tyr–Gly–Phe–Leu
Met-enkephalin	Tyr–Tyr–Gly–Phe–Met

There are also extended versions of [Met]enkephalin (neo-endorphins) and [Leu]enkephalin (dynorphins), and truncated forms of β-endorphin (α-, γ-, δ-endorphin) and dynorphin. A number of related peptides exist in lower organisms, some of which contain D amino acids, e.g. deltorphin (Tyr–DMet–Phe–His–Leu–Met–Asp–NH2).

Distribution and synthesis

Three precursor polypeptides give rise to endorphins. Preproenkephalin A contains six copies of [Met]-enkephalin and one of [Leu]-enkephalin; preproenkephalin B (preprodynorphin) contains three copies of [Leu]-enkephalin and one copy of dynorphin; pro-opiomelanocortin (also known as corticotrophin-β–lipoprotein precursor) contains one copy of [Met]-enkephalin and is also the precursor of α-, β- and γ-melanocyte-stimulating hormone, corticotrophin and β-endorphin.

Opioid peptides are present in the central and peripheral nervous systems, and are also found in pituitary gland, gastrointestinal tract and adrenal medulla.

Reviews

1 Goldstein, A. (1987) Binding selectivity profiles for ligands of multiple receptor types: focus on opioid receptors. *Trends Pharmacol. Sci.* **8**, 456–459.

2 Loh, H.H. and Smith, A.P. (1990) Molecular characterisation of opioid receptors. *Annu. Rev. Pharmacol. Toxicol.* **30**, 123–147

3 Rees, D.C. and Hunter, J. (1990) Opioid receptors. *Comprehensive Medicinal Chemistry*, vol. 3 (Hansch, C., Sammes, P.G. and Taylor, J.B., eds) Pergamon Press, Oxford, pp. 805–847.

4 Takemori, Y. and Portoghese, P.S. (1992) Selective naltrexone-derived opioid receptor antagonists. *Annu. Rev. Pharmacol. Toxicol.* **32**, 239–271.

[5] Zimmerman, D.M. and Leander, J.D. (1990) Selective opioid agonists and antagonists: research tools and potential therapeutic agents. *J. Med. Chem.* **33**, 895–902.

OPIOID RECEPTORS

The term opioid is applied to a substance which produces its effects through an interaction with the major classes of opioid receptors and whose action is reversed by naloxone or naltrexone.

Naloxone: R = CH₂CH=CH₂
Naltrexone: R = CH₂CH(CH₂)₂

Morphine

μ-, δ- and κ-opioid receptors

These are all G-protein coupled and are described below.

The placenta opioid receptor

A protein has been cloned from human placenta which binds opioid ligands with moderate affinity but whose pharmacology does not correspond with the above classification. Acceptance of this protein as an opioid receptor is provisional. It is described below.

THE μ-OPIOID RECEPTOR

β-Endorphin is the most potent endogenous ligand, inducing activation in low nanomolar concentrations; dynorphin A is slightly more potent than Met and Leu-enkephalin.

Distribution

In the CNS, the μ-opioid receptor is found in cerebral cortex, thalamus, hypothalamus, periaqueductal grey, interpeduncular nucleus and median raphe. In the periphery, it is found in myenteric plexus and in certain smooth muscles e.g. mouse vas deferens (there is species variation). μ-Opioid receptors are believed to mediate analgesia, miosis, bradycardia, hypothermia, respiratory depression, constipation, nausea, euphoria and physical dependence.

Pharmacology

Agonists: DAMGO and PL017 have selectivities of greater than 100-fold; morphine has a selectivity of ~50-fold.

Antagonists: CTOP and CTAP (pA2 6.4–7.9) have selectivities of greater than 1000-fold; naloxonazine has low selectivity. β-Funaltrexamine [1] is an irreversible antagonist.

Radioligands: [³H]DAMGO has a selectivity of more than 100-fold.

Structures:

CTAP	DPhe–Cys–Tyr–DTrp–Arg–Thr–Pen–Thr–NH2
CTOP	DPhe–Cys–Tyr–DTrp–Orn–Thr–Pen–Thr–NH2
DAMGO	Tyr–DAla–Gly–[N–MePhe]–NH(CH2)2–OH
PL017	[N-MePhe3, DPro4]-morphiceptin

Naloxonazine

β-Funaltrexamine

Predominant effector pathways

Activation of K+ channels and inhibition of adenylyl cyclase through a pertussis-toxin-sensitive G-protein, most likely of the G_i/G_o class [2].

Comment

The existence of μ_1- and μ_2-opioid receptor subtypes has been proposed. Naloxonazine is a selective agonist at μ_1- receptors and many δ-opioid receptor agonists also have high affinity at this receptor subtype.

References
[1] Portoghese, P.S. *et al.* (1980) *J. Med. Chem.* **23**, 233–234.
[2] Ueda, H. *et al.* (1988) *Proc. Natl Acad. Sci. USA.* **85**, 7013–7017.

THE δ-OPIOID RECEPTOR

The endogenous opioid peptides all have similar potencies at δ-opioid receptors and induce activation in nanomolar concentrations.

Distribution

The δ-opioid receptor has a more discrete distribution in the CNS relative to μ- and κ-opioid receptors. It is found in cerebral cortex, amygdala, nucleus accumbens, olfactory tubercle and pontine nucleus. It is also found in certain smooth muscles, e.g. hamster vas deferens (there is species variation) and in cell lines, e.g. NG 108-15 neuroblastoma cells. δ-Opioid receptors mediate analgesia.

Pharmacology

Agonists: The metabolically stable peptide analogues of Leu-enkephalin, DPDPE, DSBULET and DADLE are selective. [DAla^2]-deltorphin is also highly selective.

Antagonists: Naltrindole [1] (pA$_2$ 9.7) has a selectivity of more than 100-fold; ICI 174864 (pA$_2$ 7.5) is also selective. DALCE and naltrindole 5′-isothiocyanate are selective, irreversible antagonists [2].

Radioligands: [^{3}H]DPDPE (K_d 5 nM) has a selectivity of more than 100-fold.

Structures:

DADLE	[DAla^2, DAla, DLeu^5]-enkephalin
DALCE	[DAla^2,Leu^5,Cys^6]-enkephalin
DPDPE	[DPen^2, DPen]-enkephalin
DSBULET	Tyr–DSer(OtBu)Gly–Phe–Leu–Thr
ICI 174864	N-N-diallyl–Tyr–Aib–Aib–Phe–Leu

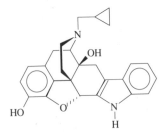

Naltrindole

Predominant effector pathways
Inhibition of adenylyl cyclase and activation of K$^+$ channels through a pertussis-toxin-sensitive G-protein, most likely of the G$_i$/G$_o$ class.

Amino acid (rat) [3]

```
                                                          TM 1
  1 MELVPSARAE LQSSPLVNLS DAFPSAFPSA GANASGSPGA RSASSLALAI
                                   TM 2
 51 AITALYSAVC AVGLLGNVLV MFGIVRYTKL KTATNIYIFN LALADALATS
                          TM 3
101 TLPFQSAKYL METWPFGELL CKAVLSIDYY NMFTSIFTLT MMSVDRYIAV
                          TM 4
151 CHPVKALDFR TPAKAKLINI CIWVLASGVG VPIMVMAVTQ PRDGAVVCML
                   TM 5
201 QFPSPSWYWD TVTKICVFLF AFVVPILIIT VCYGLMLLRL RSVRLLSGSK
                 TM 6                                  TM 7
251 EKDRSLRRIT RMVLVVVGAF VVCWAPIHIF VIVWTLVDIN RRDPLVVAAL

301 HLCIALGYAN SSLNPVLYAF LDENFKRCFR QLCRTPCGRQ EPGSLRRPRQ

351 ATTRERVTAC TPSDGPGGGA AA
```

Amino acids	372
Molecular weight	40 561
Glycosylation	Asn18*, Asn32*
Disulfide bonds	Cys121 – Cys198*
Palmitoylation	Cys328*, Cys333*

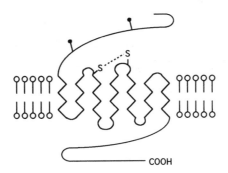

Database accession numbers

	PIR	SWISSPROT	EMBL/GENBANK	REFERENCES
Rat			L07271	3

Gene

The organization of the gene is not known. Multiple transcripts have been identified in Northern blots and these may reflect distinct genes or alternative splicing [3].

Comment

The existence of δ_1- and δ_2-opioid receptors has been proposed. The agonists DPDPE and DADLE and the antagonist DALCE are selective to δ_1-opioid receptors; the agonist DAla[2]-deltorphin and the antagonists naltrindole 5'-isothiocyanate [2] and naltrindole [4] are selective to δ_2-opioid receptors.

References

1 Portoghese, P.S. *et al.* (1988) *Eur. J. Pharmacol.* **146**, 185–186.
2 Jiang, Q. *et al.* (1991) *J. Pharmacol Ther.* **257**, 1069–1075.
3 Evans, C.J. *et al.* (1992) *Science* **258**, 1952–1955.
4 Sofuoglu, M. *et al.* (1991) *J. Pharmacol. Ther.* **257**, 676–680.

THE κ-OPIOID RECEPTOR

Dynorphin is the most potent of the endogenous ligands inducing activation in low nanomolar concentrations; β-endorphin is slightly more potent than Leu- and Met-enkephalin.

Distribution

In the CNS, the κ-opioid receptor is found in cerebral cortex, substantia nigra, interpeduncular nucleus, striatum and hippocampus (marked species variation exists). In the periphery, it is found in myenteric plexus of the guinea-pig ileum and in certain smooth muscles, e.g. rabbit vas deferens. κ-Opioid receptors are believed to mediate analgesia, sedation, miosis and diuresis.

Pharmacology

Agonists: U69593 and the structural derivative CI977 [1] have selectivities of more than 100-fold and induce activation in high nanomolar concentrations.

Antagonists: nor-Binaltorphimine (pA$_2$ 10.3) has a selectivity of more than 50-fold [2].

Radioligands: [^{3}H]U69593 [3] (K_d 3 nM) and [^{3}H]CI977 (K_d 5 nM) have selectivities of more than 100-fold.

Structures

CI977

U50488

nor-Binaltorphimine

Predominant effector pathways

Inhibition of voltage-dependent Ca^{2+} channels via a pertussis-toxin-sensitive G-protein, most likely of the G_i/G_o class [4].

Comment

The existence of κ_1, κ_2 and κ_3 receptor subtypes has been proposed. Arylacetamides, e.g. U69593, bind to a subset of the sites (named κ_1) labelled by benzomorphans; subtypes of κ_1 receptors may also exist [5]. Ethylketocyclazocine binds to κ_1 and κ_2 sites. There is no known functional correlate for the κ_2 receptors. Under κ-selective labelling conditions, naloxone benzoylhydrazone labels a further population of sites named κ_3 which possess a similar pharmacology to μ-opioid receptors [6].

References

[1] Halfpenny, P.R. et al. (1990) J. Med. Chem. **33**, 286–291.
[2] Portoghese, P.S. et al. (1987) Life Sci. **40**, 1287–1292.
[3] Lahti, R.A. et al. (1985) Eur. J. Pharmacol. **109**, 281–284.
[4] Attali, B. et al. (1989) J. Biol. Chem. **264**, 347–353.
[5] Clark, J.A. et al. (1989) J. Pharmacol. Ther. **251**, 461–468.
[6] Jiang, Q. (1991) J. Pharmacol. Ther. **257**, 1069–1075.

THE PLACENTA OPIOID RECEPTOR

A protein has been cloned from human placenta which binds opioid ligands with moderate affinity but does not correspond with the above classification. In particular, it does not display selectivity for κ-ligands despite the high levels of κ-opioid receptors in this tissue. Acceptance of this protein as an opioid receptor is provisional.

Pharmacology

Agonists: Certain agonists selective for μ-opioid receptors (DAMGO), δ receptors (DPDPE) and κ receptors (U50488) bind with K_i values in the nanomolar range [1].

Amino acid sequence [1]

```
  1 MASPAGNLSA WPGWGWPPPA ALRNLTSSPA PTASPSPAPS WTPSPRPGPA
                TM 1                                        TM 2
 51 HPFLQPPWAV ALWSLAYGAV VAVAVLGNLV VIWIVLAHKR MRTVTNSFLV
                                               TM 3
101 NLAFADAAMA ALNALVNFIY ALHGEWYFGA NYCRFQNFFP ITAVFASIYS
                                               TM 4
151 MTAIAVDRYM AIIDPLKPRL SATATRIVIG SIWILAFLLA FPQCLYSKIK
                          TM 5
201 VMPGRTLCYV QWPEGSRQHF TYHMIVIVLV YCFPLLIMGI TYTIVGITLW
                          TM 6
251 GGEIPGDTCD KYQEQLKAKR KVVKMMIIVV VTFAICWLPY HIYFILTAIY
          TM 7
301 QQLNRWKYIQ QVYLASFWLA MSSTMYNPII YCCLNKRFRA GFKRAFRWCP

351 FIHVSSYDEL ELKATRLHPM RQSSLYTVTR MESMSVVFDS NDGDSARSSH

401 QKRGTTRDVG SNVCSRRNSK STSTTASFVS SSHMSVEEGS
```

Amino acids	440
Molecular weight	49 422
Glycosylation sites	Asn7*, Asn24*
Disulfide bond	Cys133–Cys208*
Palmitoylation	Cys349*

The receptor is 93% homologous to the human NK$_3$ tachykinin receptor in the transmembrane regions but does not bind tachykinins.

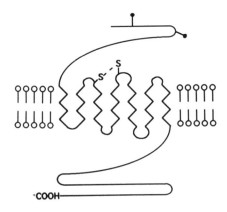

Database accession numbers

	PIR	SWISSPROT	EMBL/GENBANK	REFERENCE
Human			M84605	1

Gene

The organization of the gene is not known.

References

1 Xie, G.-X. *et al.* (1992) *Proc. Natl Acad. Sci. USA* **89**, 4124–4128.

Opsins

INTRODUCTION

Opsins are the photoreceptors of animal retinas [1-8]. Vertebrate rhodopsin is found in rod cells and is involved in vision in dim light. Blue, green and red opsins are found in cone cells and are involved in colour vision.

Opsins differ from other G-protein linked receptors in forming a covalent link with their ligand, 11-*cis*-retinal (vitamin A aldehyde). This covalent link is in the form of a protonated Schiff base between the aldehyde of retinal and the ε-amino group of a lysine residue located in the seventh transmembrane region.

Bovine rhodopsin was the first member of the G-protein linked receptor family to be purified and sequenced because of the relatively high amounts found in retina, where it makes up more than 90% of membrane protein. Much of our understanding of G-protein linked receptors is based on information derived from studies on rhodopsin.

Distribution

In vertebrates, the visual photoreceptors are the only known system to contain the opsin family of proteins. Rhodopsin is found in rod cells, and blue, green and red opsins are found in cone cells. In certain bacteria retinal-based light transducing proteins also exist, e.g. bacteriorhodopsin and halorhodopsin.

Activation

11-*cis*-Retinal is bound via a Schiff base linkage and is converted to all-*trans*-retinal on receiving a single photon of light. This isomerization leads to a conformational change in the receptor and activation of the associated G-protein, transducin. The polarity of the amino acid environment of the retinal binding pocket influences the wavelength of maximal absorption giving each visual pigment its own absorption spectrum.

11-*cis*-retinal   all-*trans*-retinal

Desensitization

Exposure to light allows the receptor protein to interact with rhodopsin kinase. Once bound, rhodopsin kinase phosphorylates specific serine and threonine residues, although this does not completely block the ability of the receptor to activate transducin. Phosphorylation, however, allows the protein arrestin to bind to the surface of photo-excited rhodopsin and thereby induce complete desensitization.

Reviews

1 Applebury, M.L. and Hargrave, P.A. (1986) Molecular biology of the visual pigments. *Vision Res.* **26**, 1881–1895.
2 Chabre, M. and Deterre, P. (1989) Molecular mechanism of visual transduction. *Eur. J. Biochem.* **179**, 255–266.
3 Hargrave, P.A. and McDowell, J.H. (1992) Rhodopsin and phototransduction: a model system for G protein-linked receptors. *FASEB J.* **6**, 2323–2331.
4 Khorana, H.G. (1992) Rhodopsin, photoreceptor of the rod cell. *J. Biol. Chem.* **267**, 1–4.
5 Liebman, P.A. *et al.* (1987) The molecular mechanism of visual excitation and its relation to the structure and composition of the rod outer segment. *Annu. Rev. Physiol.* **49**, 765–791.
6 Saibil, H. (1990) Cell and molecular biology of photoreceptors. *Seminars Neurosci.* **2**, 15–23.
7 Stryer, L. (1988) Molecular basis of visual excitation. *Cold Spring Harbour Symp. Quant. Biol.* **53**, 283–294.
8 Stryer, L. (1991) Visual excitation and recovery. *J. Biol. Chem.* **266**, 10711–10714.

OPSIN RECEPTORS

Rhodopsin; blue, green and red opsins

These are all seven transmembrane proteins and are discussed below.

Squid and octopus rhodopsins

Squid [1] and octopus [2] rhodopsins have been sequenced and extensively characterized.

Drosophila melanogaster rhodopsin

Four rhodopsins have been identified in *Drosophila melanogaster* [3–8].

Bacteriorhodopsin and related proteins

A related subset of light-sensitive proteins are found in bacteria. These include bacteriorhodopsin and halorhodopsin which are proton and chloride pumps, respectively. In this family of proteins light catalyses the conversion of all-*trans*- to 13-*cis*-retinal, thereby bringing about an alteration in protein conformation. These proteins do not interact with G-proteins and are functionally unrelated to the opsins.

Bacteriorhodopsin is the only seven transmembrane protein for which electron crystallography and 3-D reconstruction has been performed [9,10].

References

1 OuchinnikovY.A. *et al.* (1988) *FEBS Lett.* **232**, 69–72.
2 Hall, D. *et al.* (1991) *Biochem. J.* **274**, 35–40.

3 Fryxell, K.J. and Meyerowitz, E.M. (1987) *J. Neurosci.* **7**, 1550–1557.
4 Zuker, C.S. *et al.* (1985) *Cell* **40**, 851–858.
5 Montell, C. *et al.* (1987) *J. Neurosci.* **7**, 1558–1566.
6 Neufeld, T.P. *et al.* (1991) *Proc. Natl Acad. Sci. USA* **88**, 10203–10207.
7 Fortini, M.E. and Rubin, G.M. (1987) *Genes Dev.* **4**, 444–463.
8 Carulli, J.P. and Hartl, D.L. (1992) *Genetics* **132**, 193–204.
9 Henderson, R. and Unwin, P.N.T. (1975) *Nature* **273**, 443–446.
10 Henderson, R. *et al.* (1990) *J. Mol. Biol.* **213**, 899–929.

BLUE OPSIN

Blue-sensitive opsin is involved in colour vision and is found in cone photoreceptors. It has an absorption maximum at 420 nm. The ratio of blue cones to rods is ~ 1:200.

Deficiency in blue opsin causes tritan colour blindness.

Effector pathway
Activation of transducin leading to stimulation of cGMP-phosphodiesterase.

Amino acid sequence [1]

```
                                          TM 1
    1 MRKMSEEEFY LFKNISSVGP WDGPQYHIAP VWAFYLQAAF MGTVFLIGFP
                             TM 2
   51 LNAMVLVATL RYKKLRQPLN YILVNVSFGG FLLCIFSVFP VFVASCNGYF
                 TM 3
  101 VFGRHVCALE GFLGTVAGLV TGWSLAFLAF ERYIVICKPF GNFRFSSKHA
      TM 4
  151 LTVVLATWTI GIGVSIPPFF GWSRFIPEGL QCSCGPDWYT VGTKYRSESY
      TM 5
  201 TWFLFIFCFI VPLSLICFSY TQLLRALKAV AAQQQESATT QKAEREVSRM
      TM 6                                      TM 7
  251 VVVMVGSFCV CYVPYAAFAM YMVNNRNHGL DLRLVTIPSF FSKSACIYNP

  301 IIYCFMNKQF QACIMKMVCG KAMTDESDTC SSQKTEVSTV SSTQVGPN
```

Amino acids	348
Chromosome	7
Molecular weight	39 135
Glycosylation	Asn14*
Disulfide bonds	Cys107–Cys182*
Palmitoylation	Cys313*
Retinal binding	Lys293

Alternative transmembrane regions have been published [2].

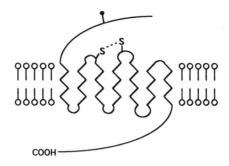

Database accession numbers

	PIR	SWISSPROT	EMBL/GENBANK	REFERENCES
Human	A03156	P03999	M13295	*1*
			M13296	
			M13297	
			M13298	
			M13299	
Chick		P28682	M92037	*3*

Gene (human) *1*

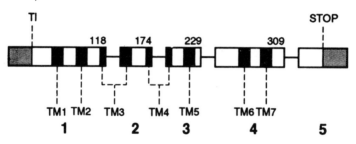

References

1. Nathans, J. *et al.* (1986) *Science* **232**, 193–202.
2. Hargrave, P.A. and McDowell, J.H. (1992) *FASEB J.* **6**, 2323–2331.
3. Okano, T. *et al.* (1992) *Proc. Natl Acad. Sci USA.* **89**, 5932–5936.

GREEN OPSIN

Green-sensitive opsin is involved in colour vision and is found in cone photoreceptors. It has an absorption maximum at 530 nm. The ratio of green and red cones to rods is ~ 1:30.

Deficiency of green opsin causes deutan colour blindness.

Effector pathway

Activation of transducin leading to stimulation of cGMP-phosphodiesterase.

Amino acid sequence [1]

```
  1 MAQQWSLQRL AGRHPQDSYE DSTQSSIFTY TNSNSTRGPF EGPNYHIAPR
       TM 1                                     TM 2
 51 WVYHLTSVWM IFVVIASVFT NGLVLAATMK FKKLRHPLNW ILVNLAVADL
                                       TM 3
101 AETVIASTIS VVNQVYGYFV LGHPMCVLEG YTVSLCGITG LWSLAIISWE
                                     TM 4
151 RWMVVCKPFG NVRFDAKLAI VGIAFSWIWA AVWTAPPIFG WSRYWPHGLK
                                     TM 5
201 TSCGPDVFSG SSYPGVQSYM IVLMVTCCIT PLSIIVLCYL QVWLAIRAVA
                                     TM 6
251 KQQKESESTQ KAEKEVTRMV VVMVLAFCFC WGPYAFFACF AAANPGYPFH
       TM 7
301 PLMAALPAFF AKSATIYNPV IYVFMNRQFR NCILQLFGKK VDDGSELSSA

351 SKTEVSSVSS VSPA
```

Amino acids	364
Chromosome	X
Molecular weight	40 584
Glycosylation	Asn34*
Disulfide bonds	Cys126–Cys203*
Retinal binding	Lys312

Alternative transmembrane regions have been published [2].

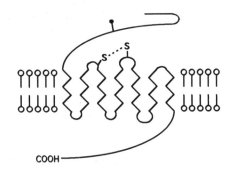

Database accession numbers

	PIR	SWISSPROT	EMBL/GENBANK	REFERENCES
Human	A30158	P04001	M13306	1
			K03491	
			K03492	
			K03493	
			K03494	
			K03495	
			K03496	
			K03497	
			K03490	
Chick		P28683	M92038	3

Gene (human) [1]

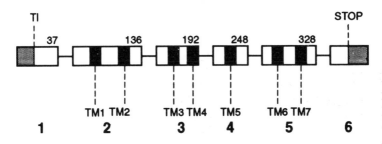

References

1 Nathans, J. et al. (1986) Science **232**, 193–202.
2 Hargrave, P.A. and McDowell, J.H. (1992) FASEB J. **6**, 2323–2331.
3 Okano, T. et al. (1992) Proc. Natl Acad. Sci. USA **89**, 5932–5936.

RED OPSIN

Red-sensitive opsin is involved in colour vision and is found in cone photoreceptors. It has an absorption maximum at 560 nm.

Deficiency in red opsin causes protan colour blindness.

Effector pathway

Activation of transducin leading to stimulation of cGMP-phosphodiesterase.

Amino acid sequence [1]

```
  1 MAQQWSLQRL AGRHPQDSYE DSTQSSIFTY TNSNSTRGPF EGPNYHIAPR
    TM 1                                           TM 2
 51 WVYHLTSVWM IFVVTASVFT NGLVLAATMK FKKLRHPLNW ILVNLAVADL
                                     TM 3
101 AETVIASTIS IVNQVSGYFV LGHPMCVLEG YTVSLCGITG LWSLAIISWE
                  TM 4
151 RWLVVCKPFG NVRFDAKLAI VGIAFSWIWS AVWTAPPIFG WSRYWPHGLK
```

```
                        TM 5
201 TSCGPDVFSG SSYPGVQSYM IVLMVTCCII PLAIIMLCYL QVWLAIRAVA
                        TM 6
251 KQQKESESTQ KAEKEVTRMV VVMIFAYCVC WGPYTFFACF AAANPGYAFH
       TM 7
301 PLMAALPAYF AKSATIYNPV IYVFMNRQFR NCILQLFGKK VDDGSELSSA

351 SKTEVSSVSS VSPA
```

Amino acids	364
Chromosome	10
Molecular weight	40 572
Glycosylation	Asn34*
Disulfide bonds	Cys126–Cys203*
Retinal binding	Lys312

Alternative transmembrane regions have been published [2].

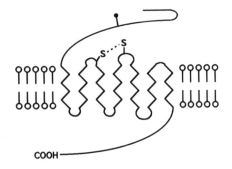

Database accession numbers

	PIR	SWISSPROT	EMBL/GENBANK	REFERENCE
Human	A03157	P04000	M13300	1
			M13301	
			M13302	
			M13303	
			M13304	
			M13305	

Gene (human) [1]

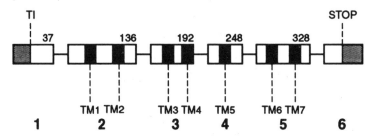

References
[1] Nathans, J. et al. (1986) Science **232**, 193–202.
[2] Hargrave, P.A. and McDowell, J.H. (1992) FASEB J. **6**, 2323–2331.

RHODOPSIN

Rhodopsin mediates vision in dim light and is found in rod cells. It has an absorption maximum at 495 nm.

Effector pathway
Activation of transducin leading to stimulation of cGMP-phosphodiesterase.

Amino acid sequence [1]

```
                                                    TM 1
    1 MNGTEGPNFY VPFSNATGVV RSPFEYPQYY LAEPWQFSML AAYMFLLIVL
                              TM 2
   51 GFPINFLTLY VTVQHKKLRT PLNYILLNLA VADLFMVLGG FTSTLYTSLH
                 TM 3
  101 GYFVFGPTGC NLEGFFATLG GEIALWSLVV LAIERYVVVC KPMSNFRFGE
           TM 4
  151 NHAIMGVAFT WVMALACAAP PLAGWSRYIP EGLQCSCGID YYTLKPEVNN
           TM 5
  201 ESFVIYMFVV HFTIPMIIIF FCYGQLVFTV KEAAAQQQES ATTQKAEKEV
         TM 6                                   TM 7
  251 TRMVIIMVIA FLICWVPYAS VAFYIFTHQG SNFGPIFMTI PAFFAKSAAI

  301 YNPVIYIMMN KQFRNCMLTT ICCGKNPLGD DEASATVSKT ETSQVAPA
```

Amino acids	348
Chromosome	3
Molecular weight	38 892
Glycosylation	Asn2*, Asn15*
Disulfide bonds	Cys110–Cys187*
Palmitoylation	Cys322, Cys323
Retinal binding	Lys296

Palmitoylation has been demonstrated [2,3]. Alternative transmembrane regions have been published [4].

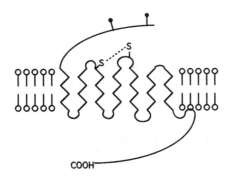

COOH

Database accession numbers

	PIR	SWISSPROT	EMBL/GENBANK	REFERENCES
Bovine	A03154	P02699	K00502	5-7
			M21606	
Chick			P22328	8
			D00702	
Human	A36235	P08100	K02281	1
	A36537			
Ovine	A30407			9
	A03155			

Gene (human) [1]

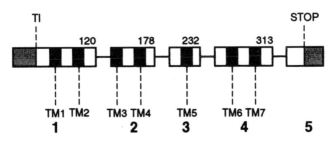

References

[1] Nathans, J. and Hogness, D.S. (1984) *Proc. Natl Acad. Sci. USA* **81**, 4851–4855.
[2] Ovchinnikov, Y.A. *et al.* (1988) *FEBS Lett.* **230**, 1–5.
[3] Papac, D.I. *et al.* (1992) *J. Biol. Chem.* **267**, 16889–16894.
[4] Hargrave, P.A. and McDowell, J.H. (1992) *FASEB J.* **6**, 2323–2331.
[5] Ovchinnikov, Y.A. *et al.* (1982) *Bioorg. Khim.* **8**, 1011–1014.
[6] Hargrave, P.A. *et al.* (1983) *Biophys. Struct. Mech.* **9**, 235–244.
[7] Nathans, J. and Hogness, D.S. (1983) *Cell* **34**, 807–814.
[8] Takao, M. *et al.* (1988) *Vision Res.* **28**, 471–480.
[9] Brett, M. and Findley, J.B.C. (1983) *Biochem. J.* **211**, 611–670.

Orphan receptors

A number of seven transmembrane receptors have been cloned but their endogenous ligands are not known. They have been termed orphan receptors.

EDG1

The cDNA sequence was cloned from human umbilical vein endothelial cells, and is also expressed in vascular smooth muscle cells, fibroblasts, melanocytes and cells of epitheloid origin [1].

Amino acid sequence [1]

```
                                                             TM 1
  1 MGPTSVPLVK AHRSSVSDYV NYDIIVRHYN YTGKLNISAD KENSIKLTSV
                                    TM 2
 51 VFILICCFII LENIFVLLTI WKTKKFHRPM YYFIGNLALS DLLAGVAYTA
                       TM 3
101 NLLLSGATTY KLTPAQWFLR EGSMFVALSA SVFSLLAIAI ERYITMLKMK
               TM 4
151 LHNGSNNFRL FLLISACWVI SLILGGLPIM GWNCISALSS CSTVLPLYHK
            TM 5
201 HYILFCTTVF TLLLLSIVIL YCRIYSLVRT RSRRLTFRKN ISKASRSSEN
               TM 6                                  TM 7
251 VALLKTVIIV LSVFIACWAP LFILLLLDVG CKVKTCDILF RAEYFLVLAV

301 LNSGTNPIIY TLTNKEMRRA FIRIMSCCKC PSGDSAGKFK RPIIAGMEFS

351 RSKSDNSSHP QKDEGDNPET IMSSGNVNSS S
```

Amino acids	381
Molecular weight	42 695
Glycosylation	Asn30*, Asn36*
Disulfide bonds	None
Palmitoylation	Cys327*

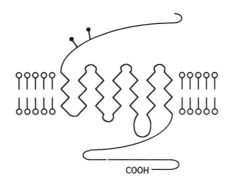

COOH

Database accession numbers

	PIR	SWISSPROT	EMBL/GENBANK	REFERENCE
Human	A35300	P21453	M31210	1

Comment

A novel cDNA (AGR16) cloned from rat aortic smooth with 51% sequence homology to EDG1 has been identified[2]. The mRNA encodes a 352 amino acid protein and is expressed in high levels in heart, lung, stomach and intestine, but is also present in many other tissues including brain.

References

1 Hla, T. and Maciag, M. (1990) *J. Biol. Chem.* **265**, 9308–9313.
2 Okazaki, H. *et al.* (1993) *Biochem. Biophys. Res. Commun.* **190**, 1104–1109.

G10d

G10d was isolated from a rat genomic library and a liver cDNA library. It has a widespread distribution and is found in high levels in lung, liver and adrenal gland, but is also present in kidney, aorta, heart, spinal cord, gut and testis.

Amino acid sequence (rat) [1]

```
  1 MSVIPSSRPV STLAPDNDFR EIHNWTELLH LFNQTFSDCH MELNENTKQV
       TM 1                                    TM 2
 51 VLFVFYLAIF VVGLVENVLV ICVNCRRSGR VGMLNLYILN MAVADLGIIL
                            TM 3
101 SLPVWMLEVM LEYTWLWGSF SCRFIHYFYL ANMYSSIFFL TCLSIDRYVT
                 TM 4
151 LTNTSPSWQR HQHRIRRAVC AGVWVLSAII PLPEVVHIQL LDGSEPMCLF
                 TM 5
201 LAPFETYSAW ALAVALSATI LGFLLPFPLI AVFNILSACR LRRQGQTESR
       TM 6                                    TM 7
251 RHCLLMWAYI VVFAICWLPY HVTMLLLTLH TTHIFLHCNL VNFLYFFYEI

301 IDCFSMLHCV ANPILYNFLS PSFRGRLLSL VVRYLPKEQA RAAGGRASSS

351 SSTQHSIIIT KEGSLPLQRI CTPTPSETCR PPLCLRTPHL HSAIP
```

Amino acids	395
Molecular weight	45 168
Glycosylation	Asn24*, Asn32*
Disulfide bonds	Cys122–Cys198*

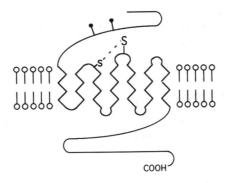

Database accession numbers

	PIR	SWISSPROT	EMBL/GENBANK	REFERENCE
Rat		-	L09249	1
			L04672	

Gene
The gene is intronless.

References
1 Harrison, J.K. *et al.* (1993) *FEBS Lett.* **318**, 17–22.

LCR1

LCR1 was isolated from a locus coeruleus library and its mRNA is also present in cerebellum, pons, dorsal raphe, thalamus and substantia nigra. It was originally proposed to encode for the neuropeptide YY$_3$ receptor.

Protein sequence (bovine)1

```
                                                        TM 1
   1 MEGIRIFTSD NYTEDDLGSG DYDSMKEPCF REENAHFNRI FLPTVYSIIF
                                      TM 2
  51 LTGIVGNGLV ILVMGYQKKL RSMTDKYRLH LSVADLLFVL TLPFWAVDAV
              TM 3
 101 ANWYFGKFLC KAVHVIYTVN LYSSVLILAF ISLDRYLAIV HATNSQKPRK
          TM 4
 151 LLAEKVVYVG VWLPAVLLTI PDLIFADIKE VDERYICDRF YPSDLWLVVF
        TM 5                                      TM 6
 201 QFQHIVVGLL LPGIVILSCY CIIISKLSHS KGYQKRKALK TTVILILTFF
                                                  TM 7
 251 ACWLPYYIGI SIDSFILLEI IQQGCEFEST VHKWISITEA LAFFHCCLNP

 301 ILYAFLGAKF KTSAQHALTS VSRGSSLKIL SKGKRGGHSS VSTESESSSF

 351 HSS
```

Amino acids	353
Molecular weight	39 938
Glycosylation	Asn11*
Disulfide bonds	Cys110–Cys187*
Palmitoylation	None

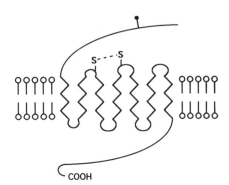

Database accession numbers

	PIR	SWISSPROT	EMBL/GENBANK	REFERENCE
Bovine		P25930	M86739	*1*

References
[1] Rimland, J. *et al.* (1991) *Mol. Pharmacol.* **40**, 869–875.

THE *MAS* ONCOGENE

The *mas* oncogene was discovered following co-transfection with DNA isolated from a human epidermoid carcinoma. It efficiently induces tumorigenicity and has weak focus-inducing activity in NIH 3T3 cells. It is the only oncogene to have been sequenced which encodes for a seven transmembrane spanning protein.

It has been claimed that *mas* is a receptor for angiotensin [1], but this has been challenged [2]. *Mas* has little sequence homology with the AT_1 receptor.

Distribution

In the CNS high levels of the *mas* oncogene transcript are present in cerebral cortex, with lower amounts in hippocampus and cerebellum. In the periphery, it is expressed in low levels in kidney, adrenals and liver.

Amino acid sequence [3]

```
                                    TM 1
 1 MDGSNVTSFV VEEPTNISTG RNASVGNAHR QIPIVHWVIM SISPVGFVEN
                        TM 2
51 GILLWFLCFR MRRNPFTVYI THLSIADISL LFCIFILSID YALDYELSSG
```

```
         TM 3                                               TM 4
101 HYYTIVTLSV TFLFGYNTGL YLLTAISVER CLSVLYPIWY RCHRPKYQSA
                                             TM 5
151 LVCALLWALS CLVTTMEYVM CIDREEESHS RNDCRAVIIF IAILSFLVFT
                        TM 6
201 PLMLVSSTIL VVKIRKNTWA SHSSKLYIVI MVTIIIFLIF AMPMRLLYLL
          TM 7
251 YYEYWSTFGN LHHISLLFST INSSANPFIY FFVGSSKKKR FKESLKVVLT

301 RAFKDEMQPR RQKDNCNTVT VETVV
```

Amino acids   325
Molecular weight   37 465
Glycosylation   Asn5*, Asn16*, Asn22*
Disulfide bind   None
Palmitoylation   Cys316*

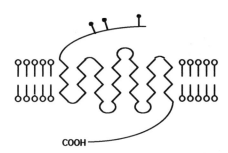

COOH

Database accession numbers

	PIR	SWISSPROT	EMBL/GENBANK	REFERENCES
Human	A01375	P04201	M13150	1
Rat	A31816	P12526	J03823	4

Comment

A novel cDNA (RTA) cloned from rat thoracic aorta with 34% sequence homology to *mas* has been identified [4]. The mRNA encodes a 343 amino acid protein and is expressed in high levels in gut, vas deferens, uterus, aorta and cerebellum, but is also present in many other tissues. There is one intron interrupting the N-terminal region of the sequence.

References

[1] Jackson, T. *et al.* (1988) *Nature* **335**, 437–440.
[2] Catt, K. and Abbott, A. (1991) *Trends Pharmacol. Sci.* **12**, 279–281.
[3] Young, W.S. *et al.* (1986) *Cell* **45**, 711–719.
[4] Ross, P.C. *et al.* (1990) *Proc. Natl Acad. Sci. USA* **87**, 3052–3056.

RDC1

It was originally claimed that the human homologue of RDC1 codes for a VIP receptor [1], but this is no longer thought to be correct.

Distribution

mRNA is present in the CNS, intestine, spleen, heart, lung and kidney.

Amino acid sequence [1]

```
                                                         TM 1
  1 MDLHLFDYAE PGNFSDISWP CNSSDCIVVD TVMCPNMPNK SVLLYTLSFI
                                   TM 2
 51 YIFIFVIGMI ANSVVVWVNI QAKTTGYDTH CYILNLAIAD LWVVLTIPVW
                           TM 3
101 VVSLVQHNQW PMGELTCKVT HLIFSINLFS GIFFLTCMSV DRYLSITYFT
                     TM 4
151 NTPSSRKKMV RRVVCILVWL LAFCVSLPDT YYLKTVTSAS NNETYCRSFY
                   TM 5
201 PEHSIKEWLI GMELVSVVLG FAVPFSIIAV FYFLLARAIS ASSDQEKHSS
          TM 6                                      TM 7
251 RKIIFSYVVV FLVCWLPYHV AVLLDIFSIL HYIPFTCRLE HALFTALHVT

301 QCLSLVHCCV NPVLYSFINR NYRYELMKAF IFKYSAKTGL TKLIDASRVS

351 ETEYSALEQN AK
```

Amino acids	362
Molecular weight	41 473
Glycosylation	Asn13*, Asn22*, Asn39*
Disulfide bonds	Cys117–Cys196*
Palmitoylation	None

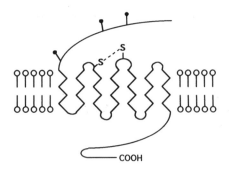

Database accession numbers

	PIR	Swissprot	EMBL/Genbank	Reference
Human			M64749	*1*
Canine	A30341 S12821	P11613	X14048	*3*

Gene

The complete coding sequence is contained in one exon; an intron divides the 5' untranslated region [2].

References

[1] Sreedharan, S. *et al.* (1991) *Proc. Natl Acad. Sci. USA* **88**, 4986–4990.
[2] Cook, J.S. *et al.* (1992) *FEBS Lett.* **300**, 149–152.
[3] Libert, F. *et al.* (1989) *Science* **244**, 569–572.

R334

R334 was isolated from a rat pituitary library and is found in discrete regions of the brain, pituitary and testis but is absent in other tissues.

Amino acid sequence (rat) [1]

```
                                                             TM 1
  1 MNEDPKVNLS GLPRDCIEAG TPENISAAVP SQGSVVESEP ELVVNPWDIV
                                  TM 2
 51 LCSSGTLICC ENAVVVLIIF HSPSLRAPMF LLIGSLALAD LLAGLGLIIN
              TM 3
101 FVFAYLLQSE ATKLVTIGLI VASFSASVCS LLAITVDRYL SLYYALTYHS
        TM 4                                          TM 5
151 GEDPHLYLCM LVMLWGTSTC LGLLLYGLEL PEGRVHLQRG QTLTKNNAAI
                                                       TM 6
201 LSISFLFMFA LMLQLYIQIC KIVMRHAHQI ALQHHFLATS HYVTTRKGIS
                                  TM 7
251 TLALILGTFA ACWMPFTLYS LIADYTYPSI YTYATLLPAT YNSIINPVIY

301 AFRNQEIQKA PLPHLLWVHP
```

Amino acids	320
Molecular weight	35 369
Glycosylation	Asn8*, Asn24*
Disulfide bonds	None
Palmitoylation	None

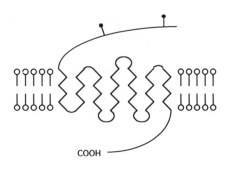

COOH

Database accession numbers

	PIR	SWISSPROT	EMBL/GENBANK	REFERENCE
Rat			X61496	1

Reference
1 Eidne, K.A. *et al.* (1991) *FEBS Lett.* **292**, 243–248.

Parathyroid hormone PTH

PTH is involved in calcium homeostasis within the body in combination with calcitonin and vitamin D [1,2]. PTH is released in response to hypocalcaemia and stimulates a rise in blood calcium; the converse is the case for calcitonin. The principle targets for PTH are bone and kidney. In bone, it acts on osteoblasts, possibly decreasing matrix production and releasing cytokines which are involved in osteoclast recruitment and activation. In kidney, PTH increases reabsorption of calcium and decreases reabsorption of phosphate in renal tubules and stimulates increased production of 1,25dihydroxyvitamin D3 which causes increased absorption of calcium in the intestine.

Antagonists at the PTH receptor are of potential clinical use in the treatment of hyperparathyroidism and short-term hypercalcaemic states.

Structure (human)

PTH   SVSEIQLMHN LGKHLNSMER VEWLRKKLQD VHNFVALGAP LAPRDAGSQR
             PRKKEDNVLV ESHEKSLGEA DKADVNVLTK AKSQ

PTH-related peptide shares 8 of the last 13 N-terminal amino acids in PTH. It is synthesized by cancer cells and causes the hypercalcaemia associated with malignancy syndrome.

Distribution and synthesis

PTH is made in the parathyroid gland and circulates as the full length peptide together with a mixture of C-terminal and N-terminal fragments. A single gene on chromosome 11, containing two introns, encodes for a large precursor protein, preproPTH, which is processed to proPTH and subsequently to PTH.

Reviews
1 Epand, R.M. and Caulfield, M.P. (1990) Calcitonin and parathyroid hormone receptors. In Comprehensive Medicinal Chemistry, Vol.3 (Membranes and Receptors) (Hansch, C., Sammes, P.G. and Taylor, J.B. eds) Pergamon Press, Oxford, pp. 1023–1045.
2 Muff, R. *et al.* (1992) Parathyroid hormone receptors in control of proximal tubule function. *Annu. Rev. Physiol.* **54**, 67–81.

THE PTH RECEPTOR
Distribution

In addition to its presence in bone and kidney, the receptor is found in lower levels in blood vessels where it mediates vasodilation. It is also found on UMR-106 cells.

Pharmacology

Agonists: PTH and PTH-related peptide are approximately equipotent and induce receptor activation in subnanomolar concentrations. Full activity resides in the N-terminal sequence of PTH, PTH$_{1-34}$.

Antagonists: [DTrp^{12},Tyr^{34}]bovinePTH$_{7-34}$NH$_2$ (K_i 5 nM)[1] and [Tyr^{34}] bovinePTH$_{7-34}$NH$_2$ (K_i 75 nM)[2] are claimed to be competitive antagonists.

Radioligands: [^{125}I]PTH$_{1-34}$ (K_d 0.5–4 nM)[3] and [^{125}I]PTH$_{1-36}$ (K_d 0.5–4 nM) bind to a single site.

Predominant effector pathways

The principle second messenger pathway is activation of adenylyl cyclase through G_s. In addition, PTH stimulates phosphoinositide metabolism on the expressed receptor[3].

Amino acid sequence (rat)[4]

```
  1 MGAARIAPSL ALLLCCPVLS SA
                             YALVDADD VFTKEEQIFL LHRAQAQCDK

 51 LLKEVLHTAA NIMESDKGWT PASTSGKPRK EKASGKFYPE SKENKDVPTG

101 SRRRGRPCLP EWDNIVCWPL GAPGEVVAVP CPDYIYDFNH KGHAYRRCDR
                                         TM 1
151 NGSWEVVPGH NRTWANYSEC LKFMTNETRE REVFDRLGMI YTVGYSMSLA
                             TM 2
201 SLTVAVLILA YFRRLHCTRN YIHMHMFLSF MLRAASIFVK DAVLYSGFTL
                         TM 3                              TM 3
251 DEAERLTEEE LHIIAQVPPP PAAAAVGYAG CRVAVTFFLY FLATNYYWIL
                             TM 4
301 VEGLYLHSLI FMAFFSEKKY LWGFTIFGWG LPAVFVAVWV GVRATLANTG
              TM 5
351 CWDLSSGHKK WIIQVPILAS VVLNFILFIN IIRVLATKLR ETNAGRCDTR
       TM 6                                          TM 7
401 QQYRKLLRST LVLVPLFGVH YTVFMALPYT EVSGTLWQIQ MHYEMLFNSF

451 QGFFVAIIYC FCNGEVQAEI RKSWSRWTLA LDFKRKARSG SSSYSYGPMV

501 SHTSVTNVGP RAGLSLPLSP RLPPATTNGH SQLPGHAKPG APATETETLP

551 VTMAVPKDDG FLNGSCSGLD EEASGSARPP PLLQEEWETV M
```

The first 22 amino acids are thought to be a signal peptide.

Amino acids	591
Molecular weight	66 260
Glycosylation	Asn151*, Asn161*, Asn166*, Asn177*
Disulfide bonds	None

The N-terminus has two hydropholic domains which are considered to be extracellular by comparison with other G-protein linked receptors.

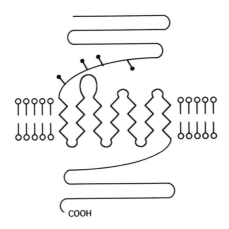

COOH

Database accession numbers

	PIR	SWISSPROT	EMBL/GENBANK	REFERENCES
Opossum			M74445	4
Rat		P25961	M77184	3

References
1 Goldman, M.E. *et al.* (1988) *Endocrinology* **123**, 2597–2597.
2 Horiuchi, N. *et al.* (1983) *Science* **220**, 1053–1055.
3 Abou-Samra, A-B. *et al.* (1992) *Proc. Natl Acad. Sci. USA* **89**, 2732–2736.
4 Jüppner, H. *et al.* (1991) *Science* **254**, 1024–1026.

PAF is an important mediator in allergic and inflammatory conditions stimulating microvascular leakage, vasodilation, endothelial adhesion, platelet aggregation, chemotaxis in neutrophils, macrophages and eosinophils, and smooth muscle contraction [1-9].

PAF antagonists are potential anti-inflammatory and anti-asthmatic agents.

PAF

PAF is 1-O-alkyl-2-acetyl-sn-glycero-3-phosphocholine. The alkyl residue attached at position 1 may be hexadecyl or octadecyl.

Distribution and synthesis

PAF is generated and released from activated neutrophils, macrophages, eosinophils, mast cells and basophils. It is derived from sn-2 fatty acyl-PAF by the action of phospholipase A$_2$ followed by acetylation (remodelling pathway) and from 1-O-alkyl-2-acetyl-sn-glycerol by the action of a cholinephosphotototransferase (*de novo* pathway).

Book
[1] Snyder, F. (1987) Platelet-activating Factor and Related Lipid Mediators, Plenum Press, New York.

Reviews
[2] Braquet, P. *et al.* (1987) Perspectives in platelet activating factor research *Pharmacol. Rev.* **39**, 97–145.
[3] Hosford, D. and Braquet, P. (1990) Antagonists of platelet-activating factor: Chemistry, pharmacology and clinical applications. *Prog. Med. Chem.* **27**, 325–380.
[4] Hwang, S-B. (1990) Specific receptors of platelet-activating factor, receptor heterogeneity, and signal transduction mechanisms. *J. Lipid Mediators* **2**, 123–158.
[5] Koltai, M. *et al.* (1991) Platelet activating factor (PAF). A review of its effects, antagonists and possible future clinical implications Part I and II. *Drugs* **42**, 9–29; *ibid.* 174-204.
[6] Prescott, S.M. *et al.* (1990) Platelet activating factor. *J. Biol. Chem.* **265**, 17381–17384.
[7] Shimizu, T. *et al.* (1992) Platelet-activating factor receptor and signal transduction. *Biochem. Pharmacol.* **44**, 1001–1008.

8 Whittaker, M. (1992) PAF receptor antagonists: recent advances. *Curr. Opinion Ther. Patents* **3**, 583–623.

9 Shukla, S.D. (1992) Platelet-activating factor receptor and signal transduction mechanisms. *FASEB J.* **6**, 2296–2301.

THE PAF RECEPTOR

PAF receptors are distributed widely in peripheral tissues and are found in high levels in platelets, leukocytes, macrophages and eosinophils; they are also present in smooth muscle, lung, endothelial cells and liver. PAF receptors have been described in brain and on cell lines, e.g. HL60.

Pharmacology

Agonists: PAF acts in subnanomolar concentrations. 1-*O*-Hexadecyl-2-*N*-methylcarbamyl-*sn*-glycero-3-phosphocholine (C-PAF) is resistant to hydrolysis by acetylhydrolase.

Antagonists: A large number of antagonists have been described. Examples from different structural classes are WEB 2086 (pA$_2$ 7.3); CV-6209 (pA$_2$ 9.5); L-652731 (pA$_2$ 6.7); SRI 63-072 (pA$_2$ 6.4) and BB-823 (K_i 8 pM).

Radioligands: The antagonist [3H]WEB 2086 (K_d 9 nM) labels a single binding site; there are several other antagonists ligands, e.g. [3H]SR 27417. [3H]PAF (K_d 0.5–2 nM) has relatively high nonspecific binding.

Structures:

CV-6209

SRI 63-072

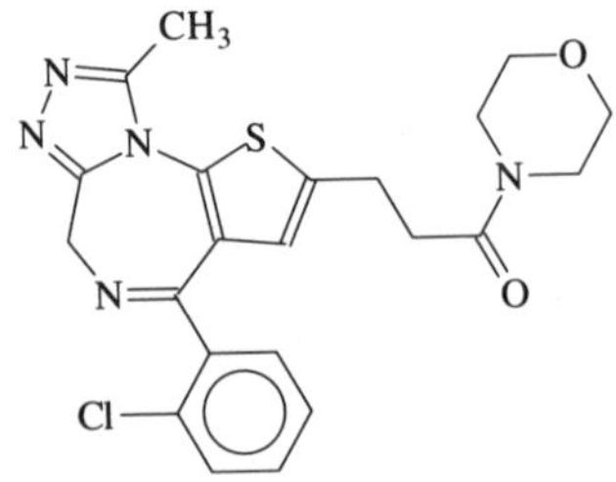

WEB 2086

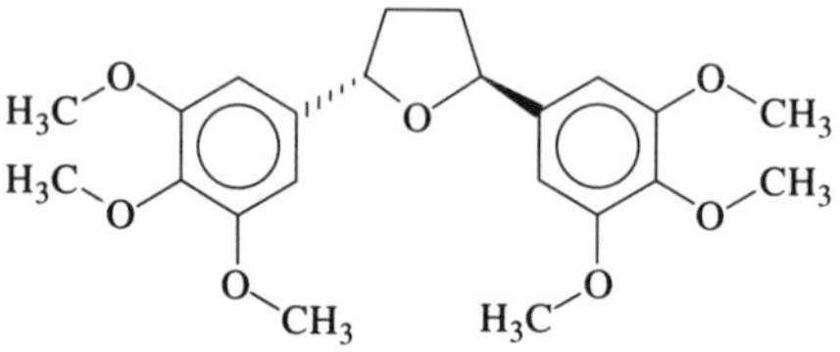

L-652731

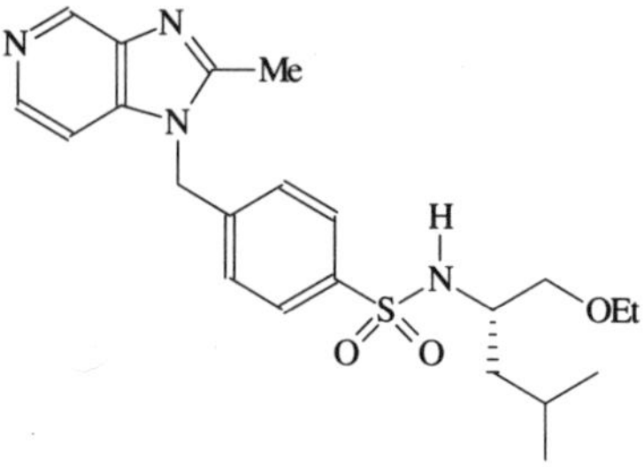

BB-823

Predominant effector pathways

G-protein activation of the phosphoinositide pathway; in a limited number of cases this is pertussis-toxin-sensitive [1].

Amino acid sequence [2-4]

```
                    TM 1
   1 MEPHDSSHMD SEFRYTLFPI VYSIIFVLGV IANGYVLWVF ARLYPCKKFN
       TM 2                                        TM 3
  51 EIKIFMVNLT MADMLFLITL PLWIVYYQNQ GNWILPKFLC NVAGCLFFIN
                                          TM 4
 101 TYCSVAFLGV ITYNRFQAVT RPIKTAQANT RKRGISLSLV IWVAIVGAAS
```

```
                                          TM 5
151 YFLILDSTNT VPDSAGSGNV TRCFEHYEKG SVPVLIIHIF IVFSFFLVFL
                                          TM 6
201 IILFCNLVII RTLLMQPVQQ QRNAEVKRRA LWMVCTVLAV FIICFVPHHV
                         TM 7
251 VQLPWTLAEL GFQDSKFHQA INDAHQVTLC LLSTNCVLDP VIYCFLTKKF

301 RKHLTEKFYS MRSSRKCSRA TTDTVTEVVV PFNQIPGNSL KN
```

Amino acids	342
Molecular weight	39 203
Glycosylation	None
Disulfide bonds	Cys90–Cys173*
Palmitoylation	Cys316*

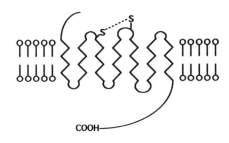

Accession numbers

	PIR	SWISSPROT	EMBL/GENBANK	REFERENCES
Human	A40191	P25105	M80436	2–4
			M76674	
Guinea-pig	S13638	P21556	X56736	5

Gene (human)

The gene organization in human is not known. Northern analysis has revealed the presence of two additional species of PAF receptor message in guinea-pig which are not found in CNS [5].

Comment

Binding studies suggest that differences exist between receptors in a number of tissues, notably neutrophils and platelets. This may, however, represent species differences or non-receptor sites.

WEB 2086 appears to distinguish between two PAF receptor subtypes or affinity states in guinea-pig eosinophils [6]. A novel receptor has been proposed to exist on macrophages on the basis of affinity and mode of antagonism by CV 6209 and other antagonists [7].

References

[1] Shukla, S.D. (1991) *Lipids* **26**, 1028–1033.
[2] Nakamura, M. *et al.* (1991) *J. Biol. Chem.* **266**, 20400–20405.

3 Ye, R.D. *et al.* (1991) *Biochem. Biophys. Res. Commun.* **180**, 105–111.
4 Kunz, D. *et al.* (1992) *J. Biol. Chem.* **267**, 9101–9106.
5 Honda, Z.I. *et al.* (1991) *Nature* **349**, 342–346.
6 Kroegel, C. *et al.* (1989) *Biochem. Biophys. Res. Commun.* **162**, 511–521.
7 Stewart, A. (1989) *Clin. Exp. Pharmacol. Physiol.* **16**, 813–820.

Prostanoids

INTRODUCTION

Prostanoids (prostaglandins [PG] and thromboxanes [TX]) mediate a wide variety of actions and have important physiological roles in the cardiovascular and immune systems and in pain sensation in peripheral tissues [1-6]. However, a lack of selective agonists/antagonists and marked species heterogeneity in response has, in many cases, hampered identification of the physiological roles of individual prostanoids and their receptors.

The opposing actions of PGI_2 and TXA_2 are involved in the regulation of the interaction of platelets with the vascular endothelium, while PGE_2, PGI_2 and PGD_2 are powerful vasodilators and potentiate the action of various autocoids to induce plasma extravasation and pain sensation, e.g. bradykinin. PGE_2 has marked effects on the immune system, inhibiting secretion of cytokines, superoxide generation and macrophage activation. It is also found in high levels in CSF during infection and may mediate the increase in temperature induced by agents such as interleukin 1. PGE_2 is also involved in water reabsorption in the kidney.

Nonsteroidal anti-inflammatory drugs, e.g. aspirin, mediate their clinical effects by preventing prostanoid synthesis through inhibition of cyclooxygenase. It has also been suggested that glucocorticoids mediate some of their clinical actions by preventing prostanoid generation through synthesis of the protein lipocortin which is proposed to inhibit phospholipase A_2 (PLA_2). Inhibition of prostanoid formation is associated with decreased inflammation, analgesia and reduced fever. Clinically, prostanoid analogues are used to induce uterine contractility, bronchodilation, vasodilation or to inhibit gastric acid secretion. There is also evidence that 3-series prostanoids, derived from eicosapentaenoic acid and found in high levels in fish-eating communities, e.g. eskimos, confer resistance to atherosclerosis and thrombotic disease.

Structures

Several species of TXs and PGs (termed A–J) have been identified. Structures of the physiologically most important ones are shown:

PGD$_2$

PGE₂

PGF₂α

PGI₂

Thromboxane A₂

Synthesis and distribution

Prostanoids are not stored within cells but are synthesized *de novo* in response to an appropriate stimulus. Prostanoids are derived from C_{20} fatty acids, principally arachidonate which is esterified in the 2-position of membrane phospholipids. Arachidonate gives rise to the 2-series of prostanoids (the major types found within cells) by the action of the enzyme cyclooxygenase. Arachidonate is liberated from phospholipids by the action of PLA_2 which is regulated by several pathways including Ca^{2+} and possibly a G-protein. In addition, arachidonate can be liberated by the combined actions of phospholipase C and diacylglycerol lipase. Many cell surface receptors induce mobilization of Ca^{2+} and therefore liberation of prostanoids. Prostanoids are also generated in response to cell damage.

Although prostanoids are found in nearly every tissue within the body, there is a great deal of heterogeneity in the individual prostanoids that are synthesized. For example, TXA_2 synthase is found in platelets and white blood cells (and also many other tissues, e.g. spleen, bronchial epithelium, liver, etc.) but is not found in the vascular endothelium; in contrast, PGI_2 is synthesized in the endothelium but not in platelets. PGE_2 is the predominant prostanoid in inflammatory sites and is also synthesized in many other tissue types while PGD_2 is synthesized in high amounts in mast cells and in brain. The synthetic pathway of $PGF_{2\alpha}$ is poorly characterized.

Reviews

[1] Coleman, R.A. (1988) Prostaglandin receptors. *Prostaglandin Perspectives* **4**, 33–35

[2] Coleman, R.A. *et al.* (1990) Prostanoids and their receptors. *In Comprehensive Medicinal Chemistry, vol. 3, Membranes and Receptors* (Hansch, C., Sammes, P.G. and Taylor, J.B. eds), Pergamon Press, Oxford, pp. 643–714.

[3] Giles, H. (1990) More selective ligands of eicosanoid receptor subtypes improve prospects in inflammatory and cardiovascular research. *Trends Pharmacol. Sci.* **11**, 301–304.

[4] Giles, H. and Leff, P. (1988) *Prostaglandins* **35**, 277–300.

[5] Halushka, P. *et al.* (1989) Thromboxane, prostaglandin and leukotriene receptors. *Annu. Rev. Pharmacol. Toxicol.* **29**, 213–239.

[6] Smith, W.L. (1989) The eicosanoids and their biochemical mechanisms of action. *Biochem. J.* **259**, 315–324.

PROSTANOID RECEPTORS

Evidence for the existence of at least five classes of prostanoid receptor has been obtained primarily from studies of rank orders of potencies of agonists and antagonists in peripheral tissues, with subtypes of the EP receptor class. This scheme of classification acknowledges that an endogenous prostanoid is at least one order of magnitude more potent at one receptor class than the other endogenous prostanoids [1,2].

Identification of receptor subtypes and their distribution, however, is hampered by expression of more than one receptor within a tissue and the poor selectivity of available agonists and antagonists. In addition, many endogenous prostanoids undergo rapid metabolism, in particular TXA$_2$ (the metabolically stable analogue, U-46619 is frequently used as a mimetic of TXA$_2$).

References
1 Kennedy, I. *et al.* (1982) *Prostaglandins* **24**, 667–689.
2 Coleman, R.A. and Kennedy, I. (1985) *Prostaglandins* **29**, 363–375.

THE DP RECEPTOR

PGD$_2$ is ~ tenfold more potent than other naturally occurring prostanoids and induces activation in nanomolar concentrations. The DP receptor has not been cloned.

Distribution

DP receptors have a limited distribution. They mediate relaxation in vascular, gastrointestinal and uterine smooth muscle in human and in a few other species; they inhibit platelet activation in some species, including human, and modify release of hypothalamic and pituitary hormones.

Pharmacology

Agonists: BW 245C has ~tenfold higher potency than PGD$_2$ and a selectivity of more than 1000-fold [1].

Antagonists: BW A868C (pA$_2$ 9.3) has a selectivity of more than 1000-fold [2].

Radioligands: [^{3}H]PGD$_2$ has weak selectivity.

Structures:

BW 245C

BW A868C

Predominant effector pathways

Activation of adenylyl cyclase through G_s.

Comment

Subtypes of DP receptor have been proposed to account for potency differences of PGD_2 analogues on the pressor response in sheep and inhibition of platelet aggregation in man [3]. These differences, however, may reflect species homologues of the DP receptor.

References

[1] Town M.H. *et al.* (1983) *Prostaglandins* **25**, 13–28.
[2] Giles, H. *et al.* (1989) *Br. J. Pharmacol.* **96**, 291–300.
[3] Jones, R.L. *et al.* (1984) In *IUPHAR 9th International Congress of Pharmacology, London* (Paton, Mitchell and Turner eds), Macmillan, London, pp. 293–301.

THE EP$_1$ RECEPTOR

The rank order of potency of prostanoids is: PGE_2 > $PGF_{2\alpha}$ $\simeq$ PGI_2 > PGD_2 > U-46619. PGE_2 induces activation in low nanomolar concentrations. It has not been cloned.

Distribution

EP$_1$ receptors mediate contraction of gastrointestinal smooth muscles in various species, and relaxation of airway and uterine smooth muscles, particularly in rodents.

Pharmacology

Agonists: 17-Phenyl-ω-trinor-PGE$_2$ has weak selectivity and induces activation in nanomolar concentrations.

Antagonists: AH 6809 (pA$_2$ 6.8) is ~ tenfold selective relative to DP and TP receptors; SC 19220 (pA$_2$ 5.6) has increased selectivity but is of low potency and solubility.

Radioligands: [^{3}H]PGE$_2$ (K_d 3 nM) has weak selectivity to EP receptors.

Structures:

AH 6809   SC 19220

Predominant effector pathways
Activation of the phosphoinositide pathway through a pertussis-toxin-insensitive G-protein most likely of the G_q/G_{11} class.

THE EP$_2$ RECEPTOR

The rank order of potency of prostanoids is: $PGE_2 > PGF_{2\alpha} \simeq PGI_2 \simeq PGD_2 >$ U-46619. PGE_2 induces activation in low nanomolar concentrations. It has not been cloned.

Distribution

EP$_2$ receptors mediate vasodilation, bronchodilation and relaxation of intestinal smooth muscle. They stimulate fluid secretion in the intestine.

Pharmacology

Agonists: Butaprost has a selectivity of more than 100-fold and acts in low micromolar concentrations; AH 13205 has a similar agonist profile but is slightly less potent.

Antagonists: There are no selective antagonists.

Radioligands: [^{3}H]PGE$_2$ (K_d 3 nM) has weak selectivity to EP receptors.

AH 13205   Butaprost

Predominant effector pathways
Activation of adenylyl cyclase through G_s.

THE EP$_3$ RECEPTOR

The rank order of potency of prostanoids is: $PGE_2 > PGI_2 > PGF_{2\alpha} \gg PGD_2 >$ U-46619. PGE_2 induces activation in low nanomolar concentrations. Two isoforms of the EP$_3$ receptor exist, EP$_{3\alpha}$ and EP$_{3\beta}$, derived by alternative splicing.

Distribution

EP$_3$ receptors mediate contraction in a wide range of smooth muscle including gastrointestinal and uterine. They also inhibit neurotransmitter release in central and autonomic nerves through a presynaptic action, and inhibit secretion in glandular tissues, e.g. acid secretion from gastric mucosa, and sodium and water reabsorption in kidney. mRNA is found in high levels in kidney and uterus, and in lower levels in brain, thymus, lung, heart, stomach and spleen.

Pharmacology

Agonists: Enprostil, GR 63799 and MB 28767 are active in low nanomolar concentrations and have selectivities of more than 30-fold.

Antagonists: There are no selective antagonists.

Radioligands: [^{3}H]PGE$_2$ (K_d 3 nM) has weak selectivity to EP receptors.

Structures:

Enprostil   MB 28767

GR 63799

Predominant effector pathways

Inhibition of adenylyl cyclase through an uncharacterized G-protein most likely of the G_i/G_o class. Stimulation of the phosphoinositide pathway has been reported although this has not been observed with the cloned receptor.

Amino acid sequence (mouse) [1]

```
                                          TM 1
  1 MASMWAPEHS AEAHSNLSST TDDCGSVSVA FPITMMVTGF VGNALAMLLV
                    TM 2
 51 SRSYRRRESK RKKSFLLCIG WLALTDLVGQ LLTSPVVILV YLSQRRWEQL
              TM 3
101 DPSGRLCTFF GLTMTVFGLS SLLVASAMAV ERALAIRAPH WYASHMKTRA
         TM 4
151 TPVLLGVWLS VLAFALLPVL GVGRYSVQWP GTWCFISTGP AGNETDPARE
              TM 5
201 PGSVAFASAF ACLGLLALVV TFACNLATIK ALVSRCRAKA AVSQSSAQWG
                 TM 6
251 RITTETAIQL MGIMCVLSVC WSPLLIMMLK MIFNQMSVEQ CKTQMGKEKE
         TM 7
                                          IRDHT NYASSSTSLP
301 CNSFLIAVRL ASLNQILDPW VYLLLRKILL RKFCQ
                                          MMNNL KWTFIAVPVS
351 CPGSSALMWS DQLER EP₃α

    LGLRISSPRE G EP₃β
```

Amino acids	365
Molecular weight	40 077
Glycosylation	Asn16*, Asn193*
Disulfide bonds	Cys107–Cys184*
Palmitoylation	Cys334*

Two isoforms of the EP₃ receptor derived by alternative splicing have been described [2]. A 30 amino acid hydrophilic sequence in the C-terminal of EP₃α is replaced by a 26 amino acid, hydrophobic sequence.

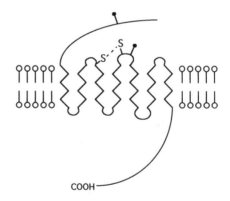

COOH

Comment

Three orders of magnitude lower concentrations of PGE₂ are required to activate the EP₃α receptor relative to the EP₃β receptor, presumably because of differing coupling profiles to the Gᵢ/Gₒ family of G-proteins [2]. The relative abundance of mRNA levels for the EP₃α receptor is ~ 1.5–5 times greater than that for the EP₃β receptor within tissues.

Database accession numbers

	PIR	SWISSPROT	EMBL/GENBANK	REFERENCES
Mouse EP$_{3\alpha}$			D10204	1
Mouse EP$_{3\beta}$			D13321	2

Gene
The organization of the gene is not known.

References
1 Sugimoto, Y. et al. (1992) *J. Biol. Chem.* **267**, 6463–6466.
2 Sugimoto, Y. et al. (1993) *J. Biol. Chem.* **268**, 2712–2718.

THE FP RECEPTOR

The rank order of potency of prostanoids is: PGF$_{2\alpha}$ > PGD$_2$ > PGE$_2$ $\simeq$ U-46619 > PGI$_2$. PGF$_{2\alpha}$ induces activation in nanomolar concentrations.

Distribution

FP receptors mediate contraction in a wide range of smooth muscle including intraocular, myometrial and bronchial tissues, and are potent stimulants of luteolysis.

Pharmacology

Agonists: Fluprostenol has a selectivity of more than 1000-fold and induces activation in low nanomolar concentrations.

Antagonists: There are no specific antagonists.

Radioligands: [³H]PGF$_{2\alpha}$ has weak selectivity.

Structures:

Fluprostenol

Predominant effector pathways
Activation of phosphoinositide pathway through a pertussis-toxin-insensitive G-protein, most likely of the G$_q$/G$_{11}$ class.

THE IP RECEPTOR

The rank order of potency of prostanoids is: $PGI_2 > PGD_2 \sim PGE_2 \sim PGF_{2\alpha} >$ U-46619. PGI_2 induces activation in nanomolar concentrations.

Distribution

IP receptors induce relaxation in a range of smooth muscles including blood vessels and potently inhibit platelet activation. They are expressed in NCB-20 neuronal hybrid cells.

Pharmacology

Agonists: the stable agonist cicaprost has a selectivity of more than 1000-fold and induces activation in low nanomolar concentrations [1]; iloprost is a potent agonist at IP receptors but is a partial agonist at EP_1 receptors.

Antagonists: There is no specific antagonist.

Radioligands: [^{3}H]iloprost (K_d 15 nM) is selective to IP and EP_1 sites.

Structures:

Cicaprost

Iloprost

Predominant effector pathways

Activation of adenylyl cyclase through G_s.

Reference

[1] Sturzebecher, C.-St. *et al.* (1986) *Prostaglandins* **31**, 95–109.

THE TP RECEPTOR

The rank order of potency of prostanoids is: U-46619 >> PGI_2 ≃ PGD_2 ≃ PGE_2 ≃ $PGF_{2\alpha}$. TXA_2 induces activation in nanomolar concentrations, although U-46619 is usually preferred because of its metabolic stability.

Distribution

TP receptors cause vasoconstriction, platelet aggregation and bronchoconstriction. They are also found in a range of other smooth muscle preparations and tissue types, e.g. thymus, placenta, spleen and lung.

Pharmacology

Agonists: Several metabolically stable agonists with a selectivity of more than 1000-fold have been described, including U-46619 which is approximately equipotent with TXA_2; STA_2 and I-BOP are further examples.

Antagonists: Several antagonists have a selectivity of more more than 1000-fold, including GR 32191 [1] (pA2 8.8) and SQ 29548 (pA2 8.7).

Radioligands: The antagonists [^{3}H]SQ 29548 (K_d 5 nM) and [^{125}I]SAP (K_d 0.2–1 nM) have a selectivity of more than 1000-fold. The agonists [^{3}H]U-46619 (K_d 10 nM) and [^{125}I]BOP (K_d 1 nM) are also selective.

Structures:

I-BOP

U-46619

STA2

GR 32191

SQ 29548

Predominant effector pathways

Activation of phosphoinositide pathway through a pertussis-toxin-insensitive G-protein, most likely of the G_q/G_{11} class.

Amino acid sequence [2]

```
                                    TM 1
  1 MWPNGSSLGP CFRPTNITLE ERRLIASPWF AASFCVVGLA SNLLALSVLA
            TM 2
 51 GARQGGSHTR SSFLTFLCGL VLTDFLGLLV TGTIVVSQHA ALFEWHAVDP
        TM 3                                          TM 4
101 GCRLCRFMGV VMIFFGLSPL LLGAAMASER YLGITRPFSR PAVASQRRAW
                                                       TM 5
151 ATVGLVWAAA LALGLLPLLG VGRYTVQYPG SWCFLTLGAE SGDVAFGLLF
                                                       TM 6
201 SMLGGLSVGL SFLLNTVSVA TLCHVYHGQE AAQQRPRDSE VEMMAQLLGI
                                                 TM 7
251 MVVASVCWLP LLWFIAQTVL RNPPAMSPAG QLSRTTEKEL LIYLRVATWN

301 QILDPWVYIL FRRAVLRRLQ PRLSTRPRSL SLQPQLTQRS GLQ
```

Amino acids	343
Molecular weight	37 518
Glycosylation	Asn4*, Asn16*
Disulfide bonds	Cys106–Cys183*
Palmitoylation	None

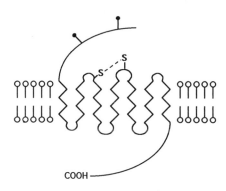

Database accession numbers

	PIR	SWISSPROT	EMBL/GENBANK	REFERENCES
Human	S13647	P21731		2
Mouse			D10849	3

Gene

The organization of the gene is not known.

References

1 Lumley, P. *et al.* (1989) *Br. J. Pharmacol.* **97**, 783–794.
2 Hirata, M. *et al.* (1991) *Nature* **349**, 617–620.
3 Namba, T. *et al.* (1992) *Biochem. Biophys. Res. Commun.* **184**, 1197–1203.

Somatostatin

INTRODUCTION

Somatostatin (SS) is a neurotransmitter/hormone with a wide spectrum of biological actions [1-3]. It has an important role in the neuroendocrine system and inhibits secetion of growth hormone and prolaction in the anterior pituitary. Somatostatin also inhibits secretion in intestine (including gastric acid in stomach), pancreatic acinar cells and pancreatic β-cells, stimulates absorption in intestine and modulates smooth muscle contractility. In the CNS, it is a neurotransmitter activating a hyperpolarizing K^+ current and inhibiting Ca^{2+} influx, and is believed to have important roles in regulating locomotor activity and cognitive function.

The somatostatin analogue, SMS 201-995 (octreotide), is used clinically in the treatment of certain tumours, carcinoid syndrome and glucagonoma. A reduction in cortical somatostatin levels has been reported in Alzheimer's disease and Parkinson's disease.

Structure

$$S\text{————————————————}S$$

SS-14   Ala–Gly–Cys–Lys–Asn–Phe–Phe–Trp–Lys–Thr–Phe–Thr–Ser–Cys

There are two secreted forms of somatostatin, SS-14 and SS-28, with the entire SS-14 sequence present in the C-terminus of SS-28. The structure of SS-14 is conserved in mammals, but not in lower organisms.

Distribution and synthesis

SS-14 and SS-28 are derived from a larger, 8 kDa precursor, preprosomatostatin, which is the product of a single gene in human or rat. Tissue processing results in considerable variation in the proportions of the mature peptides.

Somatostatin is distributed widely in neurons within the CNS. It is found in high levels in hypothalamus and is also present in cerebral cortex, hippocampus, thalamus, brainstem and spinal cord. In the periphery, it is found in D cells of the pancreas, gastrointestinal tract, neurons in autonomic ganglia and in a diverse range of other cells, e.g. thyroid, adrenal, placenta, urinary tract, etc. The highest levels of somatostatin are in gut and brain. Although SS-14 and SS-28 are co-expressed, their distribution is distinct.

Reviews

1 Bell, G.I. and Reisine, T. (1993) Molecular biology of somatostatin receptors. *Trends Neurosci.* **16**, 34–38.
2 Epelbaum, J. (1986) Somatostatin in the central nervous system: physiology and pathological modifications. *Prog. Neurobiol.* **27**, 63–100.
3 Rens-Domiano, S. and Reisine, T.R. (1992) Biochemical and functional properties of somatostatin receptors. *J. Neurochem.* **58**, 1987–1996.

SOMATOSTATIN RECEPTORS

Receptor subtypes were originally proposed on the basis of functional and radioligand binding studies but their full characterization was prevented by the lack of selective ligands. As a consequence, no clear scheme of classification became established. Cloning studies have confirmed the existence of receptor subtypes, but, in many cases, it is unclear how these correspond to earlier schemes of classification.

SS_1, SS_2, SS_3 and SS_4 receptors

Four subtypes of somatostatin receptors have been identified in cloning studies [1,2] and named in the order in which they were isolated. They were originally named $SSTR_1$–$SSTR_4$. All are seven transmembrane proteins and are described below.

Earlier schemes of classification

Functional and radioligand binding studies had previously provided evidence for subtypes of somatostatin receptor. Binding sites for iodinated analogues of SS-14 and SS-28 have distinct distributions in brain, and SS-14 and SS-28 have reverse rank orders of potency and opposing actions on certain biological responses [1]. SMS 201-995 (DPhe–Cys–Phe–DTrp–Lys–Thr–Cys–Thr-ol) induces a biphasic displacement of [^{125}I]SS-14 binding to brain membranes [2,3]. The agonist MK 678 has high affinity for $SRIF_1$ receptors and low affinity for $SRIF_2$ receptors [4].

References
1 Wang, H.L. *et al.* (1989) *Proc. Natl Acad. Sci. USA* **86**, 9616–9620.
2 Tran, V. *et al.* (1985) *Science* **228**, 492–495.
3 Reubi, J.C. (1984) *Neurosci. Lett.* **49**, 259–263.
4 Raynor, K. and Reisine, T. (1989) *J. Pharmacol. Exp. Ther.* **251**, 510–517.

THE SS_1 RECEPTOR

SS-28 and SS-14 have similar affinities and induce activation in low nanomolar concentrations. It is thought to correspond to the $SRIF_2$ receptor [1].

Distribution

In human, high levels of mRNA for the SS_1 receptor are found in jejunum and stomach, with lower levels in pancreas, colon and kidney; mRNA is absent in brain [2]. In contrast, in rodent tissue, high levels of mRNA are present in brain but are absent in peripheral tissues [3].

Pharmacology

Agonists: There are no selective agonists.

Antagonists: No antagonist has been described.

Radioligands: [^{125}I-Tyr^{11}]SS-14 (K_d 0.1 nM) is nonselective.

Effector pathway

The binding of agonists to recombinant receptor expressed in CHO cells is not regulated by GTP and is unaffected by pertussis toxin [1].

Amino acid sequence [2]

```
  1 MFPNGTASSP SSSPSPSPGS CGEGGGSRGP GAGAADGMEE PGRNASQNGT
            TM 1                                      TM 2
 51 LSEGQGSAIL ISFIYSVVCL VGLCGNSMVI YVILRYAKMK TATNIYILNL
                                     TM 3
101 AIADELLMLS VPFLVTSTLL RHWPFGALLC RLVLSVDAVN MFTSIYCLTV
                         TM 4
151 LSVDRYVAVV HPIKAARYRR PTVAKVVNLG VWVLSLLVIL PIVVFSRTAA
                    TM 5
201 NSDGTVACNM LMPEPAQRWL VGFVLYTFLM GFLLPVGAIC LCYVLIIAKM
                    TM 6
251 RMVALKAGWQ QRKRSERKIT LMVMMVVMVF VICWMPFYVV QLVNVFAEQD
        TM 7
301 DATVSQLSVI LGYANSCANP ILYGFLSDNF KRSFQRILCL SWMDNAAEEP

351 VDYYATALKS RAYSVEDFQP ENLESGGVFR NGTCTSRITT L
```

Amino acids	391
Molecular weight	42 657
Glycosylation	Asn4*, Asn44*, Asn48*
Disulfide bonds	Cys130–Cys208*
Palmitoylation	Cys339*
Phosphorylation (PKA)	Thr172*, Ser265*

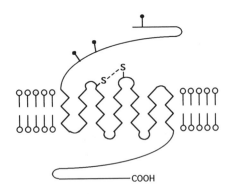

Database accession numbers

	PIR	SWISSPROT	EMBL/GENBANK	REFERENCES
Human			M81829	2
Mouse			M81831	2
Rat			M97656	3

The rat SS$_1$ receptor was initially identified as an orphan receptor [4].

Gene
The receptor is encoded by an intronless gene [2].

References
1 Rens-Domiano, S. *et al.* (1992) *Mol. Pharmacol.* **42**, 28–34.
2 Yamada, Y. *et al.* (1992) *Proc. Natl Acad. Sci. USA* **89**, 251–255.
3 Li, X.-J. *et al.* (1992) *J. Biol. Chem.* **267**, 21307–21312.
4 Meyerhof, W. *et al.* (1992) *DNA Cell Biol.* **10**, 689–694.

THE SS$_2$ RECEPTOR

SS-14 and SS-28 have similar affinities and induce activation in low nanomolar concentrations. It appears to correspond to the SRIF$_1$ receptor [1] and is also known as the SS$_{2A}$ receptor.

A novel form of the SS$_2$ receptor, derived by alternative splicing, and named SS$_{2B}$, has been identified in mouse (see Comment).

Distribution

In human, mRNA for the SS$_2$ receptor is found in high levels in human brain, kidney and pituitary, with lower levels in jejunum, colon, pancreas and liver [2].

Pharmacology

Agonists: MK 678 has a selectivity of more than 100-fold. SMS 201-995 is selective to SS$_2$ and SS$_4$ receptors.

Antagonists: None has been described.

Radioligands: [^{125}I]MK 678 (K_d 0.15 nM) [2] is selective; [^{125}I-Tyr^3]SMS 201-995 (K_d 0.5 nM) binds to SS$_2$ and SS$_4$ receptors. [^{125}I-Tyr^{11}]SS-14 (K_d 0.1 nM) is nonselective.

Structures:

MK 678	cyc(n-Met–Ala–Tyr–DTrp–Lys–Val–Phe)
SMS 201-995	cyc(DPhe–Cys–Phe–DTrp–Lys–Thr–Cys–Thr-ol)

Predominant effector pathways
It has been shown to be coupled to G$_o$ and G$_{i3}$ but not to G$_{i1}$; it is linked to inhibition of adenylyl cyclase.

Amino acid sequence [1]

```
                                                             TM 1
  1 MDMADEPLNG SHTWLSIPFD LNGSVVSTNT SNQTEPYYDL TSNAVLTFIY
                                  TM 2
 51 FVVCIIGLCG NTLVIYVILR YAKMKTITNI YILNLAIADE LFMLGLPFLA
                 TM 3
101 MQVALVHWPF GKAICRVVMT VDGINQFTSI FCLTVMSIDR YLAVVHPIKS
            TM 4
151 AKWRRPRTAK MITMAVWGVS LLVILPIMIY AGLRSNQWGR SSCTINWPGE
          TM 5
201 SGAWYTGFII YTFILGFLVP LTIICLCYLF IIIKVKSSGI RVGSSKRKKS
        TM 6                                     TM 7
251 EKKVTRMVSI VVAVFIFCWL PFYIFNVSSV SMAISPTPAL KGMFDFVVVL

301 TYANSCANPI LYAFLSDNFK KSFQNVLCLV KVSGTDDGER SDSKQDKSRL

351 NETTETQRTL LNGDLQTSI
```

Amino acids	369
Molecular weight	41 305
Glycosylation	Asn9*, Asn22*, Asn29*, Asn32*
Disulfide bonds	Cys115–Cys193*
Palmitoylation	Cys328*
Phosphorylation	Ser250*

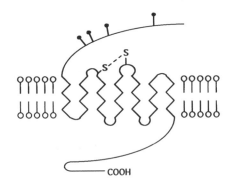

Database accession numbers

	PIR	SWISSPROT	EMBL/GENBANK	REFERENCES
Human			M81830	1
Mouse			M81832	1
Rat			M93273	2

Gene

The genomic sequence has only been partially determined and contains at least one intron [2].

Comment

An SS_{2B} receptor derived by alternative splicing of the SS_2 gene has been identified in mouse [3]. It has nearly complete identity with the SS_{2A} receptor up to amino acid 332 (leucine and threonine residues replace isoleucine and serine at positions 180 and 305, respectively), but has 23 fewer amino acids and differs in the last 15 C-terminal residues. Somatostatin has a similar affinity for both forms of the receptor. mRNA encoding the SS_{2B} receptor is present in higher levels in mouse brain than that for the SS_{2A} receptor.

References

[1] Yamada, Y. *et al.* (1992) *Proc. Natl Acad. Sci. USA* **89**, 251–255.
[2] Kluxen, F.W. *et al.* (1992) *Proc. Natl Acad. Sci. USA* **89**, 4618–4622.
[3] Vanetti, M. *et al.* (1992) *FEBS Lett.* **311**, 290–294.

THE SS_3 RECEPTOR

SS-28 and SS-14 have similar affinities and induce activation in low nanomolar concentrations. It has also been called the SSR-28 receptor.

Distribution

mRNA for the SS_3 receptor is widely distributed in mouse brain with high levels in forebrain, hippocampus and amygdala; moderate levels are also present in substantia nigra [1].

Pharmacology

Agonists: There is no selective agonist. CGP 23996 has similar affinity at SS_2, SS_3 and SS_4 receptors.

Antagonists: None has been described.

Radioligands: [^{125}I]CGP 23996 (K_d 2 nM) binds with high affinity to SS_2 and SS_3 receptors. [^{125}I-Tyr^{11}]SS-14 (K_d 0.1 nM) is nonselective.

Structures:

CGP 23966   des-Ala^1,Gly^2-[desamino-Cys^3,Tyr^{11}]-3,14 dicarbasomatostatin-14

Predominant effector pathways

Inhibition of adenylyl cyclase through a pertussis-toxin-sensitive G-protein, most likely of the G_i/G_o class [1].

Amino acid sequence [1]

```
                                                     TM 1
   1 MDMLHPSSVS TTSEPENASS AWPPDATLGN VSAGPSPAGL AVSGVLIPLV
                                       TM 2
  51 YLVVCVVGLL GNSLVIYVVL RHTASPSVTN VYILNLALAD ELFMLGLPFL
                        TM 3
 101 AAQNALSYWP FGSLMCRLVM AVDGINQFTS IFCLTVMSVD RYLAVVHPTR
              TM 4
 151 SARWRTAPVA RTVSAAVWVA SAVVVLPVVV FSGVPRGMST CHMQWPEPAA
            TM 5
 201 AWRAGFIIYT AALGFFGPLL VICLCYLLIV VKVRSAGRRV WAPSCQRRRR
            TM 6                                   TM 7
 251 SERRVTRMVV AVVALFVLCW MPFYVLNIVN VVCPLPEEPA FFGLYFLVVA

 301 LPYANSCANP ILYGFLSYRF KQGFRRVLLR PSRRVRSQEP TVGPPEKTEE

 351 EDEEEEDGEE SREGGKGKEM NGRVSQITQP GTSGQERPPS RVASKEQQLL

 401 PQEASTGEKS STMRISYL
```

Amino acids	418
Molecular weight	48 855
Glycosylation	Asn17*, Asn30*
Disulfide bonds	Cys116–Cys191*
Palmitoylation	None
Phosphorylation (PKA)	Ser251*, Thr256*
Phosphorylation (PKC)	Ser151*, Ser251*, Ser317*, Ser332*, Thr412*

The SS$_3$ receptor has a glutamic acid-rich region in the carboxyl domain, unique among G-protein coupled receptors.

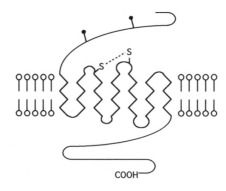

Database accession numbers

	PIR	SWISSPROT	EMBL/GENBANK	REFERENCES
Mouse			M91000	2
Human			S53994	1
			M96738	
Rat			X63574	3

Gene

The receptor is encoded by an intronless gene [1].

References

1 Yasuda, K. *et al.* (1992) *J. Biol. Chem.* **28**, 20422–20428.
2 Yasuda, K. *et al.* (1992) *Mol. Endocrinol.* **6**, 2136–2142.
3 Meyerhof, W. *et al.* (1992) *Proc. Natl Acad. Sci. USA* **89**, 10267–10271.

THE SS₄ RECEPTOR

SS-28 is one order of magnitude more potent that SS-14 and induces activation in high picomolar concentrations.

Distribution

mRNA is present in high levels in pituitary but is absent from brain and peripheral tissues [1].

Pharmacology

Agonists: SMS 201-995 is selective to SS₄ and SS₂ receptors. MK 678 has high affinity for SS₄ receptors.

Antagonists: None has been described.

Radioligands: [^{125}I-Tyr^3]SMS 201-995 (K_d 0.5 nM) binds to SS₂ and SS₄ receptors. [^{125}I-Tyr^{11}]SS-14 (K_d 0.1 nM) is nonselective.

Predominant effector pathways

Inhibition of adenylyl cyclase through a pertussis-toxin-sensitive G-protein of the G_i/G_o class [1].

Amino acid sequence (rat) 1

```
                                      TM 1
  1 MEPLSLASTP SWNASAASSG NHNWSLVGSA SPMGARAVLV PVLYLLVCTV
                          TM 2
 51 GLSGNTLVIY VVLRHAKMKT VTNVYILNLA VADVLFMLGL PFLATQNAVV
              TM 3
101 SYWPFGSFLC RLVMTLDGIN QFTSIFCLMV MSVDRYLAVV HPLRSARWRR
        TM 4                                      TM 5
151 PRVAKMASAA VWVFSLLMSL PLLVFADVQE GWGTCNLSWP EPVGLWGAAF
```

```
                                                              TM 6
201 ITYTSVLGFF GPLLVICLCY LLIVVKVKAA GMRVGSSRRR RSEPKVTRMV
                                              TM 7
251 VVVVLVFVGC WLPFFIVNIV NLAFTLPEEP TSAGLYFFVV VLSYANSCAN

301 PLLYGFLSDN FRQSFRKVLC LRRGYGMEDA DAIEPRPDKS GRPQAHCPHA

351 AARPMGSCRP AGFECPCNTL GVLQASTVLS SGI
```

Amino acids	383
Molecular weight	~48 000
Glycosylation	Asn13*, Asn186*
Disulfide bonds	Cys110–Cys185*
Palmitoylation	Cys320*

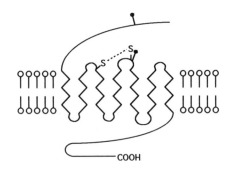

Database accession numbers

	PIR	SWISSPROT	EMBL/GENBANK	REFERENCE
Rat			L04535	1

Gene
The organization of the gene is not known.

References
1 O'Carroll, A.-M. *et al.* (1992) *Mol. Pharmacol.* **42**, 939–946.

Tachykinins

INTRODUCTION

Tachykinins are a family of peptide neurotransmitters which share a common C-terminal amino acid sequence Phe–X–Gly–Leu–Met–NH$_2$ [1–10]. Full biological activity resides in this C-terminal portion with the N-terminal sequence having a modulatory role. Substance P is the best known member of the family.

Tachykinins have important physiological roles in a number of pathways. In the periphery they stimulate smooth muscle contraction and are involved in peristalsis and micturition; they stimulate glandular secretion, induce activation of cells of the immune system and activate peripheral nerves. In the CNS they regulate dopaminergic neurons in the basal ganglia and mesolimbic areas of brain and are involved in the transmission of sensory information including noxious stimuli to the spinal cord.

Structures

Substance P (SP)	Arg–Pro–Lys–Pro–Gln–Gln–Phe–Phe–Gly–Leu–Met–NH$_2$
Neurokinin A (NKA)	His–Lys–Thr–Asp–Ser–Phe–Val–Gly–Leu–Met–NH$_2$
Neurokinin B (NKB)	Asp–Met–His–Asp–Phe–Phe–Val–Gly–Leu–Met–NH$_2$

Neuropeptide γ and neuropeptide K are N-terminally extended forms of NKA present in mammalian tissues. NKA was previously called substance K, neurokinin α or neuromedin L; NKB was previously named neurokinin β or neuromedin K. Additional members of the family exist in lower organisms, e.g. eledoisin, physalaemin, kassinin and sychliorhinin II.

Distribution and synthesis

SP and NKA have a similar distribution, distinct from that of NKB. In brain, SP and NKA are concentrated in basal ganglia, dorsal horn, hypothalamus and limbic forebrain; NKB is found in high levels in the dorsal horn. SP and NKA are present in peripheral nerves, including enteric neurons and capsaicin-sensitive primary afferent neurons, but NKB is absent.

Two genes encode for mammalian tachykinins. The preprotachykinin (PPT) A gene transcript is alternatively spliced to three precursors: α-PPT encodes for SP; β-PPT encodes for SP, NKA and neuropeptide K; γ-PPT encodes for SP, NKA and neuropeptide γ. The preprotachykinin B gene transcript is alternatively spliced to two precursors, both of which give NKB.

Reviews

1 Guard, S. and Watson, S.P. (1991) Tachykinin receptor types: classification and membrane signalling mechanisms. *Neurochem. Int.* **18**, 149–165.
2 Helke, C.J. *et al.* (1989) Diversity in mammalian tachykinin peptidergic neurons: multiple peptides, receptors, and regulatory mechanisms. *FASEB J* **4**, 1606–1615.
3 Holzer, P. (1988) Local effector functions of capsaicin-sensitive sensory nerve endings: involvement of tachykinins, calcitonin gene-related peptide and other neuropeptides. *Neuroscience* **24**, 739–768.
4 Maggi, C.A. *et al.* (1993) Tachykinin receptors and tachykinin receptor antagonists. *J. Auton. Pharmacol.* **13**, 23–93.

5 Maggio, J.E. (1988) Tachykinins. *Annu. Rev. Neurosci.* **11**, 13–28.
6 Nakanishi, S. (1987) Substance P precursor and kininogen: Their structures, gene organizations and regulation. *Physiol. Rev.* **67**, 1117–1142.
7 Nakanishi, S. (1991) Mammalian tachykinin receptors. *Annu. Rev. Neurosci.* **14**, 123–136.
8 Pernow, B. (1984) Substance P *Pharmacol. Rev.* **35**, 86–141.
9 Regoli, D. *et al.* (1988) New selective agonists for neurokinin receptors: pharmacological tools for receptor classification. *Trends Pharmacol. Sci.* **9**, 290–295.
10 Watling, K.J. and Krause, J.E. (1993) The rising sun shines on substance P and related peptides. *Trends Pharmacol. Sci.* **14**, 81–84.

TACHYKININ RECEPTORS

NK$_1$, NK$_2$ and NK$_3$ receptors

All three are G-protein linked and described below.

Drosophila locustatachykinin receptor

A *Drosophila* cDNA encoding for a protein, NKD, with 38% homology to the NK$_3$ receptor has been described [1]. Stable cell lines expressing this protein respond to locustatachykinins, peptides isolated from locust and which exhibit a limited degree of sequence homology with tachykinins, but not to members of the tachykinin family.

N-terminal receptors/binding sites

Binding and functional studies indicate the presence of sites in the CNS and adrenal medulla that are recognized by N-terminal fragments of substance P, although their physiological significance is not established [2,3].

References
1 Monnier, D. *et al.* (1992) *J. Biol. Chem.* **267**, 1298–1302.
2 Geraghty, D.P. *et al.* (1990) *Neurosci. Lett.* **112**, 276–281.
3 Igwe, O.J. *et al.* (1990) *J. Neurosci.* **10**, 3653–3663.

THE NK$_1$ TACHYKININ RECEPTOR

SP is the most potent mammalian tachykinin, inducing activation in low nanomolar concentrations; NKA and NKB are 5–50-fold less potent, with NKA having the higher affinity. The receptor was previously named the substance P or SP-P receptor.

Distribution

In brain, high levels of NK$_1$ binding sites are found in striatum, olfactory bulb, dendate gyrus, locus coeruleus and spinal cord. In peripheral tissues, NK$_1$ receptors are found in smooth muscle, e.g. ileum and bladder, enteric neurons, secretory glands, e.g. parotid, cells of the immune system and vascular endothelium.

Pharmacology

Agonists: SP methyl ester, [Sar^9,Met(O$_2$)^{11}]SP and [L-Pro^9]SP have a selectivity of at least 1000-fold and are active in nanomolar concentrations.

Antagonists: CP-96,345 **[1]** (pA$_2$ 9.0) and RP 67580 **[2]** (pA$_2$ 7.5) are highly selective, nonpeptide antagonists (see Comment). The tripeptide FR 113680 (pA$_2$ 7.5) is also highly selective. Various peptide antagonists have lower affinities and are less selective, e.g. L-668169 (pA$_2$ 7.0) and GR 82334 (pA$_2$ = 7.4).

Radioligands: The nonpeptide antagonists [^{3}H]CP96345 (K_d 0.22 nM) and [^{125}I]L-703606 (K_d 0.3 nM) are highly selective (see Comment). Selective agonist ligands include [^{3}H]-[Sar^9,Met(O$_2$)^{11}]SP, [^{3}H]-[Pro^9]SP or [^{125}I]SP (K_d 0.1–1 nM).

Structures:

L-703606

CP 96345

RP 67580

FR 113680 N^α-(N^α-[N^α-acetyl-L-threonyl]-N^{in}-formyl-D-tryptophyl)-N-methyl-N-phenylmethyl-L-phenylalaninamide

GR 82334   [DPro^9(spiro-γ-lactam),Leu^{10}]physaelamin

L-668169   cyc(Gln–DTrp–[N-MePhe(R)]–Gly[ANC-2]–Leu–Met)

Predominant effector pathways

Activation of the phosphoinositide pathway through a pertussis-toxin-insensitive G-protein.

Amino acid sequence [3–6]

```
                                              TM 1
  1 MDNVLPVDSD LSPNISTNTS EPNQFVQPAW QIVLWAAAYT VIVVTSVVGN
                 TM 2
 51 VVVMWIILAH KRMRTVTNYF LVNLAFAEAS MAAFNTVVNF TYAVHNEWYY
           TM 3                                         TM 4
101 GLFYCKFHNF FPIAAVFASI YSMTAVAFDR YMAIIHPLQP RLSATATKVV
                                                         TM 5
151 ICVIWVLALL LAFPQGYYST TETMPSRVVC MIEWPEHPNK IYEKVYHICV
                                                        TM 6
201 TVLIYFLPLL VIGYAYTVVG ITLWASEIPG DSSDRYHEQV SAKRKVVKMM
                                          TM 7
251 IVVVCTFAIC WLPFHIFFLL PYINPDLYLK KFIQQVYLAI MWLAMSSTMY

301 NPIIYCCLND RFRLGFKHAF RCCPFISAGD YEGLEMKSTR YLQTQGSVYK

351 VSRLETTIST VVGAHEEEPE DGPKATPSSL DLTSNCSSRS DSKTMTESFS

401 FSSNVLS
```

Chromosome location	2
Amino acids	407
Molecular weight	46 248
Glycosylation	Asn14*, Asn18*
Disulfide bonds	Cys105–Cys180*
Palmitoylation	Cys322*, Cys333*

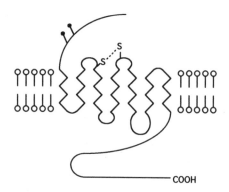

Database accession numbers

	PIR	SWISSPROT	EMBL/GENBANK	REFERENCES
Human	JQ1274	P25103	M74290	3–6
	S21188			
Mouse			X69934	7
Rat	A34357	P14600	J05097	8,9
			M31477	

A shorter form of the human NK_1 receptor is produced by alternative splicing [5]. Substance P has an ~10 times lower affinity on the shorter form which has a truncated C-terminal amino acid sequence of 7 amino acids. The physiological significance of the shorter form is unclear.

Gene (rat) [8]

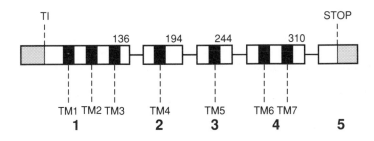

Comment

Marked species differences in the pharmacology of NK_1 receptors has been demonstrated. For example, the antagonist CP 96345 is ~two orders of magnitude less potent on rat and mouse NK_1 receptors relative to NK_1 receptors in other species, including human [1]. The converse is true for RP 67580 [2]. The molecular basis for the difference in affinity of CP 96345 is primarily due to the presence of isoleucine at position 290 in the human receptor in contrast to serine in the equivalent position in the rat receptor [10,11].

References

1 Snider, R.M. et al. (1991) Science **251**, 435–439.
2 Garret, C. et al. (1991) Proc. Natl Acad. Sci. USA **88**, 10208–10212.
3 Takeda, Y. et al. (1991) Biochem. Biophys. Res. Commun. **179**, 1232–1240.
4 Hopkins, B. et al. (1991) Biochem. Biophys. Res. Commun. **180**, 1110–1117; ibid. **180**, 1514.
5 Fong, T.M. et al. (1992) Mol. Pharm. **41**, 24–30.
6 Takahashi, K. et al. (1992) Eur. J. Biochem. **204**, 1025–1033.
7 Sundelin, J.B. et al. (1992) Eur. J. Biochem. **203**, 625–631.
8 Yokata, Y. et al. (1989) J. Biol. Chem. **264**, 17649–17652.
9 Hershey, A.D. et al. (1990) Science **247**, 958–962.
10 Fong, T.M. et al. (1992) J. Biol. Chem. **267**, 25668–25671.
11 Sachias, B.S. et al. (1993) J. Biol. Chem. **268**, 2319–2323.

THE NK₂ TACHYKININ RECEPTOR

NKA is the most potent mammalian tachykinin, inducing activation in low nanomolar concentration; NKB and SP are approximately one and two orders of magnitude less potent, respectively. The receptor was previously named the substance K receptor.

Distribution

In the periphery, NK₂ receptors are found in smooth muscle of the respiratory, gastrointestinal and urogenital systems. There is little firm evidence for their presence in CNS.

Pharmacology

Agonists: [Lys⁵,MeLeu⁹,Nle¹⁰]NKA4–10, [Nle¹⁰]NKA4–10, [β-Ala⁸]NKA4–10 and GR 64349 have a selectivity of at least 100-fold and are active in nanomolar concentrations.

Antagonists: The nonpeptide SR 48968 **1** (pA₂ 10.5) is highly selective. Peptide antagonists have lower affinities and are less selective, e.g. MEN 10207 (pA₂ 6.0–8.0), L 659877 (pA₂ 6.8–7.9) and R 396 (pA₂ 5.4–7.6).

Radioligands: [³H]SR 48968 is selective. [³H] or [¹²⁵I] analogues of NKA (K_d 0.1–1 nM), including [¹²⁵I]neuropeptide γ, have a reasonable degree of selectivity but bind to other tachykinin receptors when expressed in high levels.

Structures:

SR 48968

GR 64349	Lys–Asp–Ser–Phe–Val–Gly–[R-γ-lactam]
GR 94800	PhCO–Ala–Ala–DTrp–Phe–DPro–Pro–Nle–NH₂
L-659877	cyc(Gln–Trp–Phe–Gly–Leu–Met)
MEN 10207	[Tyr⁵,DTrp⁶,⁸,⁹,Arg¹⁰]NKA4–10
R 396	Ac–Leu–Asp–Gln–Trp–Phe–Gly–NH₂

Predominant effector pathways

Activation of the phosphoinositide pathway through a pertussis-toxin-insensitive G-protein most likely of the G_q/G_{11} class.

Amino acid sequence [2,3]

```
                                           TM 1
  1 MGTCDIVTEA NISSGPESNT TGITAFSMPS WQLALWAPAY LALVLVAVTG
                      TM 2
 51 NAIVIWIILA HRRMRTVTNY FIVNLALADL CMAAFNAAFN FVYASHNIWY
         TM 3                                          TM 4
101 FGRAFCYFQN LFPITAMFVS IYSMTAIAAD RYMAIVHPFQ PRLSAPSTKA
                                                      TM 5
151 VIAGIWLVAL ALASPQCFYS TVTMDQGATK CVVAWPEDSG GKTLLLYHLV
                                                      TM 6
201 VIALIYFLPL AVMFVAYSVI GLTLWRRAVP GHQAHGANLR HLQAKKKFVK
                                      TM 7
251 TMVLVVLTFA ICWLPYHLYF ILGSFQEDIY CHKFIQQVYL ALFWLAMSST

301 MYNPIIYCCL NHRFRSGFRL AFRCCPWVTP TKEDKLELTP TTSLSTRVNR

351 CHTKETLFMA GDTAPSEATS GEAGRPQDGS GLWFGYGLLA PTKTHVEI
```

Amino acids	398
Molecular weight	44 425
Chromosome location	10
Glycosylation	Asn11*, Asn19*
Disulfide bonds	Cys106–Cys181*
Palmitoylation	Cys324*,Cys325*

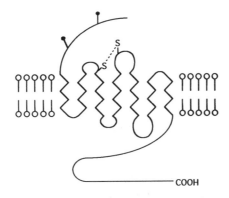

Database accession numbers

	PIR	SWISSPROT	EMBL/GENBANK	REFERENCES
Bovine	S00516	P05363	X06295	4
Human	A23658		M57414	2,3
	JQ1059		J05680	
Mouse			X62933	5
Rat	A36737	P16610	M31838	6

Gene (human) [3]

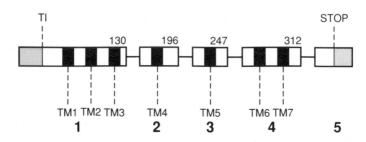

Comment

The potency order of NK_2 receptor antagonists in certain smooth muscles, e.g. rabbit pulmonary artery, MEN 10207 > L-659877 > R 396, is distinct from that in other smooth muscles, e.g. hamster trachea, L-659877 > R 396 > MEN 10207 [7]. [Leu$^8\psi$(CH$_2$NH),Leu^{10}]NKA$_{4-10}$ (MDL 28564) is a full agonist in the former but a partial agonist in the latter [8]. These results appear to reflect species homologues of the NK_2 receptor rather than distinct subtypes.

References

[1] Advenier, C. et al. (1992) Br. J. Pharmacol. **105**, 77P.
[2] Gerard, N.P. et al. (1990) J. Biol. Chem. **179**, 20455–20462; errata, ibid. **266**, 1354.
[3] Graham, A. et al. (1991) Biochem. Biophys. Res. Commun. **177**, 8–16.
[4] Masu, Y. et al. (1987) Nature **329**, 836–838.
[5] Sundelin, J.B. et al. (1992) Eur. J. Biochem. **203**, 625–631.
[6] Sasai, Y. et al. (1989) Biochem. Biophys. Res. Commun. **165**, 695–702.
[7] Maggio, C.A. et al. (1990) Eur. J. Pharmacol. **166**, 435–440.
[8] Buck, S.H. et al. (1990) Life Sci. **47**, 37–41.

THE NK3 TACHYKININ RECEPTOR

NKB is the most potent mammalian tachykinin, inducing activation in nanomolar concentrations; NKA and SP are approximately one and two orders of magnitude less potent, respectively. The receptor was previously named the neurokinin B receptor.

Distribution

NK_3 receptors are distributed widely throughout the rat CNS and are found in high levels in cerebral cortex, basal ganglia and dorsal horn of the spinal cord. They have a very limited distribution in peripheral tissues and are found in ganglia, e.g. myenteric plexus, kidney and in a limited number of smooth muscles, e.g. rat portal vein.

Pharmacology

Agonists: Senktide (succinyl-[Asp^6,MePhe^8]SP$_{6-11}$) has a selectivity of more than 10 000-fold; other highly selective agonists include [MePhe^7]NKB and [Pro^7]NKB. All are active in nanomolar concentrations.

Antagonists: [Trp^7,β-Ala^8]NKA$_{4-10}$ (pA$_2$ 7.5) is an antagonist at NK$_3$ receptors but an agonist at NK$_1$ and NK$_2$ receptors. No other class of compound has been shown to be an antagonist at the NK$_3$ receptor.

Radioligands: [^{3}H]Senktide [1] (K_d 1–10 nM) is highly selective; [^{125}I]analogues of eledoisin and sychliorhinin II are less selective and bind to other tachykinin receptor subtypes when expressed in high levels.

Predominant effector pathways
Activation of the phosphoinositide pathway through a pertussus-toxin-insensitive G-protein most likely of the G$_q$/G$_{11}$ class.

Amino acid sequence [2]

```
  1 MATLPAAETW IDGGGGVGAD AVNLTASLAA GAATGAVETG WLQLLDQAGN
                                          TM 1
 51 LSSSPSALGL PVASPAPSQP WANLTNQFVQ PSWRIALWSL AYGVVVAVAV
                          TM 2
101 LGNLIVIWII LAHKRMRTVT NYFLVNLAFS DASMAAFNTL VNFIYALHSE
                          TM 3
151 WYFGANYCRF QNFFPITAVF ASIYSMTAIA VDRYMAIIDP LKPRLSATAT
        TM 4                                          TM 5
201 KIVIGSIWIL AFLLAFPQCL YSKTKVMPGR TLCFVQWPEG PKQHFTYHII
                                                      TM 6
251 VIILVYCFPL LIMGITYTIV GITLWGGEIP GDTCDKYHEQ LKAKRKVVKM
                          TM 7
301 MIIVVMTFAI CWLPYHIYFI LTAIYQQLNR WKYIQQVYLA SFWLAMSSTM

351 YNPIIYCCLN KRFRAGFKRA FRWCPFIKVS SYDELELKTT RFHPNRQSSM

401 YTVTRMESMT VVFDPNDADT TRSSRKKRAT PRDPSFNGCS RRNSKSASAT

451 SSFISSPYTS VDEYS
```

Amino acids	465
Molecular weight	52 201
Glycosylation	Asn23*, Asn50*, Asn73*
Disulfide bonds	Cys158–Cys233*
Palmitoylation	Cys374*

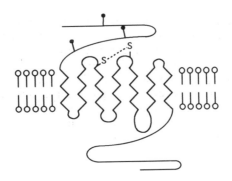

Database accession numbers

	PIR	SWISSPROT	EMBL/GENBANK	REFERENCES
Human		P29371	M89473	2, 3
Rat	A34916	P16177	J05189	4

Gene

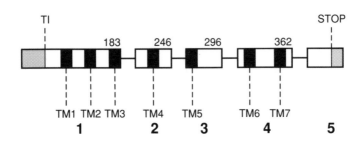

References
1 Guard, S. *et al.* (1990) *Br. J. Pharmacol.* **99**, 767–773.
2 Buell, G. *et al.* (1992) *FEBS Lett.* **299**, 90–95.
3 Huang, R.-R.C. *et al.* (1992) *Biochem. Biophys. Res. Commun.* **184**, 966–972.
4 Shigemoto, R. *et al.* (1990) *J. Biol. Chem.* **265**, 623–628.

Taste receptors

The molecular basis of taste is likely to be related to other types of sensory signalling such as phototransduction and olfactory transduction. Thus, stimuli such as sweet and bitter are believed to activate receptors in taste buds of the tongue leading to the generation of intracellular messengers. Consistent with this, a novel G-protein, gustucin, and a large gene family encoding more than 60 seven transmembrane proteins are expressed in lingual epithelium but are absent in other tissues [1]. The identification of this large multigene receptor family provides a molecular explanation for the diversity and specificity of taste detection in the tongue, although future studies are required to establish its role in taste sensation.

Reference
[1] Abe, K. *et al.* (1993) *FEBS Lett.* **316**, 253–256.

Thrombin

Thrombin is a serine protease (EC 3.4.21.5) with a pivotal role in blood clotting [1-5]. It cleaves a number of substrates involved in coagulation and activates cell surface receptors through a novel proteolytic action.

Thrombin stimulates aggregation and secretion in blood platelets at the site of vascular injury, and also has inflammatory and reparative actions, stimulating chemotaxis in monocytes, proliferation of fibroblasts and lymphocytes, and inducing endothelium-dependent relaxation of blood vessels.

Thrombin activates a number of substrates which are involved in the coagulation process. It cleaves fibrinogen to fibrin and activates coagulation factor XIII, a transglutaminase which cross-links the soluble fibrin into insoluble fibrin. It also activates factors V and VIII. On the other hand, when bound to thrombomodulin it activates plasma protein C which, in concert with protein S, inactivates factors Va and VIIIa leading to a decrease in thrombin formation.

Distribution

Prothrombin circulates in the blood and is converted to thrombin as part of the coagulation process.

Thrombin inhibitors

In vivo the enzymic activity of thrombin is terminated by various plasma proteins, e.g. heparin and antithrombin III.

Hirudin, a 65 amino acid peptide from leech saliva, binds to the anion exosite in thrombin, blocking its serine protease action and ability to induce receptor activation. Synthetic peptide mimics of hirudin have been designed. Other selective inhibitors of the serine protease action of thrombin include DPhe–Pro–Arg-chloromethylketone and Nα-[2-naphthylsulfonyl-Gly]-4-amidino-Phe piperidine.

Books
1 Machovich (ed.) The Thrombins, vols I and II, CRC Press, Boca Raton, USA.
2 Walz *et al.* (ed.) Bioregulatory functions of thrombin. *Ann. N.Y. Acad. Sci.* **485**, 1–490.

Reviews
3 Coughlin, S. *et al.* (1992) Perspectives. Characterisation of the cloned platelet thrombin receptor: issues and opportunities. *J. Clin. Invest.* **89**, 351–355.
4 Couglin, S. *et al.* (1993) Thrombin receptor structure and function. In *Cold Spring Harbor Symposium: The Cell Surface,* **57**, in press.
5 Fenton, J.W. (1988) Regulation of thrombin generation and functions. *Seminars Thromb. Hemostas.* **14**, 234–240.

THE THROMBIN RECEPTOR

Thrombin acts at low nanomolar concentrations.

Distribution

The receptor is expressed in high levels in platelets, vascular endothelial cells and various cell lines, e.g. HEL and Dami megakaryocyte cells, CCL39 hamster lung fibroblasts, etc.

Pharmacology

Agonists: Peptides which mimic the N-terminal sequence of the thrombin receptor after the site of cleavage (see below) induce activation in micromolar concentrations, e.g. SFLLRN [1]. A number of serine proteases also induce activation, e.g. trypsin.

Antagonists: Serine protease inhibitors can be used to block the action of thrombin (see below). Peptide inhibitors have been described which block receptor activation but not thrombin's enzymic activity [2]; they have weak partial agonist activity.

Radioligands: [^{125}I]Thrombin binds to high-(K_d 0.3 nM), moderate (K_d 11 nM) and low-(K_d 3000 nM) affinity sites in platelets, although their relationship to the thrombin receptor is not established [3].

Predominant effector pathways

Activation of phosphoinositide metabolism through a pertussis-toxin-insensitive G-protein and inhibition of adenylyl cyclase through a pertussis-toxin-sensitive G-protein.

Amino acid sequence [3]

```
  1 MGPRRLLLVA ACF SLCGPLL SARTRA

                     RRPE SKATNATLDP RSFLLRNPND
                                     ^
 51 KYEPFWEDEE KNESGLTEYR LVSINKSSPL QKQLPAFISE DASGYLTSSW
    TM 1                                        TM 2
101 LTLFVPSVYT GVFVVSLPLN IMAIVVFILK MKVKKPAVVY MLHLATADVL
                         TM 3
151 FVSVLPFKIS YYFSGSDWQF GSELCRFVTA AFYCNMYASI LLMTVISIDR
                         TM 4
201 FLAVVYPMQS LSWRTLGRAS FTCLAIWALA IAGVVPLVLK EQTIQVPGLN
                         TM 5
251 ITTCHDVLNE TLLEGYYAYY FSAFSAVFFF VPLIISTVCY VSIIRCLSSS
                         TM 6
301 AVANRSKKSR ALFLSAAVFC IFIICFGPTN VLLIAHYSFL SHTSTTEAAY
    TM 7
351 FAYLLCVCVS SISSCIDPLI YYYASSECQR YVYSILCCKE SSDPSSYNSS

401 GQLMASKMDT CSSNLNNSIY KKLLT
```

The first 26 amino acids are thought to be a signal sequence.

Amino acids	425
Molecular weight	47 410
Glycosylation	Asn35*, Asn62*, Asn75*, Asn259*
Disulfide bonds	Cys175–Cys254*
Palmitoylation	Cys337*, Cys338*
Λ	Site of cleavage by thrombin

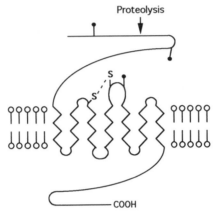

Mechanism of receptor activation

Thrombin cleaves the receptor between Arg41 and Ser42 liberating a new N-terminus which is believed to interact with the remaining portion of the receptor to induce activation. Certain peptides which mimic the new N-terminal sequence are potent agonists at the thrombin receptor. Carboxyl to the cleavage site is a sequence rich in aromatic and acidic residues which resembles the carboxyl tail of hirudin. This sequence has been shown to bind to thrombin's anion binding exosite and is an important mediator of thrombin/receptor interaction [5].

Database accession numbers

	PIR	SWISSPROT	EMBL/GENBANK	REFERENCES
Hamster			M80612	6
Human	A37912	P25116	M62424	4
Rat		P26824	M81462	7

Gene (human) [3]

The gene has an open reading frame encoding the receptor.

References

1 Vassallo, R.R. et al. (1992) J. Biol. Chem. **267**, 6081–6085.
2 Ruda, E.W. et al. (1988) Biochem. Pharmacol. **37**, 2417–2426.
3 Harmon, J.T. and Jamieson, G.A. (1986) J. Biol. Chem. **261**, 15928–15933.
4 Vu, T. et al. (1991) Cell **64**, 1057–1068.
5 Vu, T. et al. (1991) Nature **353**, 674–677.
6 Rasmussen, U.B. et al. (1991) FEBS Lett. **288**, 123–128.
7 Zhong , C. et al. (1992) J. Biol. Chem. **267**, 16975–16979.

Thyrotrophin-releasing hormone

The principle function of TRH (thryoliberin) is to stimulate synthesis and release of thyroid-stimulating hormone in the anterior pituitary [1,2]. TRH also stimulates synthesis and release of prolactin, although the physiological importance of this is uncertain.

In the CNS, TRH stimulates a number of behavioural and pharmacological actions, including increased turnover of catecholamines in the nucleus accumbens.

Structure

TRH   pyro–Glu–His–Pro–NH$_2$

Distribution and synthesis

In human, TRH is derived from a precursor which contains six copies of the sequence Lys–Arg–Gln–His–Pro–Gly–Lys(Arg)–Arg. TRH is found in highest amounts in median eminence and paraventricular nucleus of the hypothalamus, although other brain areas contain significant amounts, e.g. septum, medulla oblongata, nucleus accumbens, ventral spinal cord. In the periphery, TRH is found in pancreas and gastrointestinal tract.

Reviews

1 Bjøro, T. *et al.* (1990) The mechanisms by which vasoactive intestinal peptide (VIP) and thyrotropin releasing hormone (TRH) stimulate prolactin release from pituitary cells. *Biosci. Rep.* **10**, 189–199.
2 Gershengorn, M. (1986) Mechanism of thyrotropin releasing hormone stimulation of pituitary hormone secretion. *Annu. Rev. Physiol.* **48**, 515–526.

THE TRH RECEPTOR

Distribution

TRH receptors are found in high levels in anterior pituitary, and are also found in retina and certain areas of the brain including hypothalamus, hippocampus, amygdala, basal ganglia and cranial nerves [1]. They are expressed in high levels in GH$_3$ and GH$_4$C$_1$ rat pituitary cells.

Pharmacology

Agonists: TRH stimulates receptor activation in nanomolar concentrations. A number of structurally related analogues have been described.

Antagonists: The benzodiazepine, chlordiazepoxide, is a weak, competitive antagonist ($K_i \sim 100\ \mu$M) [2].

Radioligands: [^{3}H]TRH (K_d 10 nM).

Structures:

Chlordiazepoxide

Predominant effector pathways

Activation of phosphoinositide metabolism through a pertussis-toxin-insensitive G-protein, most likely of the G_q/G_{11} class.

Amino acid sequence (rat) [3,4]

```
                                    TM 1
  1 MENETVSELN QTELPPQVAV ALEYQVVTIL LVVVICGLGI VGNIMVVLVV
         TM 2                                          TM 3
 51 MRTKHMRTAT NCYLVSLAVA DLMVLVAAGL PNITDSIYGS WVYGYVGCLC
                                                     TM 4
101 ITYLQYLGIN ASSCSITAFT IERYIAICHP IKAQFLCTFS RAKKIIIFVW
                                                     TM 5
151 AFTSIYCMLW FFLLDLNIST YKDAIVISCG YKISRNYYSP IYLMDFGVFY

201 VMPMILATVL YGFIARILFL NPIPSDPKEN SKTWKNDSTH QNKNMNLNTT
                             TM 6                      TM 7
251 NRCFNSTVSS RKQVTKMLAV VVILFALLWM PYRTLVVVNS FLSSPFQENW

301 FLLFCRICIY LNSAINPVIY NLMSQKFRAA FRKLCNCKQK PTEKAANYSV

351 ALNYSVIKES DRFSTELDDI TVTDTYVSTT KVSFDDTCLA SEKNGPSSCT

401 YGYSLTAKQE KI
```

Amino acids	412
Molecular weight	46 554
Glycosylation	Asn3*, Asn10*, Asn167*
Disulfide bonds	Cys98–Cys179*
Palmitoylation	Cys337*

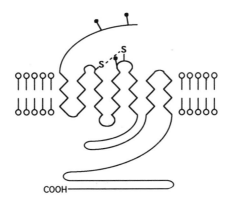

Database accession numbers

	PIR	SWISSPROT	EMBL/GENBANK	REFERENCES
Mouse	A39251	P21761	M59811	5
Rat			X64630	3,4
			M90308	

Gene
The organization of the gene is not known.

References
1 Zabavnik, J. et al. (1993) Neuroscience 53, 877–887.
2 Drummond, A. (1985) Biochem. Biophys. Res. Commun. 127, 63–70.
3 de la Pena, P. et al. (1992) Biochem. J. 284, 891–899.
4 Zhao, D. et al. (1992) Endocrinology 130, 3529–3536.
5 Straub, R.E. et al. (1990) Proc. Natl Acad. Sci. USA 87, 907–914.

INTRODUCTION

VIP is grouped with a number of structurally related peptides which share an overlapping profile of biological activity [1-5].

VIP has a wide profile of physiological actions. In the periphery, VIP induces relaxation in smooth muscle, e.g. intestine, blood vessels, trachea; inhibits secretion in certain tissues, e.g. stomach; but stimulates secretion in others, e.g. intestinal epithelium, pancreas, gall bladder; and modulates activity of cells in the immune system. In the CNS, VIP has a wide range of excitatory and inhibitory actions.

Other members of the family are involved in a more limited number of physiological pathways, e.g. secretin stimulates secretion of enzymes and ions in pancreas and intestine, but is present in only small amounts in brain, notably hypothalamus, brainstem and cerebral cortex. Growth hormone releasing factor (GRF) is found in the hypothalamus and is an important neuroendocrine agent, regulating synthesis and release of growth hormone from the anterior pituitary.

Structures (human)

GRF_{1-29} Tyr–Ala–<u>Asp</u>–Ala–Ile–<u>Phe</u>–<u>Thr</u>–Asn–Ser–<u>Tyr</u>–Arg–Lys–Val–Leu–Gly–
<u>Gln</u>–Leu–Ser–Ala–Arg–<u>Lys</u>–Leu–<u>Leu</u>–Gln–Asp–Ile–Met–Ser–Arg–NH$_2$

PACAP <u>His</u>–<u>Ser</u>–<u>Asp</u>–<u>Gly</u>–Ile–<u>Phe</u>–<u>Thr</u>–Asp–Ser–<u>Tyr</u>–<u>Ser</u>–<u>Arg</u>–Tyr–<u>Arg</u>–Lys–
<u>Gln</u>–Met–Ala–Val–<u>Lys</u>–<u>Lys</u>–<u>Tyr</u>–<u>Leu</u>–Ala–Ala–Val–Leu–NH$_2$

PHM <u>His</u>–Ala–<u>Asp</u>–<u>Gly</u>–Val–<u>Phe</u>–<u>Thr</u>–Ser–Asp–Phe–<u>Ser</u>–<u>Arg</u>–<u>Leu</u>–Leu–Gly–
<u>Gln</u>–Leu–Ser–Ala–<u>Lys</u>–<u>Lys</u>–<u>Tyr</u>–<u>Leu</u>–Glu–Ser–Leu–Ile–NH$_2$

Secretin <u>His</u>–<u>Ser</u>–<u>Asp</u>–<u>Gly</u>–Thr–<u>Phe</u>–<u>Thr</u>–Ser–Glu–Leu–Ser–<u>Arg</u>–<u>Leu</u>–<u>Arg</u>–Glu–
Gly–Ala–Arg–Leu–Gln–Arg–Leu–<u>Leu</u>–Gln–Gly–Leu–Val–NH$_2$

VIP <u>His</u>–<u>Ser</u>–<u>Asp</u>–Ala–Val–<u>Phe</u>–<u>Thr</u>–Asp–Asn–<u>Tyr</u>–Thr–<u>Arg</u>–<u>Leu</u>–<u>Arg</u>–Lys–
<u>Gln</u>–Met–Ala–Val–<u>Lys</u>–<u>Lys</u>–<u>Tyr</u>–<u>Leu</u>–Asn–Ser–Ile–Leu–Asn–NH$_2$

PACAP – pituitary adenylyl cyclase activating polypeptide; PHM – peptide histidine methionamide. Shared amino acids are underlined. There is considerable variation in structures between species and N-terminally extended forms exist. Related peptides are present in lower organisms, e.g. helodermin in lizard.

Distribution and synthesis

VIP is derived from a 20 kDa precursor which also contains the sequence for PHM. Secretin is synthesized as a precursor of 134 amino acids in S cells in the mucosa of the small intestine.

VIP immunoreactivity is found in neurons throughout the CNS and is distributed widely in peripheral organs, including gastrointestinal, genitourinary, respiratory and cardiovascular systems. VIP has also been localized to neutrophils.

Book
1 Said, S.I. and Mutt, V. (eds) (1988) Vasoactive intestinal polypeptide and related peptides. *Ann. N. Y. Acad. Sci.* **527.**

Reviews
2 Gozes, I. and Brennerman, D.E. (1989) VIP: molecular biology and neurobiological function. *Mol. Neurobiol.* **3**, 201–236.
3 Magstretti, P.J. *et al.* (1990) VIP neurons in the cerebral cortex. *Trends Pharmacol. Sci.* **11**, 250–254.
4 Rosselin, G. (1986) The receptors of the VIP family peptides. Specificities and identity. *Peptides* **7**, 89–100.
5 Said, S.I. (1988) Vasoactive intestinal peptide. *Adv. Metab. Disord.* **11**, 369–391.

VIP RECEPTOR FAMILY

Receptors have been named according to the most potent naturally occurring ligand. All receptors activate adenylyl cyclase, suggesting that they are G-protein coupled.

GRF receptor

GRF has the highest affinity of the endogenous members of the family. The GRF receptor is found in high levels in the anterior pituitary.

The helodermin-preferring receptor

Identified in several cell lines, e.g. human SUP-T1 lymphoma cells and NCJ-H345 lung carcinoma.

PACAP receptor

Identified in rat CNS, pituitary, pancreas and adrenal, and also in cell lines, e.g. rat pancreatic acinar AR4-2J, human neuroblastoma OK. The receptor can be labelled with $[^{125}I]$PACAP-27. VIP has low affinity.

Two forms of the receptor have been proposed. $PACAP_A$ receptors recognize PACAP-27 and PACAP-38 with high affinity (0.3 nM); $PACAP_B$ receptors recognise PACAP-38 with high affinity (0.3 nM) and PACAP-27 with low affinity (30 nM) [1]. $PACAP_{6-27}$ (K_i 500 nM) is an antagonist at the $PACAP_A$ receptor [1].

Secretin receptor

This is a seven transmembrane spanning receptor and is described below.

VIP receptor

This is a seven transmembrane spanning receptor and is described below.

Reference
1 Robberecht, P. *et al.* (1991) *FEBS Lett.* **286**, 133–136.

THE VIP RECEPTOR

VIP and PACAP induce activation in low nanomolar concentrations; other members of the family are less potent: VIP = PACAP > PHM >> secretin.

Distribution

VIP receptors are distributed widely in the periphery and are found throughout the gastrointestinal tract and the genitourinary system, other smooth muscles, e.g. trachea, blood vessels and in secretory glands, e.g. pancreas, gall bladder, etc. In the CNS, receptors for VIP are present in high levels in cerebral cortex, hippocampus, striatum and thalamus. They are present in cell lines, e.g. human colon adeno-carcinoma HT29 cells.

Pharmacology

Agonists: Partial sequences of VIP or other endogenous peptides have considerably lower potencies than the parent molecule or are inactive.

Antagonists: A number of analogues of VIP or its fragments have been claimed to be weak antagonists with K_d values of less than 1 μM, e.g. [Ac-Tyr^1,DPhe^2]GRF$_{1-29}$, VIP$_{10-28}$ and [4Cl-DPhe^6,Leu^{17}]VIP. All are partial agonists under certain conditions.

Radioligands: In most preparations iodinated VIP binds to two sites, a high-affinity site of low capacity (K_d 0.1–5 nM) and a low-affinity site of high capacity (K_d 10–1000 nM); [^{125}I]helodermin and [^{125}I]-[His^1,DAla^2]GRF$_{1-29}$NH$_2$ are alternative radioligands.

Predominant effector pathways

Activation of adenylyl cyclase through stimulation of G$_s$. A stimulatory effect of VIP on phosphoinositide metabolism has been described in some tissues.

Amino acid sequence (rat) [1]

```
  1 MRPPSPPHVR WLCVLAGALA CALRPAGSQA

                          ASPQHECEYL QLIEIQRQQC

 51 LEEAQLENET TGCSKMWDNL TCWPTTPRGQ AVVLDCPLIF QLFAPIHGYN
                                                         TM 1
101 ISRSCTEEGW SQLEPGPYHI ACGLNDRASS LDEQQQTKFY NTVKTGYTIG
                         TM 2
151 YSLSLASLLV AMAILSLFRK LHCTRNYIHM HLFMSFILRA TAVFIKDMAL
                         TM 3
201 FNSGEIDHCS EASVGCKAAV VFFQYCVMAN FFWLLVEGLY LYTLLAVSFF
```

```
           TM 4                                              TM 5
251 SERKYFWGYI LIGWGVPSVF ITIWTVVRIY FEDFGCWDTI INSSLWWIIK
                                                            TM 6
301 APILLSILVN FVLFICIIRI LVQKLRPPDI GKNDSSPYSR LAKSTLLLIP
                        TM 7
351 LFGIHYVMFA FFPDNFKAQV KMVFELVVGS FQGFVVAILY CFLNGEVQAE

401 LRRKWRRWHL QGVLGWSSKS QHPWGGSNGA TCSTQVSMLT RVSPSARRSS

451 SFQAEVSLV
```

The first 30 amino acids may serve as a signal sequence.

Amino acids	429
Molecular weight	
(polypeptide)	48 946
(protein)	~55 kDa
Glycosylation	Asn58*, Asn69*, Asn100*
Disulfide bonds	Cys216–Cys86*

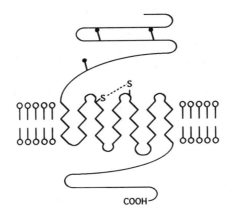

Database accession numbers

	PIR	SWISSPROT	EMBL/GENBANK	REFERENCE
Rat			M86835	1

Gene

The organization of the gene is not known. In addition to a ubiquitous 55 kb mRNA, two smaller mRNAs (1.3 and 2.4) hybridize with cloned cDNA in rat brain [1].

Comment

It has been claimed that the human homologue of the RDC1 gene [2] encodes a VIP receptor [3]; however, neither the expressed canine nor the rat protein exhibit [125I]VIP binding (see Orphan receptors, page 223).

References

1 Ishihara, T. *et al.* (1992) *Neuron* **8**, 811–819.
2 Libert, F. *et al.* (1989) *Science* **244**, 569–572.
3 Sreedharan, S.P. *et al.* (1991) *Proc. Natl Acad. Sci. USA* **88**, 4986–4990.

THE SECRETIN RECEPTOR

Secretin is the most potent endogenous ligand, inducing activation in low nanomolar concentrations. VIP has an ~1000 lower affinity.

Distribution

It is found in high levels in pancreas, stomach, heart and in certain cell lines, e.g. NG108-15 cells.

Pharmacology

Agonists: Partial sequences of secretin or other endogenous peptides have considerably lower potencies than the parent molecule or are inactive.

Antagonists: No antagonist have been described.

Radioligands: [^{125}I]Secretin binds to high-(K_d 0.5 nM) and low-affinity (K_d 20 nM) sites.

Predominant effector pathways

Activation of adenylyl cyclase through stimulation of G_s.

Amino acid sequence (rat) [1]

```
  1 MLSTMRPRLS LLLLRLLLLT KA


                         AHTVGVPP  RLCDVRRVLL EERAHCLQQL

 51 SKEKKGALGP ETASGCEGLW DNMSCWPSSA PARTVEVQCP KFLLMLSNKN
                                                TM 1
101 GSLFRNCTQD GWSETFPRPD LACGVNINNS FNERRHAYLL KLKVMYTVGY
                         TM 2
151 SSSLAMLLVA LSILCSFRRL HCTRNYIHMH LFVSFILRAL SNFIKDAVLF
              TM 3
201 SSDDVTYCDA HKVGCKLVMI FFQYCIMANY AWLLVEGLYL HTLLAISFFS
        TM 4                                    TM 5
251 ERKYLQAFVL LGWGSPAIFV ALWAITRHFL ENTGCWDINA NASVWWVIRG
        TM 6
301 PVILSILINF IFFINILRIL MRKLRTQETR GSETNHYKRL AKSTLLLIPL
                         TM 7
351 FGIHYIVFAF SPEDAMEVQL FFELALGSFQ GLVVAVLYCF LNGEVQLEVQ


401 KKWRQWHLQE FPLRPVAFNN SFSNATNGPT HSTKASTEQS RSIPRASII
```

The first 22 amino acids may serve as a signal sequence [1].

Amino acids	449
Molecular weight	48 696
Glycosylation	Asn72*, Asn100*, Asn106*, Asn128*, Asn291*
Disulfide bonds	Cys215–Cys285*
Palmitoylation	None

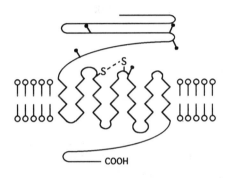

Database accession numbers

	PIR	SWISSPROT	EMBL/GENBANK	REFERENCE
Rat	S16319	P23811	X59132	1

Gene
The organization of the gene is not known.

Reference
[1] Ishihara, T. et al. (1991) EMBO J. **10**, 1635–1641.

Vasopressin and oxytocin

INTRODUCTION

Vasopressin and oxytocin are members of a family of peptides found in all mammalian species called the neurohypophyseal hormones [1-6]. They are present in high levels in the posterior pituitary.

Vasopressin has an essential role in the control of the water content of the body and acts in the kidney to increase water and sodium absorption. In higher concentrations vasopressin stimulates contraction of vascular smooth muscle, stimulates glycogen breakdown in liver, induces platelet activation and evokes release of corticotrophin from the anterior pituitary. Vasopressin or its analogues are used clinically to treat diabetes insipidus.

Oxytocin stimulates contraction of uterine smooth muscle, although the physiological signficance of this is uncertain. Oxytocin stimulates milk secretion in response to suckling by inducing contraction of myoepithelial cells in the mammary gland. It is used clinically to induce labour and to promote lactation.

Structures (human)

Oxytocin

$$S\text{———————}S$$
$$\text{Cys–Tyr–Ile–Gln–Asn–Cys–Pro–Leu–Gly–NH}_2$$

Vasopressin (VP)

$$S\text{———————}S$$
$$\text{Cys–Tyr–Phe–Gln–Asn–Cys–Pro–Arg–Gly–NH}_2$$

In guinea-pig vasopressin arginine in position 8 is replaced by lysine. A number of further neurohypophyseal hormones exist which differ by structural changes in positions 2, 3, 4 and 8, e.g. mesotocin, valitocin, arginine vasotocin, isotocin, phenylpressin.

Distribution and synthesis

They are stored along with their transport proteins, the neurophysins, in the posterior lobe of the pituitary; they are also found in the CNS.

Book
1 Gash and Boer (eds). Vasopressin: Principles and Properties, Plenum Press, New York.

Reviews
2 Hruby, V.J. et al. (1990) Conformational and structural considerations in oxytocin-receptor binding and biological activity. Annu. Rev. Pharmacol. Toxicol. **30**, 501–534.
3 Huffman, W.F. et al. (1988) Vasopressin antagonists. Annu. Rep. Med. Chem. **23**, 91–99.
4 Kinter, L.B. et al. (1988) Antagonists of the antidiuretic action of vasopressin. Am. J. Physiol. **254**, F165–F177.
5 Lászlo, F.O. et al. (1991) Pharmacology and clinical perspectives of vasopressin antagonists. Pharmacol. Rev. **43**, 73–108.

6 Manning, M. and Sawyer, W.H. (1989) Discovery, development, and some uses of vasopressin and oxytocin antagonists. *J. Lab. Clin. Med.* **144**, 617–632.

VASOPRESSIN AND OXYTOCIN RECEPTORS

The V$_{1A}$, V$_2$ and oxytocin receptors

These are seven transmembrane spanning proteins and are described below.

The V$_{1B}$ receptor

V$_{1B}$ receptors can be distinguished from V$_{1A}$ receptors by the low affinity of certain antagonists at the latter, e.g. the affinity of d(CH$_2$)$_5$[Tyr(Me)^2]vasopressin is approximately 1000-fold lower than at the V$_{1A}$ receptor. V$_{1B}$ receptors stimulate phosphoinositide metabolism and are found in anterior pituitary.

THE V$_{1A}$ VASOPRESSIN RECEPTOR

Vasopressin acts in low nanomolar concentrations and is ~tenfold more potent than oxytocin.

Distribution

In the periphery the V$_{1A}$ receptor is found in high levels in vascular smooth muscle, myometrium and bladder where it mediates contraction. It also induces activation of platelets, glycogenlysis in hepatocytes and adipocytes, induces ACTH release in pituitary (V$_{1B}$ receptor) and is found in spleen and testis. In CNS, V$_1$ sites are distributed widely and are found in lateral septal nucleus, hippocampus, superior collicular, substantia nigra and central grey matter.

Considerable species differences exist in distribution, e.g. the V$_{1A}$ receptor is not expressed in rabbit liver.

Pharmacology

Agonists: No selective agonist has been described.

Antagonists: The nonpeptide OPC 21268 (K_i 140 nM) has a selectivity of more than 1000-fold relative to V$_2$ receptors (affinity for oxytocin receptors not known) and is orally active [1]. Examples of peptide antagonists with selectivities of greater than tenfold are d(CH$_2$)$_5$[Tyr(Me)^2]AVP (pA$_2$ 8.6) and Phaa,DTyr(Me),Phe,Gln,Asn,Arg,Pro,Arg,Tyr-NH$_2$ (pA$_2$ 8.9) [2].

Radioligands: [^{125}I]Phaa,DTyr(Me),Phe,Gln,Asn,Arg,Pro,Arg,Tyr-NH$_2$ (K_d 0.06 nM) [2] and [^{3}H]d(CH$_2$)$_5$[Tyr(Me)^2]AVP (K_d 0.1 nM) are examples of selective antagonist ligands. [^{125}I]VP (K_d 0.6-6 nM) is nonselective.

Structures:

OPC 21268

Predominant effector pathways

Activation of phosphoinositide metabolism through a pertussis-toxin-insensitive G-protein, most likely of the G_q/G_{11} class.

Amino acid sequence (rat) [3]

```
  1 MSFPRGSQDR SVGNSSPWWP LTTEGSNGSQ EAARLGEGDS PLGDVRNEEL
              TM 1                                  TM 2
 51 AKLEIAVLAV IFVVAVLGNS SVLLALHRTP RKTSRMHLFI RHLSLADLAV
                        TM 3
101 AFFQVLPQLC WTSPSFRGPD WLCRVVKHLQ VFAMFASAYM LVVMTADRYI
              TM 4
151 AVCHPLKTLQ QPARRSRLMI ATSWVLSFIL STPQYFIFSV IEIEVNNGTK
              TM 5
201 TQDCWATFIQ PWGTRAYVTW MTSGVFVAPV VVLGTCYGFI CYHIWRNIRG
                                                         TM 6
251 KTASSRHSKG DKGSGEAVGP FHKGLLVTPC VSSVKSISRA KIRTVKMTFV
                                                       TM 7
301 IVSAYILCWA PFFIVQMWSV WDENFIWTDS ENPSITITAL LASLNSCCNP
351 WIYMFFSGHL LQDCVQSFPC CHSMAQKFAK DDSDSMSRKT DFLF
```

Amino acids	394
Molecular weight	44 202
Glycosylation	Asn14*, Asn27*
Disulfide bonds	Cys123*–Cys204*
Palmitoylation	Cys370*, Cys371*

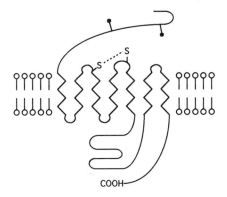

Database accession numbers

	PIR	SWISSPROT	EMBL/GENBANK	REFERENCE
Rat			Z11690	3

Gene
The organization of the gene is unknown.

References
1 Yamamura, H.J. et al. (1991) Science 252, 572–574.
2 Schmidt, R. et al. (1991) FEBS Lett. 282, 77–81.
3 Morel, F. et al. (1992) Nature 356, 523–526.

THE V₂ VASOPRESSIN RECEPTOR

Vasopressin acts at low nanomolar concentrations and is ~tenfold more potent than oxytocin.

Distribution

The V_2 receptor is found in high levels in the osmoregulatory epithelia of the terminal urinary tract where it stimulates water reabsorption. It is also present in much lower levels in endothelium and blood vessels of some species where it induces vasodilation. In the CNS binding sites are found in the subiculum with lower levels in caudate-putamen and islands of Calleja.

Pharmacology

Agonists: d[DArg^8]VP and d[Val^4]VP are selective.

Antagonists: The nonpeptide, OPC 31260 (K_i 5.1 nM), has a selectivity of more than 100-fold [1] (affinity for oxytocin receptors not known). d(CH$_2$)$_5$[DIle^2,Ile^4]AVP (pA$_2$ 8.0) is an example of a selective peptide antagonist.

Radioligands: An example antagonist radioligand is [^{3}H]d(CH$_2$)$_5$[DIle^2,Ile^4]AVP (K_i 10 nM); [^{3}H]d[DArg^8]VP (K_D 5 nM) and [^{3}H]-[Val^4]VP (K_i 5 nM) are weakly selective.

Structures:

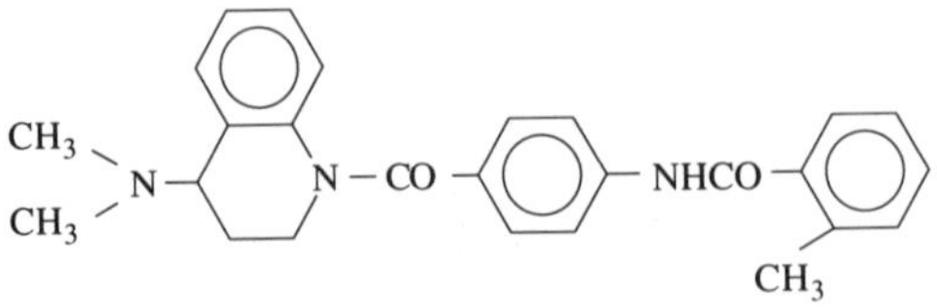

OPC 31260

Predominant effector pathways
Formation of cAMP by activation of Gs.

Amino acid sequence [2,3]

```
                                                    TM 1
  1 MLMASTTSAV PGHPSLPSLP SNSSQERPLD TRDPLLARAE LALLSIVFVA
                              TM 2
 51 VALSNGLVLA ALARRGRRGH WAPIHVFIGH LCLADLAVAL FQVLPQLAWK
                  TM 3
101 ATDRFRGPDA LCRAVKYLQM VGMYASSYMI LAMTLDRHRA ICRPMLAYRH
              TM 4
151 GSGAHWNRPV LVAWAFSLLL SLPQLFIFAQ RNVEGGSGVT DCWACFAEPW
            TM 5
201 GRRTYVTWIA LMVFVAPTLG IAACQVLIFR EIHASLVPGP SERPGGRRRG
                          TM 6
251 RRTGSPGEGA HVSAAVAKTV RMTLVIVVVY VLCWAPFFLV QLWAAWDPEA
            TM 7
301 PLEGAPFVLL MLLASLNSCT NPWIYASFSS SVSSELRSLL CCARGRTPPS

351 LGPQDESCTT ASSSLAKDTS S
```

Amino acids	371
Molecular weight	40 748
Chromosome	X
Glycosylation sites	Asn22*
Disulfide bonds	Cys112–Cys195*
Palmitoylation	Cys341*, Cys342*

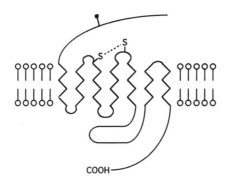

Database accession numbers

	PIR	SWISSPROT	EMBL/GENBANK	REFERENCE
Human			Z11687	2, 3

Gene

The organization of the gene is unknown.

Comment

Mutations in human gene encoding the V_2 receptor are believed to underlie the disorder nephrogenic diabetes insipidus.

References

1 Yamamura, H.J. et al. (1992) Br. J. Pharmacol. **105**, 787–791.
2 Birnbaumer, M. et al. (1992) Nature **357**, 333–335.
3 Lolait, S.J. et al. (1992) Nature **357**, 336–339.

THE OXYTOCIN RECEPTOR

Oxytocin acts in low nanomolar concentrations and has a similar or slightly greater potency than vasopressin.

Distribution

Oxytocin receptors are found in uterine smooth muscle, myoepithelial cells in mammary gland and pituitary.

Pharmacology

Agonists: [Thr⁴,Gly⁷] oxytocin is selective.

Antagonists: Examples of selective antagonists include d(CH₂)₅[Tyr(Me)²,Orn⁸] vasotocin₁₋₈ [1] (pA₂ 7.7) and cyc[DNal,Ile,DPip,Pip,DHis,Pro] (pA₂ 8.6).

Radioligands: [¹²⁵I]d(CH₂)₂[Tyr(Me)²,Thr⁴,Tyr⁹-NH₂]oxytocin (K_i 0.03 nM) has a selectivity of > 100-fold [1].

Predominant effector pathways

Activation of phosphoinositide metabolism through a pertussis-toxin-insensitive G-protein most likely of the G_q/G_{11} class.

Amino acid sequence [2]

```
                                                    TM 1
  1 MEGALAANWS AEAANASAAP PGAEGNRTAG PPRRNEALAR VEVAVLCLIL
                                 TM 2
 51 LLALSGNACV LLALRTTRQK HSRLFFFMKH LSIADLVVAV FQVLPQLLWD
                            TM 3
101 ITFRFYGPDL LCRLVKYLQV VGMFASTYLL LLMSLDRCLA ICQPLRSLRR
           TM 4
151 RTDRLAVLAT WLGCLVASAP QVHIFSLREV ADGVFDCWAV FIQPWGPKAY
          TM 5
201 ITWITLAVYI VPVIVLATCY GLISFKIWQN LRLKTAAAGA EAPEGAAAGD
251 GGRVALARVS SVKLISKAKI RTVKMTFIIV LAFIVCWTPF FFVQMWSVWD
           TM 7
301 ANAPKEASAF IIVMLLASLN SCCNPWIYML FTGHLFHELV QRFLCCSASY

351 LKGRRLGETS ASKKSNSSSF VLSHRSSSQR SCSQPSTA
```

Amino acids	388
Molecular weight	42 716
Glycosylation	Asn8*, Asn15*, Asn26*
Disulfide bonds	Cys112–Cys187*
Palmitoylation	Cys345*, Cys346*

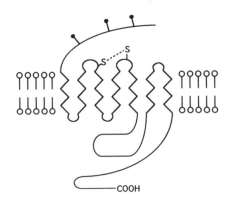

Database accession numbers

	PIR	SWISSPROT	EMBL/GENBANK	REFERENCE
Human			X64878	2

Gene
The organization of the gene is unknown. mRNAs of 6.1, 4.4, 3.6 and 2.6 kb have been identified and are thought to be derived from the same gene [2].

References
[1] Elands, J. *et al.* (1988) *Eur. J. Pharmacol.* 147, 197–207.
[2] Kimura, T. *et al.* (1992) *Nature* 356 526-529; *ibid.* 357, 176.

Viral and other nonmammalian receptors

A number of seven transmembrane receptors identified in lower organisms have no counterpart in mammalian tissues.

THE cAMP RECEPTOR

	PIR	SWISSPROT	EMBL/GENBANK	REFERENCE
Dictyostelium	A41238	P13773	M21824	1

THE OCTOPAMINE RECEPTOR

	PIR	SWISSPROT	EMBL/GENBANK	REFERENCE
Drosophilia	JH0170	P22270	M60789	2,3
	S12004		X54794	

VIRAL SEVEN TRANSMEMBRANE PROTEINS

The presence of genes encoding seven transmembrane proteins in the viral genome may provide a molecular explanation for virally induced cell transformation and proliferation. Ligands targeted to the polypeptides may represent a novel class of antiviral drugs.

CYTOMEGALOVIRUS (CMV)

Human CMV (strain AD169; EMBL/Genbank X17403) is a herpesvirus with a genome encoding about 200 genes. It causes congenital and neonatal disease and is an opportunistic pathogen in immune-deficient individuals. Three CMV genes (US27, US28 and US33) encode polypeptides which are predicted to have seven transmembrane domains.

US27

PIR	SWISSPROT	EMBL/GENBANK	REFERENCE
B27216	P09703	X17403	4
S09476		X04650	
S09941			

```
                                        TM 1
    1  MTTSTNNQTL TQVSNMTNHT LNSTEIYQLF EYTRLGVWLM CIVGTFLNVL
                    TM 2
   51  VITTILYYRR KKKSPSDTYI CNLAVADLLI VVGLPFFLEY AKHHPKLSRE
          TM 3                                            TM 4
  101  VVCSGLNACF YICLFAGVCF LINLSMDRYC VIVWGVELNR VRNNKRATCW
                                                       TM 5
  151  VVIFWILAVL MGMPHYLMYS HTNNECVGEF ANETSGWFPV FLNTKVNICG
                                        TM 6
  201  YLAPIALMAY TYNRMVRFII NYVGKWHMQT LHVLLVVVVS FASFWFPFNL
                            TM 7
  251  ALFLESIRLL AGVYNDTLQN VIIFCLYVGQ FLAYVRACLN PGIYILVGTQ
```

```
301  MRKDMWTTLR VFACCCVKQE IPYQDIDIEL QKDIQRRAKH TKRTHYDRKN

351  APMESGEEEF LL
```

US28

	PIR	SWISSPROT	EMBL/GENBANK	REFERENCE
	C27216	P09704	X17403	4
	S09477		X04650	
	S09942			

```
                                       TM 1
  1  MTPTTTTAEL TTEFDYDEDA TPCVFTDVLN QSKPVTLFLY GVVFLFGSIG
                       TM 2
 51  NFLVIFTITW RRRIQCSGDV YFINLAAADL LFVCTLPLWM QYLLDHNSLA
     TM 3                                        TM 4
101  SVPCTLLTAC FYVAMFASLC FITEIALDRY YAIVYMRYRP VKQACLFSIF
                                       TM 5
151  WWIFAVIIAI PHFMVVTKKD NQCMTDYDYL EVSYPIILNV ELMLGAFVIP
                             TM 6
201  LSVISYCYYR ISRIVAVSQS RHKGRIVRVL IAVVLVFIIF WLPYHLTLFV
                        TM 7
251  DTLKLLKWIS SSCEFERSLK RALILTESLA FCHCCLNPLL YVFVGTKFRK

301  NYTVCWPSFA SDSFPAMYPG TTA
```

UL33

	PIR	SWISSPROT	EMBL/GENBANK	REFERENCE
	S09322	P16849	X17403	4
	S09796			

```
          TM 1                                    TM 2
  1  MTGPLFAIRT TEAVLNTFII FVGGPLNAIV LITQLLTNRV LGYSTPTIYM
                             TM 3
 51  TNLYSTNFLT LTVLPFIVLS NQWLLPAGVA SCKFLSVIYY SSCTVGFATV
                        TM 4
101  ALIAADRYRV LHKRTYARQS YRSTYMILLL TWLAGLIFSV PAAVYTTVVM
                                  TM 5
151  HHDANDTNNT NGHATCVLYF VAEEVHTVLL SWKVLLTMVW GAAPVIMMTW
                        TM 6
201  FYAFFYSTVQ RTSQKQRSRT LTFVSVLLIS FVALQTPYVS LMIFNSYATT
                   TM 7
251  AWPMQCEHLT LRRTIGTLAR VVPHLHCLIN PILYALLGHD FLQRMRQCFR

301  GQLLDRRAFL RSQQNQRATA ETNLAAGNNS QSVATSLDTN SKNYNQHAKR

351  SVSFNFPSGT WKGGQKTASN DTSTKIPHRL SQSHHNLSGV
```

HERPESVIRUS SAIMIRI (HVS)

HVS is a T-lymphotropic gammaherpesvirus. It causes asymptomatic infections in its natural host, the squirrel monkey (*Saimiri sciureus*), but causes fatal lymphoproliferative diseases in other New World primates. Its genome contains one gene that encodes for a seven transmembrane protein.

ECRF3

PIR	SWISSPROT	EMBL/GENBANK	REFERENCE
		X64346	5

```
                                                       TM 1
  1  MEVKLDFSSE DFSNYSYNYS GDIYYGDVAP CVVNFLISES ALAFIYVLMF
                                      TM 2
 51  LCNAIGNSLV LRTFLKYRAQ AQSFDYLMMG FCLNSLFLAG YLLMRLLRMF
               TM 3
101  EIFMNTELCK LEAFFLNLSI YWSPFILVFI SVLRCLLIFC ATRLWVKKTL
          TM 4
151  IGQVFLCCSF VLACFGALPH VMVTSYYEPS SCIEEDGVLT EQLRTKLNTF
          TM 5                                  TM 6
201  HTWYSFAGPL FITVICYSMS CYKLFKTKLS KRAEVVTIIT MTTLLFIVFC
                                            TM 7
251  IPYYIMESID TLLRVGVIEE TCAKRSAIVY GIQCTYMLLV LYYCMLPLMF

301  AMFGSLFRQR MAAWCKTICH C
```

References

[1] Klein, P.S. *et al.* (1988) *Science* 241, 1467–1472.
[2] Sauduo, F. *et al.* (1990) *EMBO J.* 9, 3611–3617.
[3] Arakawa, S. *et al.* (1990) *Neuron* 4, 343–354.
[4] Chee, M.S. *et al.* (1990) *Nature* 344, 774–777.
[5] Nicholas, J. *et al.* (1992) *Nature* 355, 362–365.

SUPERFAMILY
OF
HETEROTRIMERIC
G-PROTEINS

Introduction: Heterotrimeric G-proteins

The first report that hormone-receptor linked signal transduction is regulated by GTP appeared in 1971 [1]. Over 20 years later we recognize that a major proportion of all neurotransmitters, hormones, local modulators as well as sensory stimuli act through similar GTP-dependent processes [2]. The basis for GTP sensitivity is now known to reflect a step early in signal transduction involving receptor-dependent activation of a class of GTPases termed G-proteins. These crucial regulatory components are located on the cytoplasmic leaflet of the plasma membrane, poised ready for interaction with agonist-bound receptors (largely members of the seven transmembrane superfamily) as well as modulation of a range of effector and second messenger systems [2-8].

STRUCTURE AND FUNCTION

G-proteins are heterotrimers

G-proteins are heterotrimeric in nature and composed of α, β, and γ components encoded by distinct genes. Each gene type represents a family exhibiting a varying degree of complexity with the most diverse encoding α-subunits. Although protein purification procedures have identified a limited molecular weight range from 39 to 52 kDa, molecular cloning has revealed the existence of at least 17 Gα genes. Some of these also have multiple protein products reflecting alternative RNA splicing. Based on predicted amino acid primary sequences G-protein α-subunits have been subdivided into four main classes termed G_s, G_i, G_q and G_{12} [9] (Fig. 1; Table 1) In addition to diversity among α-chains, there are also multiple genes encoding at least four β- and six γ-subunits of 35–36 kDa and 6–10 kDa, respectively (Figs 2 and 3; Tables 2 and 3).

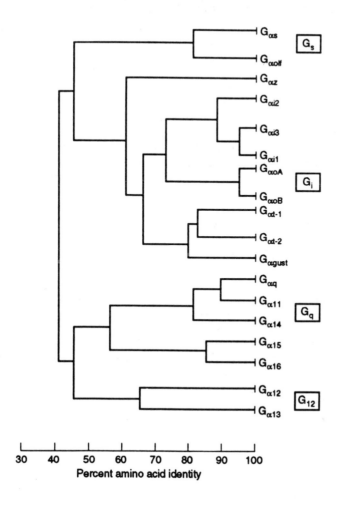

Figure 1. *Structural relationship between G-protein α-chains. Adapted from refs 9 and 10. Members of each G-protein class Gs, Gi, Gq and G12 (boxed) are grouped according to amino acid identity with branch junctions corresponding to approximate comparative values for the α-chains indicated. Percentage sequence identities were calculated using the GAP routine of the Wisconsin GCG sequence analysis software package version 7. Actual values for sequence identity comparisons between all α-chains are given in Table 1.*

Table 1. *Percentage Amino acid identity between G-protein α-chains*

	Gαs-L	Gαolf	Gαi1	Gαi2	Gαi3	GαoA	GαoB	Gαt-1	Gαt-2	Gαgust	Gαz	Gαq	Gα11	Gα14	Gα15	Gα16	Gα12	Gα13
Gαs-S	100	79	44	43	44	46	45	42	43	44	42	41	42	42	38	40	39	42
Gαs-L	Gαs-L	80	44	44	44	46	45	42	44	44	42	42	42	42	39	41	39	42
Gαolf		Gαolf	43	43	42	44	44	43	44	43	41	41	40	40	39	38	37	43
Gαi1			Gαi1	88	94	73	73	68	69	67	67	55	52	52	45	45	41	38
Gαi2				Gαi2	86	70	70	66	70	67	66	51	52	51	45	45	41	40
Gαi3					Gαi3	72	71	65	69	66	67	54	51	51	45	44	40	41
GαoA						GαoA	94	63	62	63	61	51	50	51	44	44	43	39
GαoB							GαoB	61	60	62	59	51	51	52	44	46	42	44
Gαt-1								Gαt-1	82	79	54	50	50	47	43	44	42	43
Gαt-2									Gαt-2	79	57	50	50	49	45	45	40	39
Gαgust										Gαgust	57	50	50	49	41	44	41	38
Gαz											Gαz	48	47	49	43	43	43	39
Gαq												Gαq	89	82	58	56	43	40
Gα11													Gα11	81	57	56	42	45
Gα14														Gα14	57	56	42	46
Gα15															Gα15	85	39	45
Gα16																Gα16	40	41
Gα12																	Gα12	65
Gα13																		Gα13

Where possible, comparisons were performed using human sequence data. Exceptions to this are Gαolf and Gαgust (rat) as well as Gαq, Gα11, Gα12, Gα13 Gα14 and Gα15 (mouse). Identity was calculated using the GAP routine of the Wisconsin GCG sequence analysis software package version 7.

Note: Percentage identities as calculated reflect matches over entire molecules and mask highly conserved domains important for guanine nucleotide binding and GTPase activity (regions G1–G5) *11*. Amino acid differences are clustered mainly outside these domains.

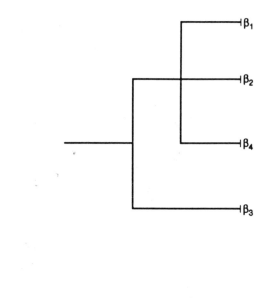

60   70   80   90   100
Percent amino acid identity

Figure 2. *Structural relationship between G-protein β-chains. β-subunits are grouped according to amino acid identity with branch junctions corresponding to approximate comparative values. Percentage sequence identities were calculated using the GAP routine of the Wisconsin GCG sequence analysis software package version 7. Actual values for sequence identity comparisons between all β-chains are given in Table 2.*

Table 2. *Percentage amino acid identities between G-protein β-chains*

	β2	β3	β4
β1	90	83	90
	β2	81	89
		β3	79
			β4

Identities were calculated using GAP routine of the Wisconsin GCG sequence analysis software package version 7. Comparisons were performed using human (β1, β2, β3) and mouse (β4) sequence data.

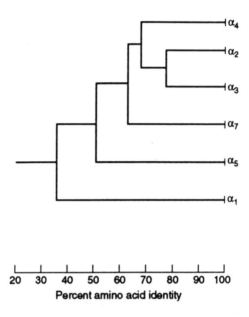

20 30 40 50 60 70 80 90 100
Percent amino acid identity

Figure 3. *Structural relationship between G-protein γ–chains. γ-subunits are grouped according to amino acid identity with branch junctions corresponding to approximate comparative values. Percentage sequence identities were calculated using the GAP routine of the Wisconsin GCG sequence analysis software package version 7. Actual values for sequence identity comparisons between all γ-chains are given in Table 3.*

Table 3. *Percentage amino acid identities between G-protein γ-chains*

	γ2	γ3	γ4[a]	γ5	γ7
γ1	37	36	30	25	35
	γ2	76	70	49	68
		γ3	64	46	57
			γ4[a]	52	55
				γ5	49
					γ7

Identities were calculated using GAP routine of the Wisconsin GCG sequence analysis software package version 7. Comparisons were performed using bovine (γ1, γ2, γ3, γ5, γ7) and mouse (γ4) sequence data.
[a]γ4 represents a partial sequence (33 amino acids).

Receptor-dependent G-protein activation

A generally accepted model for G-protein activation illustrated in Fig.4 is based largely on studies performed on hormone-stimulated adenylyl cyclase by $G_{\alpha s}$ and light-activated retinal rod cyclic GMP-PDE by transducin $(G_{\alpha t})$ [2-12]. At rest, G-proteins exist as heterotrimers with GDP bound to the α-chain. Activation follows interaction with agonist-bound receptor and dissociation of GDP to convert the "inactive" G-protein into a transiently "empty state". This complex of ligand/receptor/G-protein is believed to underlie guanine nucleotide-sensitive high-affinity agonist binding detectable in membrane fractions by radioligand binding [2,13,14]. Under normal cellular conditions GTP binds the empty guanine

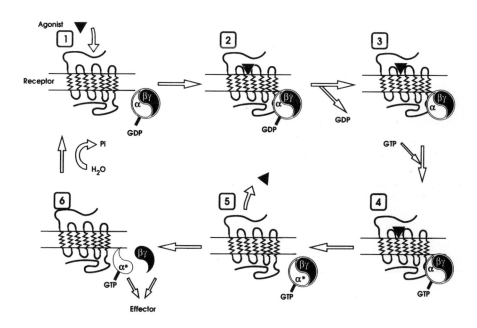

Figure 4. *G-protein activation/deactivation cycle. Inactive G-protein exists as a heterotrimer with a GDP-bound α-chain (GDP-$\alpha\beta\gamma$) (1). Binding of agonist to receptor promotes its association with heterotrimeric G-protein (2) to trigger dissociation of GDP and formation of a transient "empty" state ($\alpha\beta\gamma$) (3). This association can be disrupted by either GDP or GTP and probably corresponds to high-affinity agonist binding detectable by radioligand binding. Within the cellular environment GTP then binds (4) and in the presence of Mg^{2+} triggers a conformational change and activation of the G-protein (GTP-$\alpha*\beta\gamma$) (5). This may then be followed by dissociation into free GTP-$\alpha*$ and $\beta\gamma$ units which then regulate effector activities (e.g. adenylyl cyclases, phospholipases, ion channels) (6). Endogenous GTPase activity hydrolyses GTP, leaving GDP-α which subsequently reassociates with $\beta\gamma$ reforming heterotrimeric GDP-$\alpha\beta\gamma$. Note: antagonist binding does not trigger GDP dissociation and G-protein activation.*

nucleotide binding site to cause G-protein activation, illustrated here as a two-step process. The first step is a conformational transition from αβγ-GTP to α*βγ-GTP which appears to be Mg^{2+}-dependent and this is followed by dissociation of the α*βγ-GTP complex to form free α*-GTP and βγ dimer [15]. G-protein activation and subunit dissociation is accompanied by separation from the receptor which reverts to the low-affinity agonist binding state. Free α*-GTP and βγ dimers (see below) then modulate activities associated with a range of effector enzymes or ion channels. It should be noted that although activated G-proteins undoubtedly undergo subunit dissociation under certain experimental conditions (e.g. with non-hydrolysable GTP analogues and detergent) some kinetic and co-purification studies have led to the suggestion that undissociated yet active heterotrimeric G-protein could also regulate effector function [16,17].

G-protein deactivation by endogenous GTPase

Following G-protein activation, hydrolysis of bound GTP by an intrinsic GTPase terminates its ability to regulate effector activity and leads to α-GDP reassociation with βγ (**Fig. 4**). This interaction appears to be of high affinity [18–20]. The nature of the GTPase activity plays at least two crucial roles. Firstly, GTP hydrolysis is irreversible and makes the overall process of signal transduction unidirectional. Secondly, GTP hydrolysis is slow. K_{cat} values of 2–5 min^{-1} have been reported for $G_{\alpha s}$ or $G_{\alpha i}$[21-24] while considerably lower values have been detected for $G_{\alpha q}$ (K_{cat} 0.8 min^{-1}) [25] and $G_{\alpha z}$ (K_{cat} 0.05 min^{-1}) [26]. Slow deactivation by GTPase can allow substantial signal amplification as indicated by generation of GTP-bound transducin by photoactivated rhodopsin [27,28] and catalytic activation of $G_{\alpha s}$ or $G_{\alpha q}$ following reconstitution with receptor [21,25,29]. Interestingly, although signal amplification is an important aspect of receptor-linked transduction, endogenous GTPase activity may well be more rapid under physiological condions as there is evidence that effector interaction can augment G-protein GTPase activity considerably [30–32].

Fluoride, non-hydrolysable GTP-analogues, cholera and pertussis toxins act at sites within G-protein activation/deactivation cycle

Several biochemical tools used widely to study G-protein linked signal transduction pathways act at sites within the activation/deactivation cycle (**Fig. 4**).

1 Fluoride is a GTP-independent G-protein activator that in the presence of aluminium or beryllium ions forms complexes believed to bind the nucleotide site adjacent to the β-phosphate of GDP. This mimics the γ-phosphate of bound GTP and triggers G-protein activation [33,34].

2 Non-hydrolyzable analogues of GTP such as guanosine-5'-O-(3-thiotriphosphate) (GTPγS) and guanylyl-imidodiphosphate (Gpp(NH)p) produce G-protein activation due to their resistance to GTPase.

3 Cholera toxin from *Vibrio cholerae* transfers the ADP-ribose moiety from NAD to a conserved arginine residue within $G_{\alpha s}$ and transducin α-chains. This modification requires the presence of an additional protein, ADP-ribosylating factor (ARF), and suppresses GTPase activity, thereby locking the G-protein in the active GTP-bound conformation [35,36]. Note: Under certain activation

conditions (e.g. agonist stimulation) other α-chains such as $G_{\alpha i}$ and $G_{\alpha o}$ may also undergo cholera-toxin-dependent ADP-ribosylation [37-39].

4 Pertussis toxin (islet activating protein from *Bordetella pertussis*) ADP-ribosylates G-protein α-chains at a cysteine located four residues from the C-terminal of $G_{\alpha i1-3}$, $G_{\alpha oA}$, $G_{\alpha oB}$, $G_{\alpha t1}$ and $G_{\alpha t2}$. This modification appears dependent upon an αβγ heterotrimeric state and results in G-protein uncoupling from receptor [40-45].

Functional roles for βγ

Several crucial roles have been assigned to G-protein βγ dimers. Association of GDP-bound α-subunit with ligand-bound receptor is promoted by βγ and, together with a preference for binding α–GDP rather than α-GTP, βγ dimers appear to facilitate receptor-dependent G-protein activation [2,46-51]. βγ also stabilizes α-GDP as GDP dissociates from αβγ more slowly than from free α-subunit [50]. This is likely to diminish receptor-independent GDP dissociation and suppress signal transduction events in the absence of agonist. This increase in signal-to-noise ratio of signal transduction processes has been demonstrated for hormonal stimulated adenylyl cyclase [52]. βγ dimers may also be important for presenting α–subunits in the correct orientation for receptor activation, anchorage of G-proteins to inner surface of membrane, facilitating guanine nucleotide exchange and targeting certain kinases to enhance agonist-induced receptor phosphorylation and desensitization [48-57].

Recent data have also indicated that G-protein βγ dimers play an important role as direct mediators of signal transduction in at least three distinct systems. First, βγ subunits have been implicated in receptor control of potassium channel activity, possibly through activation of PLA_2 and generation of arachidonic acid and eicosanoids [58-61]. These observations remain controversial, however, as in the hands of other workers βγ has no stimulatory effects on potassium channels and moreover, evidence in support of a role for α-chains of G_i has been reported [2,62]. A second potential role for βγ is to mediate increased phosphoinositide hydrolysis as purified and recombinantly expressed βγ dimers activate phospholipase C (PLC) to breakdown phosphatidylinositol 4,5-bisphosphate. PLC-β2 and PLC-β3 appear particularly sensitive to activation by G-protein βγ while PLC-γ1 and PLC-δ1 are unresponsive [63-68]. βγ dimers derived from G_i or G_o, rather than α-subunits, may therefore mediate pertussis-toxin-sensitive activation of phosphoinositide hydrolysis identified in some cell types. A third functional response modulated by G-protein βγ dimers is regulation of adenylyl cyclase activity. One model for receptor-linked inhibition proposes that βγ dimers from activated G_i can associate with and promote inactivation of free stimulatory $G_{\alpha s}$ [69,70]. Others, however, dispute an important action for βγ and suggest instead a role for $G_{\alpha i}$ [71] which is also consistent with adenylyl cyclase inhibition in S49 *cyc⁻* cell membranes lacking $G_{\alpha s}$ [72]. Moreover, recombinant expression of mutationally activated $G_{\alpha i1}$, $G_{\alpha i2}$, $G_{\alpha i3}$ and $G_{\alpha z}$ leads to constitutive inhibition of adenylyl cylase supporting further an important role for the α-chain [73-76]. These apparent discrepant observations may be resolved when the true molecular identity of the adenylyl cyclase subtypes under study become clear. Hence, current cloning strategies have identified up to eight mammalian adenylyl cyclase subtypes [77] and recent data demonstrate that in the presence of activated $G_{\alpha s}$, one form of adenylyl cyclase is inhibited by βγ (type I calmodulin activated) while others can actually be activated (types II and IV) [78-80]. Interestingly, both $G_{s\alpha}$ and βγ-subunits associate and co-purify with activated adenylyl cyclase, suggesting that both G-protein components may be required to fullfil its complete physiological role [17]. It should be noted that an active signal transduction role for βγ subunits is not without precedent as in the budding yeast *Saccharomyces cerevisiae* powerful genetic evidence supports a role for βγ– subunits mediating G-protein-dependent mating responses [81-83].

SPECIFICITY OF RECEPTOR/G-PROTEIN INTERACTIONS

Precise orchestration and specificity of interaction between receptors and G-proteins is likely to be a crucial determinant of cellular responsiveness to a wide range of external signals. Indeed, although many early reconstitutions suggested substantial promiscuity [84-88], more recent studies employing highly purified and molecularly characterized components indicate a higher degree of selectivity between receptors and α-chains (Table 4).

Table 4. *Selectivity of receptor/G-protein interaction upon reconstitution*

Receptor	Selectivity upon reconstitution[a]	References
Dopamine D_2	$G_{\alpha i2}>>G_{\alpha i1},G_{\alpha i3}>>G_{\alpha o}$	[89]
Adenosine A_1	$G_{\alpha i3}>G_{i\alpha1},G_{i\alpha2},G_{o\alpha}>>G_{\alpha s}\text{-S},G_{\alpha s}\text{-L},G_{\alpha z}$	[90]
α_2-Adrenoceptor	$G_{\alpha i3}>G_{\alpha i1},G_{\alpha i2}>G_{\alpha o}>>G_{\alpha s}$	[91]
β-Adrenoceptor	$G_{\alpha s}\text{-S},\ G_{\alpha s}\text{-L}>G_{i\alpha1},G_{i\alpha3}>>G_{\alpha i2},\ G_{\alpha o},\ G_{\alpha z}$	[49, 92]
Muscarinic M_2	$G_{\alpha o}=G_{\alpha z}>G_{\alpha i1},G_{\alpha i3}>>G_{\alpha s}$	[92]
Serotonergic 5-HT_{1A}	$G_{\alpha i3}>G_{\alpha i2}>G_{\alpha i1}>>G_{\alpha o}>>G_{\alpha s}$	[93]

[a] Relative rank order for G-protein interactions with selected receptor subtypes upon reconstitution.

Despite this, heterologously expressed recombinant receptors interact with distinct G-protein systems depending on the host cell environment [94-96] and moreover, G-proteins can also exhibit strict compartmentalization of receptor interaction, specific subcellular localizations and undergo internalization in endocytic vesicles as well as intracellular trafficking to selected sites within specialized cells [97-103]. This indicates that ultimate delineation of physiologically relevant interactions between receptors and G-proteins will require studies in native membranes under conditions where compartmentalization and other subcellular constraints on protein movement are preserved. Results using three approaches to probe such interactions are presented in Table 5. These include use of site-directed antibodies to either block G-protein activation, to uncouple solubilized receptor/G-protein interactions or in immunoprecipitation. In addition, observations made with agonist-dependent cholera toxin ADP-ribosylation of G-protein α-chains (used to indicate coupling to members of the G_i family) as well as use of antisense oligonucleotides have also been included.

Table 5. *Receptor/G-protein interactions in native membranes*

Receptor (cell/tissue)	G-protein (function)	Reference
Dopaminergic D_2 (pituitary)	$G_{\alpha o}$ (inhibit calcium channels) $G_{\alpha i3}$ (activate potassium channels)	*104*[a]
α_{2B} Adrenoceptor Opioid δ (NG108-15)	$G_{\alpha i2}$ (inhibit adenylyl cyclase) $G_{\alpha o}$ (inhibit calcium channels)	*105–107*[a]
Opioid μ (SH-SY5Y) Opioid δ (SH-SY5Y)	$G_{\alpha i3}$, $G_{\alpha i2}$, $G_{\alpha o}$ (inhibit adenylyl cyclase) $G_{\alpha i1}$, $G_{\alpha i2}$, $G_{\alpha o}$	*108*[a]
Adrenergic α_2 (Platelets)	$G_{\alpha i2}$ (inhibit adenylyl cyclase)	*109*[a]
Bradykinin (liver, NG108-15) Vasopressin (1321N1 cells) Histamine	$G_{\alpha q}$, $G_{\alpha 11}$ (phospholipase C activation)	*110*[a]
TRH (GH_3 pituitary)	$G_{\alpha q}$, $G_{\alpha 11}$ (phospholipase C activation)	*119*[a]
Thromboxane A_2 (platelets)	$G_{\alpha q}$ (phospholipase C activation)	*111*[a]
Somatostatin (brain, AtT-20)	$G_{\alpha i1}$, $G_{\alpha i3}$, $G_{\alpha 0}$	*112–114*[a]
Somatostatin SS_2 (GH_4C_1)	$G_{\alpha i2}$, $G_{\alpha i3}$	*115*[a]
Muscarinic M_2 (heart atria) (heart ventricle) (cerebellum)	$G_{\alpha o}$ $G_{\alpha o}$, $G_{\alpha i2}$ $G_{\alpha o}$, $G_{\alpha i1}$, $G_{\alpha i2}$	*116*[a]
Opioid δ (NG108-15)	$G_{\alpha i2}$, $G_{\alpha i3}$, $G_{\alpha o}$	*117*[b]
formyl-Met-Leu-Phe (HL60)	$G_{\alpha i2}$	*37,118*[b]
Muscarinic M_4 (GH_3) Somatostatin (GH_3) Somatostatin (FTO-2B Adenosine A1 hepatoma) Thrombin (F9 teratocarcinoma)	$G_{\alpha oA}$ with β3γ4 (inhibit calcium channels) $G_{\alpha oB}$ with β1γ3 (inhibit calcium channels) $G_{\alpha 12}$ (inhibit adenylyl cyclase) $G_{\alpha i2}$ (inhibit adenylyl cyclase)	*120-122*[c] *183*[c] *123*[c]

Receptor/G-protein interactions detected in native membranes using specific site-directed G-protein antibodies in membrane fractions or solubilized receptor preparations [a], agonist-dependent ADP-ribosylation by cholera toxin [b] or pretreatment with antisense oligonucleotides [c]. Where parallel measurements of effector activity were made this is indicated in parenthesis as is the cell/tissue type under investigation.

MOLECULAR BASIS FOR G-PROTEIN MEMBRANE ASSOCIATION

N-terminal myristoylation

G-protein α-chains are localized predominantly to membrane fractions despite the fact that purification and primary amino acid sequences demonstrate their hydrophilic nature. One possible explanation comes from the observation that G-protein α-subunits fail to associate with phospholipid vesicles unless βγ is already incorporated, suggesting a role in anchoring G-proteins to the plasma membrane [54]. Indeed, tryptic digestion releases soluble α-chains missing an N-terminal fragment of 1–2 kDa [124] and this region also appears to be important for βγ interaction [125]. Interestingly, $G_{\alpha o}$, $G_{\alpha z}$ and all three $G_{\alpha i}$ isoforms have recently been shown to undergo co-translational addition of 14-carbon myristic acid (C14:0) on Gly2 after cleavage of the N-terminal Met (Fig. 5). Transducin undergoes similar N-terminal modification although by heterogenous fatty acids (C12:0, C14:0, C14:1 and C14:2). Increased hydrophobic interaction between fatty acid and membrane phospholipid and/or protein components such as G-protein βγ could play an important role in membrane localization as well as subunit interaction [126–134] and downstream signalling. The consensus sequence for fatty acylation is unclear but includes a prerequisite for Gly at position 2 and a preference for a hydroxyamino acid such as Ser at position 6 (MGXXXS) [130,135]. For instance, substitution of Gly2 for Ala (G2A) within both $G_{\alpha i1}$ and $G_{\alpha o}$ results in unmyristoylated α-chains localized predominantly in the cytosol [128,129].

G-protein myristoylation does not, however, account in full for G-protein membrane localization. Hence, $G_{\alpha s}$ is clearly membrane-associated although this subunit fails to undergo myristoylation [126,128]. In addition, $G_{\alpha 11}$ and $G_{\alpha q}$ appear to be membrane bound although they lack the MGXXXS consensus [136–138]. Other G-protein α-chains, including $G_{\alpha olf}$, $G_{\alpha 12}$, $G_{\alpha 13}$, $G_{\alpha 14}$, $G_{\alpha 15}$ and $G_{\alpha 16}$, are also devoid of the predicted consensus sequence for myristoylation. For these G-protein subunits some other unique post-translational modification and/or protein domain may mediate membrane anchorage. Palmitoylation of cysteine residues within the N-terminal region could be one such modification [182]. Alternatively, proteolytic removal of the 1 kDa C-terminal domain of $G_{\alpha s}$ abolishes membrane interaction [139] and chimeric α-chains show that amino acids 367–376 of $G_{\alpha s}$ induce membrane association when fused to a soluble N-terminally deleted $G_{\alpha i1}$ [140]. This suggests a role for the C-terminal region of $G_{\alpha s}$ in membrane anchorage. Notably, $G_{\alpha q}$, $G_{\alpha 11}$, $G_{\alpha 12}$ and $G_{\alpha 14}$ contain a sequence similar to $G_{\alpha s}$ within this domain ((I/T)DTENIR(R/F)VF).

γ-subunit isoprenylation and carboxymethylation.

G-protein β– and γ-subunits remain tightly associated under non-denaturing conditions and appear hydrophobic as detergent is required for membrane extraction. Although primary amino acid sequences do not predict regions of high hydrophobicity, the C-terminus of γ-subunits shares some sequence similarity to $p21^{ras}$ which ends in a CAAX motif (A=aliphatic residue, X=unspecified). As established for $p21^{ras}$, G-protein γ-subunits undergo post-translational modification of the CAAX C-terminus which increases protein hydrophobicity. First, a

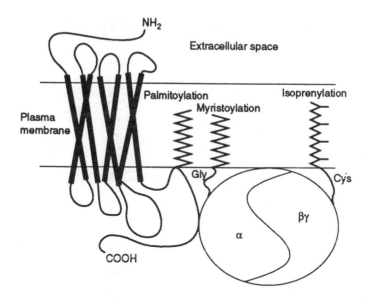

Figure 5 *Schematic representation of lipid modifications believed important for G-protein association with the plasma membrane. Some G-protein α-chains are myristoylated at an N-terminal Gly within consensus MGXXXS while the γ-subunit is isoprenylated at a C-terminal Cys within the consensus CAAX (A=aliphatic residue, X=unspecified). Both farnesylation (C-15) and geranylgeranylation (C-20) have been identified on G-protein γ-subunits. Also shown is potential palmitoylation of some seven transmembrane receptors. Lipid moieties are shown within the cell membrane lipid bilayer although interactions with additional membrane-associated proteins is not excluded.*

long-chain isoprenyl group is attached to the Cys, after which a protease cleaves the AAX residues and the Cys is then carboxymethylated. Different Gγs undergo modification with different isoprenyl groups as the rod photoreceptor γ-subunit (γ1) is modified with the 15-carbon farnesyl moeity while brain γ-subunits such as γ2 have attached the longer 20-carbon geranylgeranyl chain [141–146] (Fig. 5). These distinct isoprenylations are catalysed by at least two different enzymes and sensitivity to these is determined at least in part by the C-terminal X residue. Farnesyl transferase preferentially modifies proteins expressing a C-terminal Ala, Ser, Gln, Cys or Met whereas one of the two geranylgeranyl transferases identified prefers a C-terminal Leu. The CAAX sequences CVIS in γ1 and CAIL in γ2 therefore probably underlies their differential isoprenylation [130,151–154]. Blocking isoprenoid synthesis or mutagenesis of the crucial Cys in the γ-subunit prevents C-terminal processing and results in G-protein βγ dimer which is both localized in soluble fractions and functionally inactivated [67,130, 142,147–150].

LINEARIZED MODEL OF G-PROTEIN α-CHAINS.

Crystal structures of the GTP-binding proteins p21[H-*ras*] and bacterial EF-Tu show their guanine nucleotide binding domains to be remarkably similar [155-161]. Models of G-protein α-chains have been proposed based on these structures and allowed identification of potentially important residues in guanine nucleotide binding and the GTP hydrolysis reaction [162-164] (Fig. 6). Many amino acids crucial to these functions lie in highly conserved domains classified as regions 1–5 (G1–G5) by Bourne [11] and adhere closely to areas identified by Halliday and designated A, C, G and I [165]. Figure 7 depicts schematically conserved domains G1 to G5 and also other regions likely to be generally important in G-protein function. Experimental evidence supporting this model is summarized below.

Figure 6. *Homology model of G$_{\alpha q}$ based on p21$_{ras}$. Three-dimensional model of G$_{\alpha q}$ together with bound GTP based on homology with 1.35 Å resolution coordinates of p21^{ras}* [161]. *Modelling was performed only for G-protein primary sequence aligning with homologous regions in p21^{ras}. Model does not include N- or C-terminal of G$_{\alpha q}$. AMBER was used to parametrize the GTP and magnesium ion and dock them into G$_{\alpha q}$. Homology model was optimized using energy minimization with 5000 steps of steepest descent to obtain low energy structure of approximately –2000 kcal. (Manjula Lusti-Narasimhan, Frank Brown and Tim Wells are acknowledged for providing homology model of G$_{\alpha q}$.)*

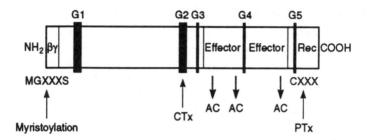

Figure 7. *Schematic representation of generalized G-protein α-chain showing domains likely to be important for interaction with receptor (Rec), effector and βγ dimers. Also indicated are sites for myristoylation and ADP-ribosylation by cholera (CTx) and pertussis toxins (PTx). Shown within effector region are three zones identified within $G_{\alpha s}$ as particularly important for interaction with adenylyl cyclase (AC). Conserved regions probably important for binding guanine nucleotides and GTPase activity are denoted by G1 to G5. Major experimental observations supporting this model are described in text.*

C-terminal is important for receptor and effector interaction

1 Pertussis-toxin-catalysed ADP-ribosylation of some G-protein α-chains at Cys located four residues from the C-terminal blocks interaction with receptor [40–45].

2 Synthetic peptides based on C-terminal sequences (311–328 and 340–350) inhibit rhodopsin association with $G_{\alpha t}$ [168], as well as mimic G-protein in stabilizing active metarhodopsin II conformation. Antibodies raised against this G-protein region or synthetic peptides from C-terminal sequences of $G_{\alpha s}$, $G_{\alpha i2}$, $G_{\alpha o}$, $G_{\alpha q}$ also block receptor interaction [105–107,111,137,166,167].

3 Mutational replacement of G-protein α chain C-terminal residues alters specificity for receptor interaction. Also, $G_{\alpha s}$ point mutation R389P (responsible for the *unc* S49 phenotype) located six residues from C-terminus results in a G-protein uncoupled from receptor [171,172].

4 Studies with a chimeric G-protein ($G_{\alpha i/s(Bam)}$) consisting of the N-terminal 60% of $G_{\alpha i2}$ and the C-terminal 40% of $G_{\alpha s}$ [$G_{\alpha i2}$ (1–212)/$G_{\alpha s}$ (235–394)] indicate that the C-terminal 40% of $G_{\alpha s}$ contains structural determinants required for interaction with receptor and effector [173,174]. Two additional chimeras ($G_{\alpha i2}$ (1–355)/$G_{\alpha s}$ (235-356)) and ($G_{\alpha s}$ (1–356)/$G_{\alpha i2}$ (320–355)) suggest that the site of effector interaction is located within a 121 amino acid core corresponding to Ile235–Arg356 and does not involve the C-terminal 38 residues [175,176]. Within this core, mutagenesis studies have identified crucial residues within three short regions of $G_{\alpha s}$ likely to activate adenylyl cyclase [177,178].

5 Peptides corresponding to residues 293–314 and 300–314 of $G_{\alpha t-1}$ mediate direct activation of cyclicGMP-PDE extracted from bovine retina. Peptides from other regions were inactive [179].

N-terminal is important for membrane anchorage, myristoylation and interaction with βγ.

1 Monoclonal antibody 4A directed against the N-terminal of Gαt causes dissociation from βα and disrupts interaction with rhodopsin [166, 167, 169].

2 Tryptic digestion releases soluble α-chains missing an N-terminal fragment of 1–2 kDa, suggesting an important role in membrane anchorage [124].

3 Proteolytic cleavage of the 20–21 N-terminal residues of Gαt and Gαo or deletion of codons 2–29 from Gαs inhibits α-chain interaction with βγ-subunits and receptor-dependent activation [43,180,181].

4 Mutational studies of the N-terminal 20 residues of Gαo suggests residues 7–10 have an important role in βγ interaction independent of myristoylation [125].

5 Gαo, Gαi1-3 and Gαz undergo co-translational addition of myristic acid chains on Gly2 after cleavage of N-terminal Met. Increased hydrophobicity associated with this modification probably contributes to membrane association of these G-protein α-chains [130].

References.

[1] Rodbell, M. et al. (1971) J. Biol. Chem. 246, 1877–1992.
[2] Birnbaumer, L. et al. (1990) Biochim. Biophys. Acta 1031, 163–224.
[3] Gilman, A. G. (1987) Annu. Rev. Biochem. 56, 615–649.
[4] Neer, E. J. and Clapham, D. E. (1988) Nature 333, 129–134.
[5] Stryer, L. and Bourne, H. R. (1986) Annu. Rev. Cell Biol. 2, 391–419.
[6] Taylor, C. W. (1990) Biochem. J. 272, 1–13.
[7] Ross, E. M. (1989) Neuron 3, 141–152.
[8] Stryer, L. (1991) J. Biol. Chem. 266, 10711–10714.
[9] Simon, M. I. et al. (1991) Science 252, 802–808.
[10] Kaziro, Y. et al. (1991) Annu. Rev. Biochem. 60, 349–400.
[11] Bourne, H. R. et al. (1991) Nature 349, 117–127.
[12] Helper, J. R. and Gilman, A. G. (1992) Trends Biochem. Sci. 17, 383–387.
[13] DeLean, A. et al. (1980) J. Biol. Chem. 255, 7108–7117.
[14] Kent, R. S. e. al. (1980) Mol. Pharmacol. 17, 14–23.
[15] Codina, J. et al. (1984) J. Biol. Chem. 259, 11408–11418.
[16] Levitzki, A. (1987) FEBS Lett. 211, 113–118.
[17] Marbach, I. et al. (1990) J. Biol. Chem. 265, 9999–10004.
[18] Phillips, W. J. and Cerione, R. A. (1991) J. Biol. Chem. 266, 11017–11024.
[19] Kohnken, R. E. and Hildebrandt, J. D. (1989) J. Biol. Chem. 264, 20688–20696.
[20] Heithier, H. et al. (1992) Eur. J. Biochem. 204, 1169–1181.
[21] Brandt, D. R. and Ross, E. M. (1986) J. Biol. Chem. 261, 1656–1664.
[22] Higashijima, T. et al. (1987) J. Biol. Chem. 262, 757–761.
[23] Linder, M. E. et al. (1990) J. Biol. Chem. 265, 8243–8251.
[24] Graziano. M. P. et al. (1989) J. Biol. Chem. 264, 409–418.
[25] Bernstein, G. et al. (1992) J. Biol. Chem. 267, 8081–8088.
[26] Casey, P. J. et al. (1990) J. Biol. Chem. 265, 2383–2390.
[27] Fung, B. K.-K. et al. (1981) Proc. Natl Acad. Sci. USA 78, 152–156.
[28] Vuong, T. M. (1984) Nature 311, 659–661.
[29] Levis, M. J. and Bourne, H. R. (1992) J. Cell Biol. 119, 1297–1307.
[30] Arshavsky, V. Y. and Bownds, M. D. (1992) Nature 357, 416–417.

[31] Bernstein, G. et al. (1992) Cell 70, 411–418.

[32] Pages, F. et al. (1992) J. Biol. Chem. 267, 22018–22021.

[33] Bigay, J. et al. (1985) FEBS Lett. 191, 181–185.

[34] Antonny, B. and Chabre, M. (1992) J. Biol. Chem. 267, 6710–6718.

[35] Van Dop, C. et al. (1984) J. Biol. Chem. 259, 696–698.

[36] Kahn, R. A. and Gilman, A. G. (1984) J. Biol. Chem. 259, 6235–6240.

[37] Gierschik, P. et al. (1989) J. Biol. Chem. 264, 21470–21473.

[38] **Milligan, G. (1988) Biochem. J. 255, 1–13.**

[39] Roerig, S. C. et al. (1992) Mol. Pharmacol. 41, 822–831.

[40] Huff, R. M. and Neer, E. J. (1986) J. Biol. Chem. 261, 1105–1110.

[41] West, R. E. et al. (1985) J. Biol. Chem. 260, 14428–14430.

[42] Watkins, P. A. et al. (1985) J. Biol. Chem. 260, 13478–13482.

[43] Navon, S. E. and Fung, B.K.-K. (1987) J. Biol. Chem. 262, 15746–15751.

[44] Van Dop, C. et al. (1984) J. Biol. Chem. 259, 23–26.

[45] Okajima, F. et al. (1985) J. Biol. Chem. 260, 6761–6768.

[46] Florio, V. A. and Sternweis, P. C. (1985) J. Biol. Chem. 260, 3477–3483.

[47] Kanaho, Y. et al. (1984) J. Biol. Chem. 259, 7378–7381.

[48] Fung, B. K.-K. et al. (1983) J. Biol. Chem. 258, 10495–10502.

[49] Rubenstein, R. C. et al. (1991) Biochemistry 30, 10769–10777.

[50] Higashijima, T. et al. (1987) J. Biol. Chem. 262, 762–766.

[51] Phillips, W. J. et al. (1992) J. Biol. Chem. 267, 17040–17046.

[52] Cerione, R. A. et al. (1986) J. Biol. Chem. 261, 9514–9520.

[53] Florio, V. A. and Sternweis, P. C. (1985) J. Biol. Chem. 260, 3477–3483.

[54] Sternweis, P. C. (1986) J. Biol. Chem. 261, 631–637.

[55] Correze, C. et al. (1987) J. Biol. Chem. 262, 15182–15187.

[56] Haga, K. and Haga, T. (1992) J. Biol. Chem. 267, 2222–2227.

[57] Pitcher, J. A. et al. (1992) Science 257, 1264–1267.

[58] Logothetis, D. E. et al. (1987) Nature 325, 321–326.

[59] Jelsema, C. L. and Axelrod, J. (1987) Proc. Natl Acad. Sci. USA 84, 3623–3627.

[60] Ito, H. et al. (1992) J. Gen. Physiol. 99, 961–983.

[61] Kim, D. et al. (1989) Nature 337, 557–560.

[62] Yatani, A. et al. (1988) Nature 336, 680–682.

[63] Camps, M. et al. (1992) Eur. J. Biochem. 206, 821–831.

[64] Blank, J. L. et al. (1992) J. Biol. Chem. 267, 23069–23075.

[65] Park, D. et al. (1993) J. Biol. Chem. 268, 4573–4576.

[66] Camps, M. et al. (1992) Nature 360, 684–686.

[67] Katz, K. et al. (1992) Nature 360, 686–689.

[68] Carozzi, A. et al. (1993) FEBS Letts. 315, 340–342.

[69] Katada, T. et al. (1984) J. Biol. Chem. 259, 3568–3577.

[70] Katada, T. et al. (1984) J. Biol. Chem. 259, 3578–3585.

[71] Hildebrandt, J. D. and Kohnken, R. E. (1990) J. Biol. Chem. 265, 9825–9830.

[72] Katada, T. et al. (1984) J. Biol. Chem. 259, 3586–3595.

[73] Wong, Y. H. et al. (1991) Nature 351, 63–65.

[74] Wong, Y. H. et al. (1992) Science 255, 339–342.

[75] Hermouet, S. et al. (1991) Proc. Natl Acad. Sci. USA 88, 10455–10459.

[76] Lowndes, J. M. et al. (1991) J. Biol. Chem. 266, 14193–14197.

[77] Tang, W.-J. and Gilman, A. G. (1992) Cell 70, 869–872.

[78] Tang, W.-J. and Gilman, A. G. (1991) Science 254, 1500–1503.

[79] Federman, A. D. et al. (1992) Nature 356, 159–161.
[80] Gao, B. and Gilman, A. G. (1991) Proc. Natl Acad. Sci. USA 88, 10178–10182.
[81] Dietzel, C. and Kurjan, J. (1987) Cell 50, 1001–1010.
[82] Miyajima, I. et al. (1987) Cell 50, 1011–1019.
[83] Whiteway, M. et al. (1989) Cell 56, 467–477.
[84] Asano, T. et al. (1984) J. Biol. Chem. 259, 9351–9354.
[85] Florio, V. A. and Sternweis, P. C. (1985) J. Biol. Chem. 260, 3477–3483.
[86] Haga, K. et al. (1986) J. Biol. Chem. 261, 10133–10140.
[87] Kurose, H. et al. (1986) J. Biol. Chem. 261, 6423–6428.
[88] Kim, M. H. and Neubig, R. R. (1987) Biochemistry 26, 3664–3672.
[89] Senogles, S. E. et al. (1990) J. Biol. Chem. 265, 4507–4514.
[90] Freissmuth, M. et al. (1991) J. Biol. Chem. 266, 17778–17783.
[91] Kurose, H. et al. (1991) Biochemistry 30, 3335–3341.
[92] Parker, E. M. et al. (1991) J. Biol. Chem. 266, 519–527.
[93] Bertin, B. et al. (1992) J. Biol. Chem. 267, 8200–8206.
[94] Vallar, L. et al. (1990) J. Biol. Chem. 265, 10320–10326.
[95] Fang Liu, Y. and Albert, P. R. (1991) J. Biol. Chem. 266, 23689–23697.
[96] Duzic, E. and Lanier, S. M. (1992) J. Biol. Chem. 267, 24045–24052.
[97] Volpp, B. D. et al. (1989) J. Immunol. 142, 3206–3212.
[98] Ali, N. et al. (1989) Biochem. J. 261, 905–912.
[99] Ercolani, L. et al. (1990) Proc. Natl Acad. Sci. USA 87, 4635–4639.
[100] Rodbell, M. (1992) Curr. Top. Cell. 32, 1–47.
[101] Lewis, J. M. et al. (1991) Cell Regulation 2, 1097–1113.
[102] Takei. Y. et al. (1992) J. Biol. Chem. 267, 5085–5089.
[103] Vogel, S. S. et al. (1990) Proc. Natl Acad. Sci. USA 88, 1775–1778.
[104] Lledo, P. M. et al. (1992) Neuron 8, 455–463.
[105] McKenzie, F. R. and Milligan, G. (1990) Biochem. J. 267, 391–398.
[106] McClue, S. J. and Milligan, G. (1990) FEBS Lett. 269, 430–434.
[107] McFadzean, I. et al. (1989) Neuron 3, 177–182.
[108] Laugwitz, K.-L. et al. (1993) Neuron 10, 233–242.
[109] Simonds, W. F. et al. (1989) Proc. Natl Acad. Sci. USA 86, 7809–7813.
[110] Gutowski, S. et al. (1991) J. Biol. Chem. 266, 20519–20524.
[111] Shenker, A. et al. (1991) J. Biol. Chem. 266, 9309–9313.
[112] Law, S. F. et al. (1991) J. Biol. Chem. 266, 17885–17897.
[113] Murray-Whelan, R. and Schlegel,W. (1992) J. Biol. Chem. 267, 2960–2965.
[114] Law, S. F. and Reisine, T. (1992) Mol. Pharmacol. 42, 398–402.
[115] Luthin, D. R. et al. (1993) J. Biol. Chem. 268, 5990–5996.
[116] Matesic, D. F. et al. (1991) Mol. Pharmacol. 40, 347–353.
[117] Roerig, S. C. et al. (1992) Mol. Pharmacol. 41, 822–831.
[118] Iiri, T. et al. (1992) J. Biol. Chem. 267, 1020–1026.
[119] Aragay, A. M. et al. (1992) J. Biol. Chem. 267, 24983–24988.
[120] Kleuss, C. et al. (1991) Nature 353, 43–48.
[121] Kleuss, C. et al. (1992) Nature 358, 424–426.
[122] Kleuss, C. et al. (1993) Science 259, 832–834.
[123] Watkins, D. C. et al. (1992) Science 258, 1373–1375.
[124] Eide, B. et al. (1987) Biochem. Biophys. Res. Commun. 148, 1398–1405.
[125] Denker, B. M. et al. (1992) J. Biol. Chem. 267, 6272–6277.
[126] Buss J. E. et al. (1987) Proc. Natl Acad. Sci. USA 84, 7493–7497.

127 Schultz, A. M. et al. (1987) Biochem. Biophys. Res. Commun. 146, 1234–1239.
128 Jones, T. L. Z. et al. (1990) Proc. Natl Acad. Sci. USA 87, 568–572.
129 Mumby, S. M. et al. (1990) Proc. Natl Acad. Sci. USA 87, 728–732
130 Spiegel, A. M. et al. (1991) Trends Biochem. Sci. 16, 338–341.
131 Linder, M. E. et al. (1991) J. Biol. Chem. 266, 4654–4659.
132 Kokame, K. et al. (1992) Nature 359, 749–752.
133 Neubert, T. A. et al. (1992) J. Biol. Chem. 267, 18274–18277.
134 Yang, Z. and Wensel, T. G. (1992) J. Biol. Chem. 267, 23197–23201.
135 Towler, D. A. et al. (1988) J. Biol. Chem. 263, 1784–1790.
136 Pang, I.-H. and Sternweis, P. C. (1990) J. Biol. Chem. 265, 18707–18712.
137 Gutowski, S. et al. (1991) J. Biol. Chem. 266, 20519–20524.
138 Wu, D. et al. (1992) J. Biol. Chem. 267, 1811–1817.
139 Audigier, Y. et al. (1990) J. Cell Biol. 111, 1427–1435.
140 Journot, L. et al. (1991) Proc. Natl Acad. Sci. USA 88, 10054–10058.
141 Yamane, H. K. et al. (1990) Proc. Natl Acad. Sci. USA 87, 5868–5872.
142 Fukada, Y. et al. (1990) Nature 346, 658–660.
143 Backlund, P. S. et al. (1990) J. Biol. Chem. 265, 15572–15576.
144 Mumby, S. M. et al. (1990) Proc. Natl Acad. Sci. USA 87, 5873–5877.
145 Fung, B. K.-K. et al. (1990) FEBS Lett. 260, 313–317.
146 Lai, R. K. et al. (1990) Proc. Natl. Acad. Sci. USA 87, 7673–7677.
147 Simonds, W. F. et al. (1991) J. Biol. Chem. 266, 5363–5366.
148 Muntz, K. H. et al. (1992) Mol. Biol. Cell 3, 49–61.
149 Iniguez-Lluhi, J. A. et al. (1992) J. Biol. Chem. 267, 23409–23417.
150 Ohguro, H. et al. (1991) EMBO J. 10, 3669–3674.
151 Yokoyama, K. et al. (1991) Proc. Natl Acad. Sci. USA 88, 5302–5306.
152 Casey, P. J. et al. (1991) Proc. Natl Acad. Sci. USA 88, 8631–8635.
153 Moores, S. L. et al. (1991) J. Biol. Chem. 266, 14603–14610.
154 Reiss, Y. et al. (1991) Proc. Natl Acad. Sci. USA 88, 732–736.
155 Jurnak, F. (1985) Science 230, 32–36.
156 La Cour, T. F. M. et al. (1985) EMBO J. 4, 2385–2388.
157 Pai, E. F. et al. (1989) Nature 341, 209–214.
158 Schlichting, I. et al. (1990) Nature 345, 309–315.
159 Milburn, M. V. et al. (1990) Science 247, 939–945.
160 Brunger, A. T. et al. (1990) Proc. Natl Acad. Sci. USA 87, 4849–4853.
161 Pai, E. F. et al. (1990) EMBO J. 9, 2351–2359.
162 Masters, S. B. et al. (1986) Protein Eng. 1, 47–54.
163 Deretic, D. and Hamm, H. E. (1987) J. Biol. Chem. 262, 10839–10847.
164 Holbrook, S. R. and Kim, S.-H. (1989) Proc. Natl Acad. Sci. USA 86, 1751–1755.
165 Halliday, K. R. (1984) Cyclic Nucl. Prot. Phos. Res. 9, 435–448.
166 Hamm, H. E. et al. (1987) J. Biol. Chem. 262, 10831–10838.
167 Mazzoni, M.R. et al. (1991) J. Biol. Chem. 266, 14072–14081.
168 Hamm, H. E. et al. (1988) Science 241, 832–835.
169 Mazzoni, M. R. and Hamm, H. E. (1989) Biochemistry 28, 9873–9880.
170 Palm, D. et al. (1990) FEBS Lett. 261, 294–298.
171 Sullivan, K. A. et al. (1987) Nature 330, 758–760.
172 Rall, T. and Harris, B. A. (1987) FEBS Lett. 224, 365–371.
173 Masters, S. B. et al. (1988) Science 241, 448–451.

174 Osawa, S. et al. (1990) Mol. Cell. Biol. 10, 2931–2940.
175 Woon, C. W. et al. (1989) J. Biol. Chem. 264, 5687–5693.
176 Osawa, S. et al. (1990) Cell 63, 697–706.
177 Berlot, C. H. and Bourne, H. R. (1992) Cell 68, 911–922.
178 Itoh, H. and Gilman, A. G. (1991) J. Biol. Chem. 266, 16226–16231.
179 Rarick, H. M. et al. (1992) Science 256, 1031–1033.
180 Neer, E. J. et al. (1988) J. Biol. Chem. 263, 8996–9000.
181 Journot, L. et al. (1991) J. Biol. Chem. 266, 9009–9015.
182 Parenti, M. et al. (1993) Biochem. J. 291, 349–353.
183 Moxham, C.W. et al. (1993) Science 260, 991–995.

G-protein α-chains

GαS

Molecular Weights

Polypeptide   44 189 (Gαs-S-A); 44 294 (Gαs-S-B)
                           45 664 (Gαs-L-A); 45 769 (Gαs-L-B)

Reduced SDS PAGE   42–52 kDa

Structure

Four molecularly distinct forms of Gαs resulting from differential splicing of a single gene have been identified [1,2]. The Gαs gene contains 13 exons and while all become incorporated into the long forms (Gαs-L), exon 3 encoding 15 amino acids is excluded from the short form (Gαs-S). Additionally, in each case two alternate splice sites are used at the 5'-end of exon 4 giving rise to mRNAs that differ in the presence or absence of three nucleotides encoding Ser at the beginning of exon 4. The two predicted forms of Gαs-S consist of 379 (Gαs-S-A) and 380 amino acids (Gαs-S-B) with Asp71-Glu72 and Asp71-Ser72 at the splice junction, respectively. The long forms consist of 394 (Gαs-L-A) and 395 amino acids (Gαs-L-B), having a 15 amino acids insert in the place of Asp71 (starting Glu) and either Gly86-Glu87 or Gly86-Ser87 at the exon 3-exon 4 splice junction. These splice variants probably account for the diverse molecular weight forms of Gαs.

Gαs Gene structure (human) [1]

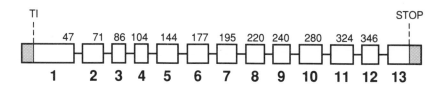

Organization of human Gαs gene adapted from ref. 1. The amino acid numbers correspond to Gαs-L-A.

Differential splicing of G$_{\alpha s}$ gene

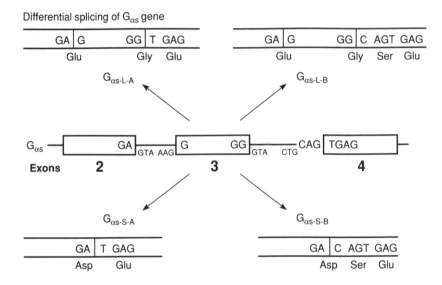

Differential splicing of G$_{\alpha s}$ gene

G$_{\alpha s\text{-L-A}}$ (394): Exons 2, 3 and 4 spliced together.
G$_{\alpha s\text{-L-B}}$ (395): Exons 2, 3 and 4 spliced together using alternative alternative TG splice site at 5'-end of exon 4.
G$_{\alpha s\text{-S-A}}$ (379): Exon 2 spliced to exon 4.
G$_{\alpha s\text{-S-B}}$ (380): Exon 2 spliced to exon 4 using alternative TG splice site at 5'-end of exon 4.

Model for alternative splicing of human G$_{\alpha s}$ gene adapted from ref. 1. The G$_{\alpha s}$ gene is shown in the centre.

Human gene location and size
Chromosome 12, >20 kb [1,3].

Tissue distribution

Ubiquitous.

Function

G$_{\alpha s}$ is firmly established as the G-protein mediating stimulation of adenylyl cyclase [4–10] although it can also activate voltage-gated calcium channels in cardiac and skeletal muscle [8,11–15]. Alternatively spliced forms appear similarly active [8–10].

Comments
In vitro phosphorylation by c-*src*, PKC and PKA has been reported [16–18].
   Oncogenic *gsp* mutations of G$_{\alpha s}$ have been identified in pituitary and thyroid tumours where somatic mutation has replaced Arg201 with Cys or His (R201C or

R201H) or Gln227 with Arg (Q227R) (numbering according to G$_{αs-L-A}$) [19,20]. These and other substitutions at identical positions are within domains conserved in all G-proteins (G-2 and G-3 regions) [21] and cause suppressed GTPase activity leading to constitutive G$_{αs}$ activation [19,20, 22–24]. An identical point mutation within G$_{αs}$ (R201H) has been identified as a somatic mutation underlying McCune-Albright syndrome [25].

Amino acid sequence

```
  1 MGCLGNSKTE DQRNEEKAQR EANKKIEKQL QKDKQVYRAT HRLLLLGAGE

                       |-Exon 3 Splice Variation-|
 51 SGKSTIVKQM RILHVNGFNG ᴰEGGEEDPQAARSNSDG[S]EKA TKVQDIKNNL

101 KEAIETIVAA MSNLVPPVEL ANPENQFRVD YILSVMNVPD FDFPPEFYEH

151 AKALWEDEGV RACYERSNEY QLIDCAQYFL DKIDVIKQAD YVPSDQDLLR

    CTx
201 CRVLTSGIFE TKFQVDKVNF HMFDVGGQRD ERRKWIQCFN DVTAIIFVVA

251 SSSYNMVIRE DNQTNRLQEA LNLFKSIWNN RWLRTISVIL FLNKQDLLAE

301 KVLAGKSKIE DYFPEFARYT TPEDATPEPG EDPRVTRAKY FIRDEFLRIS

351 TASGDGRHYC YPHFTCAVDT ENIRRVFNDC RDIIQRMHLR QYELL
```

Amino acids	395
CTx	Site of ADP-ribosylation by cholera toxin.

Database accession numbers

	PIR	SWISSPROT	EMBL/GENBANK	REFERENCES
Human	B24366	P04895	J03647	1,2,26,27
	A24366		X04408	
	A31927		X07036	
	S02122		M14631	
	A25919		M21139–42	
			M21740	
			M21741	
Rat	A27423	P04894	M12673	28,29
	C24882		M17525	

	PIR	SWISSPROT	EMBL/GENBANK	REFERENCES
Bovine	A23813	P04896	M13006	*30–33*
	A23615		M14014	
			X03404	
Hamster	S08140		X17481	*34*
	S10508		X53139	
Mouse	A25889	P08755	M13964	*35,36*
	S03075		Y00703	
Pig	S18963		X63893	

References

1 Kozasa, T. et al. (1988) Proc. Natl Acad. Sci. USA 85, 2081–2085.
2 Bray, P. et al. (1986) Proc. Natl Acad. Sci. USA 83, 8893–8897.
3 Blatt, C. et al. (1988) Proc. Natl Acad. Sci. USA 85, 7642–7646.
4 Northup, J. K. et al. (1980) Proc. Natl Acad. Sci. USA 77, 6516–6520.
5 Sternweis, P. C. et al. (1981) J. Biol. Chem. 256, 11517–11526.
6 Codina, J. et al. (1984) J. Biol. Chem. 259, 5871–5886.
7 May, D. C. et al. (1985) J. Biol. Chem. 260, 15829–15833.
8 Mattera, R. et al. (1989) Science 243, 804–807.
9 Graziano, M. P. et al. (1989) J. Biol. Chem. 264, 409–418.
10 Graziano, M. P. et al. (1987) J. Biol. Chem. 262, 11375–11381.
11 Yatani, A. et al. (1987) Science 238, 1288–1292.
12 Yatani, A. et al. (1988) J. Biol. Chem. 263, 9887–9895.
13 Yatani, A. et al. (1990) Science 249, 1163–1166.
14 Hamilton, S. L. et al. (1991) J. Biol. Chem. 266, 19528–19535.
15 Yatani, A. et al. (1992) Science 245, 71–74.
16 Hausdorff, W. P. et al. (1992) Proc. Natl Acad. Sci. USA 89, 5720–5724.
17 Pyne, N. J. et al. (1992) Biochem. J. 285, 333–338.
18 Pyne, N. J. et al. (1992) Biochem. Biophys. Res. Commun. 186, 1081–1086.
19 Landis, C. A. et al. (1989) Nature 340, 692–696.
20 Lyons, J. et al. (1990) Science 249, 655–659.
21 **Bourne, H. R. et al. (1991) Nature 349, 117–127.**
22 Freissmuth, M. and Gilman, A. G. (1989) J. Biol. Chem. 264, 21907–21914.
23 Masters, S. B. et al. (1989) J. Biol. Chem. 264, 15467–15474.
24 Graziano, M. P. and Gilman, A. G. (1989) J. Biol. Chem. 264, 15475–15482.
25 Schwindinger, W. F. et al. (1992) Proc. Natl Acad. Sci. USA 89, 5152–5156.
26 Harris, B. A. (1988) Nucleic Acids Res. 16, 3585.
27 Mattera, R. et al. (1986) FEBS Lett. 206, 36–42.
28 Jones, D. T. and Reed, R. R. (1987) J. Biol. Chem. 262, 14241–14249.
29 Itoh, H. et al. (1986) Proc. Natl Acad. Sci. USA 83, 3776–3780.
30 Nukada, T. et al. (1986) FEBS Lett. 195, 220–224.
31 Harris, B. A. et al. (1985) Science 229, 1274–1277.
32 Robishaw, J. D. et al. (1986) J. Biol. Chem. 261, 9587–9590.
33 Robishaw, J. D. et al. (1986) Proc. Natl Acad. Sci. USA 83, 1251–1255.
34 Mercken, L. et al. (1990) Nucleic Acids Res. 18, 662.
35 Rall, T. and Harris, B. A. (1987) FEBS Lett. 224, 365-371.
36 Sullivan, K. A. et al. (1986) Proc. Natl Acad. Sci. USA 83, 6687-6691.

Gαolf

Molecular weight

Polypeptide	44 308
Reduced SDS PAGE	45 kD

Tissue distribution

Gαolf is enriched in apical regions of olfactory neuroepithelium [1]. It is also expressed in CNS particularly striatum, substantia nigra, caudate putamen, nucleus accumbens , olfactory bulb and olfactory tubercle [8, 10].

Function

Gαolf can activate adenylyl cyclase [1,2]. Co-localization in olfactory neurons with large family of odorant receptors [3] and adenylyl cyclase type III [4,5] could indicate a role in odorant sensory signal transduction [6]. Gαolf is also co-localized with D1 receptors in striatonigral neurones and with adenylyl cyclase type V in corpus striatum, suggesting function mediating dopaminergic signalling [8,9].

Amino acid sequence

```
  1 MGCLGGNSKT AEDQGVDEKE RREANKKIEK QLQKERLAYK ATHRLLLLGA

 51 GESGKSTIVK QMRILHVNGF NPEEKKQKIL DIRKNVKDAI VTIVSAMSTI

101 IPPVPLANPE NQFRSDYIKS IAPITDFEYS QEFFDHVKKL WDDEGVKACF
                                          CTx
151 ERSNEYQLID CAQYFLERID SVSLVDYTPT DQDLLRCRVL TSGIFETRFQ

201 VDKVNFHMFD VGGQRDERRK WIQCFNDVTA IIYVAACSSY NMVIREDNNT

251 NRLRESLDLF ESIWNNRWLR TISIILFLNK QDMLAEKVLA GKSKIEDYFP

301 EYANYTVPED ATPDAGEDPK VTRAKFFIRD LFLRISTATG DGKHYCYPHF

351 TCAVDTENIR RVFNDCRDII QRMHLKQYEL L
```

Amino acids	381
CTx	Site of ADP-ribosylation by cholera toxin.

Database accession numbers

	PIR	*SWISSPROT*	*EMBL/GENBANK*	*REFERENCES*
Human			L10665	
Rat				*1*
Mouse	A33833		M57635	*7*
(partial)			M26743	

References

1 Jones, D. T. and Reed, R. R. (1989) Science 244, 790–795.
2 Jones, D. T. et al. (1990) J. Biol. Chem. 265, 2671–2676.
3 Buck, L. and Axel, R. (1991) Cell 65, 175–187
4 Bakalyar, H. A. and Reed, R. R. (1990) Science 250, 1403–1406.
5 Menco, B. P. M. et al. (1992) Neuron 8, 441–453.
6 Reed, R. R. (1992) Neuron 8, 205–209.
7 Strathmann et al. (1989) Proc. Natl Acad. Sci. USA 86, 7407–7409.
8 Drinnan, S. L. et al. (1991) Mol. Cell. Neurosci. 2, 66–70.
9 Glatt, C. E. and Snyder, S. H. (1993) Nature 361, 536–538.

Gαi1

Molecular weight

Polypeptide	40 345
Reduced SDS PAGE	40–41 kDa

Gαi1 Gene structure (human)

Full exon-intron organization is unavailable, although partial sequence splice sites in exons 1–3 are identical to Gαi2 and Gαi3 [1-3].

Human gene location

Chromosome 7 [4].

Tissue distribution

Wide distribution although highly expressed in central nervous system and neuronal cell lines [5-7].

Function

Unclear (see comments).

Comments

Sensitivity to pertussis toxin as well as reconstitution and heterologous expression suggest several potential roles including adenylyl cyclase inhibition [8-10], PLC-dependent phosphoinositide hydrolysis [11,12], potassium channel regulation [10,13] and inhibition of calcium channel opening [14,15].

Amino acid sequence

```
Mys
  1 MGCTLSAEDK AAVERSKMID RNLREDGEKA AREVKLLLLG AGESGKSTIV

 51 KQMKIIHEAG YSEEECKQYK AVVYSNTIQS IIAIIRAMGR LKIDFGDSAR

101 ADDARQLFVL AGAAEEGFMT AELAGVIKRL WKDSGVQACF NRSREYQLND
```

```
151 SAAYYLNDLD RIAQPNYIPT QQDVLRTRVK TTGIVETHFT FKDLHFKMFD

201 VGGQRSERKK WIHCFEGVTA IIFCVALSDY DLVLAEDEEM NRMHESMKLF

251 DSICNNKWFT DTSIILFLNK KDLFEEKIKK SPLTICYPEY AGSNTYEEAA

301 AYIQCQFEDL NKRKDTKEIY THFTCATDTK NVQFVFDAVT DVIIKNNLKD
    PTx
351 CGLF
```

Amino acids	354
Mys	Site of myristoylation.
PTx	Site of ADP-ribosylation by pertussis toxin.

Database accession numbers

	PIR	SWISSPROT	EMBL/GENBANK	REFERENCES
Human	A28318		M17219	16
Bovine	A23631	P04898	X03642	17
Rat	C27423	P10824	M17527	5

References

1 Itoh, H. et al. (1988) J. Biol. Chem. 263, 6656–6664.
2 Weinstein, L. S. et al. (1988) FEBS Letts. 232, 333–340.
3 **Kaziro, Y. et al. (1991) Annu. Rev. Biochem. 60, 349–400.**
4 Blatt, C. et al. (1988) Proc. Natl Acad. Sci. USA 85, 7642–7646.
5 Jones, D. T. and Reed, R. R. (1987) J. Biol. Chem. 262, 14241–14249.
6 Kim, S. et al. (1988) Proc. Natl Acad. Sci. USA 85, 4153–4157.
7 Brann, M. R. et al. (1988) FEBS Lett. 222, 191–198.
8 Wong, Y. H. et al. (1992) Science 255, 339–342.
9 Katada, T. et al. (1984) J. Biol. Chem. 259, 3578–3585.
10 Kobayashi, I. et al. (1990) Eur. J. Biochem. 191, 499–506.
11 Brandt, S. J. et al. (1985) Proc. Natl Acad. Sci. USA 82, 3277–3280.
12 Ohta, H. et al. (1985) J. Biol. Chem. 260, 15771–15780.
13 Yatani, A. et al. (1988) Nature 336, 680–682.
14 Ewald, D. A. et al. (1989) Neuron 2, 1185–1193.
15 Linder, M. E. et al. (1990) J. Biol. Chem. 265, 8243–8251.
16 Bray, P. et al. (1987) Proc. Natl Acad. Sci. USA 84, 5115–5119.
17 Nukada, T. et al. (1986) FEBS Lett.197, 305–310.

Gαi2

Molecular weight

Polypeptide	40 451
Reduced SDS PAGE	40–41 kD

Gene structure (human) [1,2]

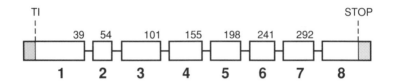

Organization of human Gαi2 gene adapted from refs 1 and 2.

Human gene location
Chromosome 3 ~ 20 kb [1, 3].

Tissue distribution

Expressed widely [4–8].

Function

Studies with site-directed G-protein antibodies and antisense oligonucleotides suggest a specific role mediating adenylyl cyclase inhibition in some cell types [15–18]. Further potential roles are unclear (see Comments).

Comments
Pertussis toxin sensitivity and functional reconstitution suggest several potential regulatory roles including PLC-dependent phosphoinositide hydrolysis [9,10], potassium channel activation [11,12] and regulation of calcium channel opening [13,14].

Gαi2 is reported to be phosphorylated by PKC [19,20].

Somatic mutation of Arg179 with His or Cys (R179H, R179C) results in an oncogenic form of Gαi2 identified in tumours of human adrenal cortex and ovary [21]. These mutations as well as the Q205L substitution result in *gip2* oncogenes which constitutively inhibit adenylyl cyclase as well as increase mitogen-activated protein (MAP) kinase activity and cause neoplastic transformation upon heterologous expression in fibroblasts [22–25].

Amino acid sequence

```
     Mys
   1 MGCTVSAEDK AAAERSKMID KNLREDGEKA AREVKLLLLG AGESGKSTIV

  51 KQMKIIHEDG YSEEECRQYR AVVYSNTIQS IMAIVKAMGN LQIDFADPSR

 101 ADDARQLFAL SCTAEEQGVL PDDLSGVIRR LWADHGVQAC FGRSREYQLN

 151 DSAAYYLNDL ERIAQSDYIP TQQDVLRTRV KTTGIVETHF TFKDLHFKMF
```

```
201 DVGGQRSERK KWIHCFEGVT AIIFCVALSA YDLVLAEDEE MNRMHESMKL

251 FDSICNNKWF TDTSIILFLN KKDLFEEKIT HSPLTICFPE YTGANKYDEA

301 ASYIQSKFED LNKRKDTKEI YTHFTCATDT KNVQFVFDAV TDVIIKNNLK
    PTx
351 DCGLF
```

Amino acids	355
Mys	Site of myristoylation.
PTx	Site of ADP-ribosylation by pertussis toxin.

Database accession numbers

	PIR	SWISSPROT	EMBL/GENBANK	REFERENCES
Human	S00618	P04899	X04828	*1,2,5,26,27*
	S02319		X07854	
	S02320		X07855	
	A29025		M20586-8	
	B28154		J03004	
Rat	B24882	P04897	M12672	*4,28*
	D27423		M17528	
Mouse	B25889	P08752	M13963	*29*
Bovine	A25888	P11015	M14207	*19*

References

1 Itoh, H. et al. (1988) J. Biol. Chem. 263, 6656–6664.
2 Weinstein, L. S. et al. (1988) FEBS Lett. 232, 333–340.
3 Blatt, C. et al. (1988) Proc. Natl Acad. Sci. USA 85, 7642–7646.
4 Jones, D. T. and Reed, R. R. (1987) J. Biol. Chem. 262, 14241–14249.
5 Beals, C. R. et al. (1987) Proc. Natl Acad. Sci. USA 84, 7886–7890.
6 Suki, W. N. et al. (1987) FEBS Lett. 220, 187–192.
7 Kim, S. et al. (1988) Proc. Natl Acad. Sci. USA 85, 4153–4157.
8 Brann, M. R. et al. (1988) FEBS Lett. 222, 191–198.
9 Brandt, S. J. et al. (1985) Proc. Natl Acad. Sci. USA 82, 3277–3280.
10 Ohta, H. et al. (1985) J. Biol. Chem. 260, 15771–15780.
11 Yatani, A. et al. (1988) Nature 336, 680–682.
12 Kobayashi, I. et al. (1990) Eur. J. Biochem. 191, 499–506.
13 Ewald, D. A. et al. (1989) Neuron 2, 1185–1193.
14 Linder, M. E. et al. (1990) J. Biol. Chem. 265, 8243–8251.
15 Simonds, W. F. et al. (1989) Proc. Natl Acad. Sci. USA 86, 7809–7813.
16 McKenzie, F. R. and Milligan, G. (1990) Biochem. J. 267, 391–398.
17 McClue, S. J. et al. (1992) Biochem. J. 284, 565–568.
18 Watkins, D. C. et al. (1992) Science 258, 1373–1375.
19 Yatomi, Y. et al. (1992) Eur. J. Biochem. 205, 1003–1009.
20 Housley, M. D. (1991) Cell. Signal. 3, 1–9.
21 J. Lyons et al. (1990) Science 249, 655–659.
22 Y. H. Wong et al. (1991) Nature 351, 63–65.
23 A. M. Pace et al. (1991) Proc. Natl Acad. Sci. USA 88, 7031–7035,

24 S. K. Gupta et al. (1992) Mol. Cell. Biol. 12, 190–197.
25 S. K. Gupta et al. (1992) J. Biol. Chem. 267, 7987–7990.
26 Michel, T. et al. (1986) Proc. Natl Acad. Sci. USA 83, 7663–7667.
27 Didsbury, J. R. et al. (1987) FEBS Lett. 211, 160–164.
28 Itoh, H. et al. (1986) Proc. Natl Acad. Sci. USA 83, 3776–3780.
29 Sullivan, K. A. et al. (1986) Proc. Natl Acad. Sci. USA 83, 6687–6691.

G$_{\alpha i3}$

Molecular weight

Polypeptide	40 532
Reduced SDS PAGE	40–41 kDa

Gene structure (human) [1]

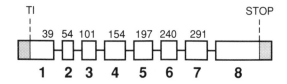

Organization of human G$_{\alpha i3}$ gene adapted from ref. 1.

Human gene location
Chromosome 1 [2].

Tissue distribution

Expressed widely [3–6].

Function

Unclear (see comments).

Comments
Pertussis toxin sensitivity as well as reconstitution and heterologous expression suggest several potential roles including adenylyl cyclase inhibition [7–9], PLC-dependent phosphoinositide hydrolysis [10,11], regulation of calcium channel opening [12] as well as sodium [13] and potassium channel activation [9,14–18].

Amino acid sequence

Mys

```
  1 MGCTLSAEDK AAVERSKMID RNLREDGEKA AKEVKLLLLG AGESGKSTIV

 51 KQMKIIHEDG YSEDECKQYK VVVYSNTIQS IIAIIRAMGR LKIDFGEAAR

101 ADDARQLFVL AGSAEEGVMT PELAGVIKRL WRDGGVQACF SRSREYQLND

151 SASYYLNDLD RISQSNYIPT QQDVLRTRVK TTGIVETHFT FKDLYFKMFD

201 VGGQRSERKK WIHCFEGVTA IIFCVALSDY DLVLAEDEEM NRMHESMKLF

251 DSICNNKWFT ETSIILFLNK KDLFEEKIKR SPLTICYPEY TGSNTYEEAA

301 AYIQCQFEDL NRRKDTKEIY THFTCATDTK NVQFVFDAVT DVIIKNNLKE
    PTx
351 CGLY
```

Amino acids	354
Mys	Site of myristoylation.
PTx	Site of ADP-ribosylation by pertussis toxin.

Database accession numbers

	PIR	SWISSPROT	EMBL/GENBANK	REFERENCES
Human	A28157	P08754	M20597-99	*1,4–6,19,20*
	A32139	P17539	M27543	
	C28154		J03005	
	S00055		J03198	
	S00078		J03238	
	S02348			
Rat	A28154	P08753	J03219	*1,3*
	E27423		M20713	

References

1 Itoh, H. et al. (1988) J. Biol. Chem. 263, 6656–6664.
2 Blatt, C. et al. (1988) Proc. Natl Acad. Sci. USA 85, 7642–7646.
3 Jones, D. T. and Reed, R. R. (1987) J. Biol. Chem. 262, 14241–14249.
4 Beals, C. R. et al. (1987) Proc. Natl Acad. Sci. USA 84, 7886–7890.
5 Suki, W. N. et al. (1987) FEBS Lett. 220, 187–192.
6 Kim, S. et al. (1988) Proc. Natl Acad. Sci. USA 85, 4153–4157.
7 Wong, Y. H. et al. (1992) Science 255, 339–342.
8 Katada, T. et al. (1984) J. Biol. Chem. 259, 3578–3585.
9 Kobayashi, I. et al. (1990) Eur. J. Biochem. 191, 499–506.
10 Brandt, S. J. et al. (1985) Proc. Natl Acad. Sci. USA 82, 3277–3280.
11 Ohta, H. et al. (1985) J. Biol. Chem. 260, 15771–15780.
12 Linder, M. E. et al. (1990) J. Biol. Chem. 265, 8243–8251.
13 Ausiello, D. A. et al. (1992) J. Biol. Chem. 267, 4759–4765.

14 Lledo, P. M. et al. (1992) Neuron 8, 455–463.
15 Yatani, A. et al. (1987) Science 235, 207–211.
16 Codina, J. et al. (1987) Science 236, 442–445.
17 Yatani, A. et al. (1988) Nature 336, 680–682.
18 Mattera, R. et al. (1989) J. Biol. Chem. 264, 465–471.

Gαo

Molecular weights

Polypeptide	40 053 (GαoA)
	40 075 (GαoB)
Reduced SDS PAGE	39–40 kDa

Structure

Two variant forms (GαoA and GαoB) both with 354 residues, result from differential RNA splicing of a single Gαo gene with the C-terminal 113 residues encoded by alternative use of duplicated exons 7 and 8 [1,2].

Gene structure (human) [1]

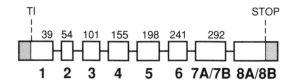

Organization of human Gαo gene adapted from ref. 1. Exons 7 and 8 (encoding amino acids 242–354) are duplicated, with 7A and 8A encoding GαoA and 7B and 8B coding for GαoB.

Tissue distribution

Gαo is expressed predominantly in central nervous system and heart. High levels in brain (~1% membrane protein [3,4]) reflect preferential expression in claustrum, endopiriform nucleus, habenula, hippocampal pyramidal cells, granule cells of the dendate gyrus, and cerebellar Purkinje cells [5–8].

Function

Inhibition of neuronal calcium channel activity [9–15]. Wider functional roles unclear (see Comments).

Comments

High G$_{\alpha o}$ levels in neuronal growth cones where activation may be promoted by interaction with another major growth cone protein, GAP-43 [16–18], and also amyloid precursor protein [19].

G$_{\alpha o}$ splice variants are regulated selectively during neuronal development, with G$_{\alpha o A}$ predominating in differentiated neurons [20–22].

G$_{\alpha o}$ splice variants show selective interaction with different receptors, β– and γ– subunits [15,23,24].

Pertussis toxin sensitivity and functional reconstitution suggest a potential functional role regulating PLC-dependent phosphoinositide hydrolysis [25,26] and potassium channel activation [27,28].

G$_{\alpha o}$ point mutation Q205L (cognate to *gsp* and *gip2* oncogenes of G$_{\alpha s}$ and G$_{\alpha i2}$) with diminished GTPase activity induces transformation in some fibroblasts [29,30].

Amino acid sequence

```
        Mys            (G)
      1 MGCTLSAEER AALERSKAIE KNLKEDGISA AKDVKLLLLG AGESGKSTIV


     51 KQMKIIHEDG FSGEDVKQYK PVVYSNTIQS LAAIVRAMDT LGIEYGDKER


    101 KADAKMVCDV VSRMEDTEPF SAELLSAMMR LWGDSGIQEC FNRSREYQLN

                          (L)
    151 DSAKYYLDSL DRIGAADYQP TEQDILRTRV KTTGIVETHF TFKNLHFRLF

                          (E)                      /NRMHESLML
    201 DVGGQRSERK KWIHCFEDVT AIIFCVALSG YDQVLHEDET T ||||||| |
                                                     \NRMHESLKL


        FDSICNNKFF IDTSIILFLN KKDLFGEKIK KSPLTICFPE YTGPNTYEDA
    251 ||||||||| |  ||||||||| ||| | |||| |||||||||| ||||       |
        FDSICNNKWF TDTSIILFLN KKDIFEEKIK KSPLTICFPE YTGPSAFTEA


        AAYIQAQFES KNRSPNKEIY CHMTCATDTN NIQVVFDAVT DIIIANNLRG
    301 |||||| || || | |||| | |||||||| ||| |||||| | ||| ||||
        VAYIQAQYES KNKSAHKEIY SHVTCATDTN NIQFVFDAVT DVIIAKNLRG


        PTx
        CGLY   GαoA
    351 ||||
        CGLY   GαoB
```

Amino acids   354
Mys   Likely site of myristoylation.
PTx   Site of ADP-ribosylation by pertussis toxin.

Human G$_{\alpha o}$ residues 16, 171 and 218 have been reported as Ser, Thr and Asp [1] or Gly, Leu and Glu, respectively [31]. Variant residues are shown in bold.

Database accession numbers

	PIR	SWISSPROT	EMBL/GENBANK	REFERENCES
Human	A40436	P09471	M60156–65	1,31
	B40436		M19182	
			M19184	
Rat	A24882	P04900	M12671	1,16,32–34
	B27423		M17526	
	C40436		M33661	
	D40436			
Mouse	A36038	P18872	M36777	35,36
		P18873	M36778	
Hamster	A42228	P17806	M33662	33
			J05476	
Bovine	A25948	P08239	M16116	37–39
	A28011		M14489	
	S00213		Y00709	
	S05600		X12924	

References

1 Tsukamoto, T. et al. (1991) Proc. Natl Acad. Sci. USA 88, 2974–2978.
2 **Kaziro, Y. et al. (1991) Annu. Rev. Biochem. 60, 349–400.**
3 Neer, E. J. et al. (1984) J. Biol. Chem. 259, 14222–14229.
4 Sternweis, P. C. and Robishaw, J. D. (1984) J. Biol. Chem. 259, 13806–13813.
5 Largent, B. L. et al. (1988) Proc. Natl Acad. Sci. USA 85, 2864–2868.
6 Worley, P. F. et al. (1986) Proc. Natl Acad. Sci. USA 83, 4561–4565.
7 Brann, M. R. et al. (1988) FEBS Lett. 222, 191–198.
8 Price, S. R. et al. (1989) Biochemistry 28, 3803–3807.
9 Heschler, J. et al. (1987) Nature 325, 445–447.
10 Ewald, D. A. et al. (1989) Neuron 2, 1185–1193.
11 Schmidt, A. et al. (1991) J. Biol. Chem. 266, 18025–18033.
12 McFadzean, I. et al. (1989) Neuron 3, 177–182.
13 Ewald, D. A. et al (1988) Proc. Natl Acad. Sci. USA 85, 3633–3637.
14 Linder, M. E. et al. (1990) J. Biol. Chem. 265, 8243–8251.
15 Kleuss, C. et al. (1991) Nature 353, 43–48.
16 Strittmatter, S. M. et al. (1990) Nature 344, 836–841.
17 Strittmatter, S. M. et al. (1991) J. Biol. Chem. 266, 22465–22471.
18 Jap Tjoen San, E. R. A. et al. (1992) Biochem. Biophys. Res. Commun. 187, 839–846.
19 Nishimoto, I. et al. (1993) Nature 362, 75–79.
20 Rouet, B. et al. (1992) Mol. Pharmacol. 41, 273–280.
21 Asano, T. et al. (1992) J. Neurochem. 58, 2176–2181.
22 Mullaney, I. and Milligan, G. (1990) J. Neurochem. 55, 1890–1898.
23 Kleuss, C. et al. (1992) Nature 358, 424–426.
24 Kleuss, C. et al. (1993) Science 259, 832–834.
25 Moriaty, T. M. et al. (1990) Nature 343, 79–82.
26 Blitzer, R. D. et al. (1993) J. Biol. Chem. 268, 7532–7537.
27 Van Dongen, A. M. J. et al. (1988) Science 242, 1433–1437.
28 Kobayashi, I. et al. (1990) Eur. J. Biochem. 191, 499–506.

29 Kroll, S. D. et al. (1992) J. Biol. Chem. 267, 23183–23188.
30 Slepak, V. Z. (1993) J. Biol. Chem. 268, 1414–1423.
31 Lavu, S. et al. (1988) Biochem. Biophys. Res. Commun. 150, 811–815.
32 Itoh, H. et al. (1986) Proc. Natl Acad. Sci. USA 83, 3776–3780.
33 Jones, D. T. and Reed, R. R. (1987) J. Biol. Chem. 262, 14241–14249.
34 Hsu, W. H. et al. (1990) J. Biol. Chem. 265, 11220–11226.
35 Strathmann, M. et al. (1990) Proc. Natl Acad. Sci. USA 87, 6477–6481.
36 Lopez, R. et al. (1992) Nature 355, 211–213.
37 Angus, C. W. et al. (1986) Proc. Natl Acad. Sci. USA 83, 5813–5816.
38 Van Meurs, K. P. et al. (1987) Proc. Natl Acad. Sci. USA 84, 3107–3111.
39 Ovchinnikov, Y. A. et al. (1987) FEBS Lett. 226, 91–95.

G$_{\alpha t}$-1

Molecular weight
Polypeptide	39 995
Reduced SDS PAGE	39–40 kDa

Gene structure (human)[1]

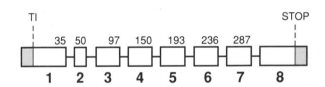

Organization of human G$_{\alpha t}$-1 gene adapted from ref. 1.

Human gene location and size
Chromosome 3; 4.9 kb [1,2]

Tissue distribution
Highly specific expression retinal rod cells [3–7].

Function
Visual signal transduction. G$_{\alpha t}$-1 mediates activation of rod-specific cyclic GMP-PDE by photoactivated rhodopsin [8,9].

Amino acid sequence of human Gαt-1

```
     FAc
   1 MGAGASAEEK HSRELEKKLK EDAEKDARTV KLLLLGAGES GKSTIVKQMK

  51 IIHQDGYSLE ECLEFIAIIY GNTLQSILAI VRAMTTLNIQ YGDSARQDDA

 101 RKLMHMADTI EEGTMPKEMS DIIQRLWKDS GIQACFERAS EYQLNDSAGY
                                CTx
 151 YLSDLERLVT PGYVPTEQDV LRSRVKTTGI IETQFSFKDL NFRMFDVGGQ

 201 RSEPKKWIHC FEGVTCIIFI AALTAYDMVL VEDDEVNRMH ESLHLFNSIC
                                  (V)
 251 NHRYFATTSI VLFLNKKDVF FEKIKKAHLS ICFPDYDGPN TYEDAGNYIK
                                        (C)                PTx
 301 VQFLELNMRR DVKEIYSHMT CATDTQNVKF VFDAVTDIII KENLKDCGLF
```

Amino acids   350
FAc   Likely site for mixed fatty acylation *11–13*.
PTx   Site of ADP-ribosylation by pertussis toxin.
. CTx   Site of ADP-ribosylation by cholera toxin.

Residues 274 and 331 of human Gαt-1 have been reported as Iso and Val *1,3* or Val and Cys, respectively *10*. Variant residues are bold in the above sequence.

Database accession numbers

	PIR	SWISSPROT	EMBL/GENBANK	REFERENCES
Human	JQ0078	P11488	X15088	*1,3,10*
	S04699		X63749	
Mouse	A33352	P20612	M25506–13	*14*
			JO4720	
Bovine	A22244	P04695	K03253	*4–7*
	A23155		K03254	
	A23156		X02440	
	JT0013			

References
1 Fong, S.-L. (1992) Nucleic Acids Res. 11, 2865–2870.
2 Blatt, C. et al. (1988) Proc. Natl Acad. Sci. USA 85, 7642–7646.
3 Lerea, C. L. et al. (1989) Neuron 3, 367–376.
4 Tanabe, T. et al. (1985) Nature 315, 242–245.
5 Medynski, D. et al. (1985) Proc. Natl Acad. Sci. USA 82, 4311–4315.
6 Yatsunami, K. and Gobind Khorana, H. (1985) Proc. Natl Acad. Sci. USA 82, 4316–4320.
7 Lerea, C. L. et al. (1986) Science 234, 77–80.
8 **Stryer, L. (1986) Annu. Rev. Neurosci. 9, 87–119.**
9 **Stryer, L. (1991) J. Biol. Chem. 266, 10711–10714.**

[10] Van Dop, C. et al. (1989) Nucleic Acids Res 17, 4887.
[11] Kokame, K. et al. (1992) Nature 359, 749–752.
[12] Neubert, T. A. et al. (1992) J. Biol. Chem. 267, 18274–18277.
[13] Yang, Z. and Wensel, T. G. (1992) J. Biol. Chem. 267, 23197–23201.
[14] Raport, C. J. et al. (1989) J. Biol. Chem. 264, 7122–7128.

G$_{\alpha t-2}$

Molecular weights

Polypeptide	40 170
Reduced SDS PAGE	39–40 kDa

Gene structure (human) [1]

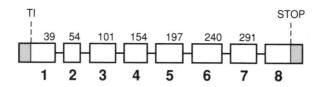

Organization of human G$_{\alpha t-2}$ gene adapted from ref. 1.

Human gene location and size
Chromosome 1; 10 kb [1,2].

Tissue distribution

Highly specific expression in retinal cone cells [3–7] although also detected in Y79 retinoblastoma, lymphocytes and some tumour cell lines [1].

Function

Visual signal transduction. G$_{\alpha t-2}$ mediates activation of cone-specific cyclic GMP-phosphodiesterase following opsin photoactivation [8,9].

Amino acid sequence

```
     FAc
  1 MGSGASAEDK ELAKRSKELE KKLQEDADKE AKTVKLLLLG AGESGKSTIV

 51 KQMKIIHQDG YSPEECLEFK AIIYGNVLQS ILAIIRAMTT LGIDYAEPSC

101 ADDGRHVNNL ADSIEEGTMP PELVEVIRRL WKDGGVQACF ERAAEYQLND
                               CTx
151 SASYYLNQLE RITDPEYLPS EQDVLRSRVK TTGIIETKFS VKDLNFRMFD
```

```
201 VGGQRSERKK WIHCFEGVTC IIFCAALSAY DMVLVEDDEV NRMHESLHLF

251 NSICNHKFFA ATSIVLFLNK KDLFEEKIKK VHLSICFPEY DGNNSYDDAG

301 NYIKSQFLDL NMRKDVKEIY SHMTCATDTQ NVKFVFDAVT DIIIKENLKD
    PTx
351 CGLF
```

Amino acids 354
FAc   Likely site for mixed fatty acylation [10-12].
PTx   Site of ADP-ribosylation by pertussis toxin.
CTx   Site of ADP-ribosylation by cholera toxin.

Database accession numbers

	PIR	SWISSPROT	EMBL/GENBANK	REFERENCES
Human	JQ0079	P19087	D10377–84	1,3
Mouse			L10666	
Bovine		P04696	M11116	13

References
1 Kubo, M. et al. (1991) FEBS Lett. 291, 245–248.
2 Blatt, C. et al. (1988) Proc. Natl Acad. Sci. USA 85, 7642–7646.
3 Lerea, C. L. et al. (1989) Neuron 3, 367–376.
4 Tanabe, T. et al. (1985) Nature 315, 242–245.
5 Medynski, D. et al (1985) Proc. Natl. Acad. Sci. USA 82, 4311–4315.
6 Yatsunami, K. and Gobind Khorana, H. (1985) Proc. Natl Acad. Sci. USA 82, 4316–4320.
7 Lerea, C. L. et al. (1986) Science 234, 77–80.
8 **Stryer, L. (1986) Annu. Rev. Neurosci. 9, 87–119.**
9 **Stryer, L. (1991) J. Biol. Chem. 266, 10711–10714.**
10 Kokame, K. et al. (1992) Nature 359, 749–752.
11 Neubert, T. A. et al. (1992) J. Biol. Chem. 267, 18274–18277.
12 Yang, Z. and Wensel, T. G. (1992) J. Biol. Chem. 267, 23197–23201.
13 Lochrie, M. A. et al. (1985) Science 228, 96–99.

GUSTUCIN (Gαgust)

Molecular weight
Polypeptide   40 294

Tissue Distribution

Highly restricted to tongue where it is enriched in circumvallate, foliate and fungiform papillae [1].

Function

Unclear (see Comments).

Comments

Perception of taste stimuli is likely to involve G-protein linked signal transduction pathways [2-7] and Gαgust could play a role within these systems.

Amino acid sequence (rat)

```
      Mys
    1 MGSGISSESK ESAKRSKELE KKLQEDAERD ARTVKLLLLG AGESGKSTIV

   51 KQMKIIHKNG YSKQECMEFK AVVYSNTLQS ILAIVKAMTT LGIDYVNPRS

  101 REDQQLLLSM ANTLEDGDMT PQLAEIIKRL WGDPGIQACF ERASEYQLND
                              CTx
  151 SAAYYLNDLD RLTAPGYVPN EQDVLHSRVK TTGIIETQFS FKDLNFRMFD

  201 VGGQRSERKK WIHCFEGVTC IIFCAALSAY DMVLVEDEEV NRMHESLHLF

  251 NSICNHKYFA TTSIVLFLNK KDLFQEKVTK VHLSICFPEY TGPNTFEDAG

  301 NYIKNQFLDL NLKKEDKEIY SHMTCATDTQ NVKFVFDAVT DIIIKENLKD
      PTx
  351 CGLF
```

Amino acids 354

Mys   Potential site for myristoylation.
CTx   Potential site for ADP-ribosylation by cholera toxin.
PTx   Potential site for ADP-ribosylation by pertussis toxin.

Database accession numbers

	PIR	SWISSPROT	EMBL/GENBANK	REFERENCE
Rat			X65747	1

References

[1] McLaughlin, S. K. et al. (1992) Nature 357, 563–569.
[2] **Kinnamon, S. C. (1988) Trends Neuro sci. 11, 491–496.**
[3] **Roper, S. D. (1989) Annu. Rev. Neurosci 12, 329–353.**
[4] Striem, B. et al. (1989) Biochem. J. 260, 121–126.
[5] Restrepo, D. (1990) Science 249, 1166–1168.
[6] Boekhoff, I. et al. (1990) EMBO J. 9, 2453–2458.
[7] Abe, K. et al. (1993) FEBS Lett. 316, 253–256.

Gαz (Gαx)

Molecular Weight

Polypeptide	40 923
Reduced SDS PAGE	41 kDa

Gene structure (human)

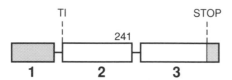

Organization of human G$_{\alpha z}$ gene adapted from ref. 1. Exon–intron organization is very different from other α-chain genes hitherto characterized. The complete gene comprises three exons although exons 2 and 3 contain the entire coding sequence. Exon 1 corresponds to 5'-noncoding region.

Human gene location and size
Chromosome 22, >60 kb [1,2].

Tissue distribution

Predominantly found in brain and neuronal cells although also found in other cells, e.g. platelets [3-5].

Function

Unknown.

Comments
Heterologously expressed wild-type G$_{\alpha z}$ as well as the Q205L point mutation (cognate to *gsp* and *gip2* oncogenes of G$_{\alpha s}$ and G$_{\alpha i2}$) mediate inhibition of adenylyl cyclase (pertussis toxin insensitive) [6].

G$_{\alpha z}$ exhibits biochemical properties distinct from other G-protein α-chains, including unusual magnesium dependence and very slow guanine nucleotide exchange (K_{diss}:GDP=0.001 min^{-1}) and GTPase activity (K_{cat}=0.05 min^{-1}) [4].

It is a target for phosphorylation by PKC [7].

Amino acid sequence

```
Mys
  1 MGCRQSSEEK EAARRSRRID RHLRSESQRQ RREIKLLLLG TSNSGKSTIV

 51 KQMKIIHSGG FNLEACKEYK PLIIYNAIDS LTRIIRALAA LRIDFHNPDR

101 AYDAVQLFAL TGPAESKGEI TPELLGVMRR LWADPGAQAC FSRSSEYHLE

151 DNAAYYLNDL ERIAAADYIP TVEDILRSRD MTTGIVENKF TFKELTFKMV

201 DVGGQRSERK KWIHCFEGVT AIIFCVELSG YDLKLYEDNQ TSRMAESLRL
```

```
251 FDSICNNNWF INTSLILFLN KKDLLAEKIR RIPLTICFPE YKGQNTYEEA

301 AVYIQRQFED LNRNKETKEI YSHFTCATDT SNIQFVFDAV TDVIIQNNLK

351 YIGLC
```

Amino acids 355
Mys   Likely site for myristoylation.
Ser27 is major site of phosphorylation by PKC [7].

Database accession numbers

	PIR	SWISSPROT	EMBL/GENBANK	REFERENCES
Human	A31339	P19086	D90150	1,8
	A36628		J03260	
			J05541	
Rat	A31316	P19627	J03773	9

References

1 Matsuoka, M. et al. (1990) J. Biol. Chem. 265, 13215–13220.
2 Blatt, C. et al. (1988) Proc. Natl Acad. Sci. USA 85, 7642–7646.
3 Hinton, D. R. et al. (1990) J. Neurosci. 10, 2763–2774.
4 Casey, P. J. et al. (1990) J. Biol. Chem. 265, 2383–2390.
5 Carlson, K. E. et al. (1989) J. Biol. Chem. 264, 13298–13305.
6 Wong, Y. H. et al. (1992) Science 255, 339–342.
7 Lounsbury, K. M. et al. (1993) J. Biol. Chem. 268, 3494–3498.
8 Fong, H. K. et al. (1988) Proc. Natl Acad. Sci. USA 85, 3066–3070.
9 Matsuoka, M. et al. (1988) Proc. Natl Acad. Sci. USA 85, 5384–5388.

G$_{\alpha q}$

Molecular weight
Polypeptide   42 153
Reduced SDS PAGE   42–43 kDa

Tissue distribution

Widely expressed in tissues and cell lines although generally lacking in T cell lines [1-6].

Function

Mediator of agonist-stimulated hydrolysis of membrane phosphoinositides by PLC-β1, PLC-β2 and PLC-β3 [6-18]. This G$_{\alpha q}$ response is insensitive to inhibition by pertussis toxin. PLC-γ1, PLC-γ2, PLC-δ1 and PLC-δ3 are not targets for activation by G$_{\alpha q}$ [6-18].

Comment

Heterologous expression of G$_{\alpha q}$ containing activating point mutations R183C and Q209L (cognate to *gsp* and *gip2* mutations in G$_{\alpha s}$ and G$_{\alpha i2}$) results in constitutive phosphoinositide hydrolysis by PLC as well as malignant transformation of fibroblasts [13,14,19,20]. Recombinant Gαq displays very low affinity for binding GTPγS [21].

Amino acid sequence (mouse)

```
  1 MTLESIMACC LSEEAKEARR INDEIERHVR RDKRDARREL KLLLLGTGES

 51 GKSTFIKQMR IIHGSGYSDE DKRGFTKLVY QNIFTAMQAM IRAMDTLKIP

101 YKYEHNKAHA QLVREVDVEK VSAFENPYVD AIKSLWNDPG IQECYDRRRE

151 YQLSDSTKYY LNDLDRVADP SYLPTQQDVL RVRVPTTGII EYPFDLQSVI

201 FRMVDVGGQR SERRKWIHCF ENVTSIMFLV ALSEYDQVLV ESDNENRMEE

251 SKALFRTIIT YPWFQNSSVI LFLNKKDLLE EKIMYSHLVD YFPEYDGPQR

301 DAQAAREFIL KMFVDLNPDS DKIIYSHFTC ATDTENIRFV FAAVKDTILQ

351 LNLKEYNLV
```

Amino acids   359

Database accession numbers

	PIR	SWISSPROT	EMBL/GENBANK	REFERENCE
Mouse	A38414	P21279	M55412	1

References
1 Strathmann, M. and Simon, M. I. (1990) Proc. Natl Acad. Sci. USA 87, 9113–9117.
2 Nakamura, F. et al. (1991) J. Biol. Chem. 266, 12676–12681.
3 Pang, I.-H. and Sternweis, P. C. (1990) J. Biol. Chem. 265, 18707–18712.
4 Wilkie, T. M. et al. (1991) Proc. Natl Acad. Sci. USA 88, 10049–10053.
5 Mitchell, F. M. et al. (1991) FEBS Lett. 287, 171–174.
6 Aragay, A. M. et al. (1992) J. Biol. Chem. 267, 24983–24988.
7 Taylor, S. J. and Exton, J. H. (1991) FEBS Lett. 286, 214–216.
8 Taylor, S. J. et al. (1991) Nature 350, 516–518.
9 Blank, J. L. et al. (1991) J. Biol. Chem. 266, 18206–18216.
10 Smrcka, A. V. et al. (1991) Science 251, 804–807.
11 Berstein, G. et al. (1992) J. Biol. Chem. 267, 8081–8088.
12 Gutawski, S. et al. (1991) J. Biol. Chem. 266, 20519–20524.
13 Wu, D. et al. (1992) J. Biol. Chem. 267, 1811–1817.
14 Conklin, B. R. et al. (1992) J. Biol. Chem. 267, 31–34.
15 Wu, D. et al. (1992) J. Biol. Chem. 267, 25798–25802.
16 Lee, C. H. et al. (1992) J. Biol. Chem. 267, 16044–16047.
17 Park, D. et al. (1992) J. Biol. Chem. 267, 16048–16055.
18 Jhon, D.-Y. et al. (1993) J. Biol. Chem. 268, 6654–6661.
19 Kalinec, G. et al. (1992) Mol. Cell. Biol. 12, 4687–4693.
20 De Vivo, M. et al. (1992) J. Biol. Chem. 267, 18263–18266.
21 Helper, J.R. et al. (1993) J. Biol. Chem. 268, 14367–14375.

Gα11 (Gα12, Gαy)

Molecular weight
Polypeptide   42 163
Reduced SDS PAGE   42–43 kDa

Tissue distribution
Widely expressed with high levels in brain. Undetectable in platelets [1-6].

Function
Mediator of agonist-stimulated hydrolysis of membrane phosphoinositides by PLC-β1, PLC-β2 and PLC-β3 [7-16]. This Gα11 response is insensitive to inhibition by pertussis toxin. PLC-γ1, PLC-γ2, PLC-δ1 and PLC-δ3 are not targets for activation by Gα11 [7-16].

Comment
Recombinant Gα11 displays very low affinity for binding GTPγS [17].

Amino acid sequence

```
  1 MTLESMMACC LSDEVKESKR INAEIEKQLR RDKRDARREL KLLLLGTGES

 51 GKSTFIKQMR IIHGAGYSEE DKRGFTKLVY QNIFTAMQAM IRAMETLKIL

101 YKYEQNKANA LLIREVDVEK VTTFEHQYVS AIKTLWEDPG IQECYDRRRE

151 YQLSDSAKYY LTDVDRIATL GYLPTQQDVL RVRVPTTGII EYPFDLENII

201 FRMVDVGGQR SERRKWIHCF ENVTSIMFLV ALSEYDQVLV ESDNENRMEE

251 SKALFRTIIT YPWFQNSSVI LFLNKKDLLE DKILYSHLVD YFPEFDGPQR

301 EPQAAREFIL KMFVDLNPDS DKIIYSHFTC ATDTENIRFV FAAVKDTILQ

351 LNLKEYNLV
```

Amino acids   359

Database accession numbers

	PIR	SWISSPROT	EMBL/GENBANK	REFERENCES
Human	A39394		M69013	3
Mouse	B38414	P21278	M55411	1
Bovine	B40891		D90336	2

References
[1] Strathmann, M. and Simon, M. I. (1990) Proc. Natl Acad. Sci. USA 87, 9113–9117.
[2] Nakamura, F. et al. (1991) J. Biol. Chem. 266, 12676–12681.
[3] Jiang, M. et al. (1991) Proc. Natl. Acad. Sci. USA 88, 3907–3911.

4 Pang, I.-H. and Sternweis, P. C. (1990) J. Biol. Chem. 265, 18707–18712.
5 Wilkie, T. M. et al. (1991) Proc. Natl Acad. Sci. USA 88, 10049–10053.
6 Mitchell, F. M. et al. (1991) FEBS Lett. 287, 171–174.
7 Taylor, S. J. and Exton, J. H. (1991) FEBS Lett. 286, 214–216.
8 Taylor, S. J. et al. (1991) Nature 350, 516–518.
9 Blank, J. L. et al. (1991) J. Biol. Chem. 266, 18206–18216.
10 Smrcka, A. V. et al. (1991) Science 251, 804–807.
11 Berstein, G. et al. (1992) J. Biol. Chem. 267, 8081–8088.
12 Wu, D. et al. (1992) J. Biol. Chem. 267, 1811–1817.
13 Wu, D. et al. (1992) J. Biol. Chem. 267, 25798–25802.
14 Aragay, A. M. et al. (1992) J. Biol. Chem. 267, 24983–24988.
15 Lee, C. H. et al. (1992) J. Biol. Chem. 267, 16044–16047.
16 Jhon, D.-Y. et al. (1993) J. Biol. Chem. 268, 6654–6661.
17 Helper, J.R. et al. (1993) J. Biol. Chem. 268, 14367–14375.

Gα14 (GαL1)

Molecular weight
Polypeptide   41 522

Tissue distribution

Limited distribution. Preferential expression in spleen, lung, kidney, uterus, testis, bone marrow stromal cells, early myeloid and progenitor B cells [1,2].

Function

Unclear, although heterologously expressed Gα14 can stimulate phosphoinositide hydrolysis by PLC-β1 [3,4].

Amino acid sequence (mouse)

```
  1 MAGCCCLSAE EKESQRISAE IERHVRRDKK DARRELKLLL LGTGESGKST

 51 FIKQMRIIHG SGYSDEDRKG FTKLVYQNIF TAMQAMIRAM DTLRIQYMCE

101 QNKENAQIIR EVEVDKVTAL SRDQVAAIKQ LWLDPGIQEC YDRRREYQLS

151 DSAKYYLTDI ERIAMPSFVP TQQDVLRVRV PTTGIIEYPF DLENIïFRMV

201 DVGGQRSERR KWIHCFESVT SIIFLVALSE YDQVLAECDN ENRMEESKAL

251 FRTIITYPWF LNSSVILFLN KKDLLEEKIM YSHLISYFPE YTGPKQDVKA

301 ARDFILKLYQ DQNPDKEKVI YSHFTCATDT ENIRFVFAAV KDTILQLNLR

351 EFNLV
```

Amino acids   355

Database accession numbers

	PIR	SWISSPROT	EMBL/GENBANK	REFERENCES
Mouse	A41534		M80631	2
Bovine	A40891		D90335	1

References

[1] Nakamura, F. et al. (1991) J. Biol. Chem. 266, 12676–12681.
[2] Wilkie, T. M. et al. (1991) Proc. Natl Acad. Sci. USA 88, 10049–10053.
[3] Lee, C. H. et al. (1992) J. Biol. Chem. 267, 16044–16047.
[4] Wu, D. et al. (1992) J. Biol. Chem. 267, 25798–25802.

Gα15

Molecular weight

Polypeptide	43 535
Reduced SDS PAGE	42 kDa

Tissue Distribution

Restricted to tissues rich in cells of haematopoetic lineage (spleen, thymus, lung, embryonic liver and bone marrow). High levels in myeloid and B-cell lineages. Not expressed in most stromal or T cell lines [1].

Function

Unclear, although heterologously expressed Gα15 can activate PLC-β1 to hydrolyze phosphoinositides [2].

Comment

Murine Gα15 is 85% identical to human Gα16 and could represent a phylogenic homologue of the same gene.

Amino acid sequence (mouse)

```
  1 MARSLTWGCC PWCLTEEEKT AARIDQEINR ILLEQKKQER EELKLLLLGP

 51 GESGKSTFIK QMRIIHGVGY SEEDRRAFRL LIYQNIFVSM QAMIDAMDRL

101 QIPFSRPDSK QHASLVMTQD PYKVSTFEKP YAVAMQYLWR DAGIRACYER

151 RREFHLLDSA VYYLSHLERI SEDSYIPTAQ DVLRSRMPTT GINEYCFSVK

201 KTKLRIVDVG GQRSERRKWI HCFENVIALI YLASLSEYDQ CLEENDQENR

251 MEESLALFST ILELPWFKST SVILFLNKTD ILEDKIHTSH LATYFPSFQG

301 PRRDAEAAKS FILDMYARVY ASCAEPQDGG RKGSRARRFF AHFTCATDTQ

351 SVRSVFKDVR DSVLARYLDE INLL
```

Amino acids	374

Database accession numbers

	PIR	SWISSPROT	EMBL/GENBANK	REFERENCES
Mouse	B41534		M80632	1

References

[1] Wilkie, T. M. et al. (1991) Proc. Natl Acad. Sci. USA 88, 10049–10053.
[2] Wu, D. et al. (1992) J. Biol. Chem. 267, 25798–25802.

Gα16

Molecular weight

Polypeptide	43 508
Reduced SDS PAGE	43 kDa

Tissue Distribution

Expression restricted to haematopoietic cells, particularly myelomonocytic and T cell lines although not in B cells [1].

Function

Unclear, although heterologously expressed, Gα16 stimulates PLC-β1, PLC-β2 and PLC-β3 to hydrolyse phosphoinositides [2-5].

Comment

Gα16 can interact with IL8 and C5a receptors to mediate ligand stimulated phosphoinositide hydrolysis.

Amino acid sequence

```
  1 MARSLTWRCC PWCLTEDEKA AARVDQEINR ILLEQKKQDR GELKLLLLGP

 51 GESGKSTFIK QMRIIHGAGY SEEERKGFRP LVYQNIFVSM RAMIEAMERL

101 QIPFSRPESK HHASLVMSQD PYKVTTFEKR YAAAMQWLWR DAGIRACYER

151 RREFHLLDSA VYYLSHLERI TEEGYVPTAQ DVLRSRMPTT GINEYCFSVQ

201 KTNLRIVDVG GQKSERKKWI HCFENVIALI YLASLSEYDQ CLEENNQENR

251 MKESLALFGT ILELPWFKST SVILFLNKTD ILEEKIPTSH LATYFPSFQG

301 PKQDAEAAKR FILDMYTRMY TGCVDGPEGS KKGARSRRLF SHYTCATDTQ

351 NIRKVFKDVR DSVLARYLDE INLL
```

Amino acids   374

Database accession numbers

	PIR	SWISSPROT	EMBL/GENBANK	REFERENCES
Human	A41096		M63904	1

References

1 Amatruda, T. T. et al. (1991) Proc. Natl Acad. Sci. USA 88, 5587–5591.
2 Wu, D. et al. (1992) J. Biol. Chem. 267, 25798–25802.
3 Lee, C. H. et al. (1992) J. Biol. Chem. 267, 16044–16047.
4 Schnabel, P. et al. (1992) Biochem. Biophys. Res. Commun. 188, 1018–1023.
5 Jhon, D.-Y. et al. (1993) J. Biol. Chem. 268, 6654–6661.

Gα12

Molecular weights

Polypeptide	44 095
Reduced SDS PAGE	45 kDa

Tissue distribution

Widely distributed, although low levels in intestine [1].

Function

Unknown.

Comment

Human Gα12 identified recently by expression cloning based on oncogenic activity [2]. Mutationally activated Gα12 facilitates arachidonic acid mobilization and is also strongly transforming when expressed in fibroblasts [3].

Amino acid sequence (mouse)

```
  1 MSGVVRTLSR CLLPAEAGAR ERRAGAARDA EREARRRSRD IDALLARERR

 51 AVRRLVKILL LGAGESGKST FLKQMRIIHG REFDQKALLE FRDTIFDNIL

101 KGSRVLVDAR DKLGIPWQHS ENEKHGMFLM AFENKAGLPV EPATFQLYVP

151 ALSALWRDSG IREAFSRRSE FQLGESVKYF LDNLDRIGQL NYFPSKQDIL

201 LARKATKGIV EHDFVIKKIP FKMVDVGGQR SQRQKWFQCF DGITSILFMV

251 SSSEYDQVLM EDRRTNRLVE SMNIFETIVN NKLFFNVSII LFLNKMDLLV

301 EKVKSVSIKK HFPDFKGDPH RLEDVQRYLV QCFDRKRRNR SKPLFHHFTT

351 AIDTENIRFV FHAVKDTILQ ENLKDIMLQ
```

Amino acids   379

Database accession numbers

	PIR	SWISSPROT	EMBL/GENBANK	REFERENCES
Mouse	A41095		M63659	1
Human			L01694	2

References
1 Strathmann, M. P. and Simon, M. I. (1991) Proc. Natl Acad. Sci. USA 88, 5582–5586.
2 Chan, A. M.-L. et al. (1993) Mol. Cell. Biol. 13, 762–768.
3 Xu, N. et al. (1993) Proc. Natl Acad Sci USA 90, 6741–6745.

Gα13

Molecular weights
Polypeptide	44 054
Reduced SDS PAGE	43 kDa

Tissue Distribution

Widely expressed, although higher levels in eye, kidney, liver, lung, uterus, testis and platelets. Undetectable in brain cortex [1,2].

Function

Unknown.

Amino acid sequence (mouse)

```
  1 MADFLPSRSV LSVCFPGCVL TNGEAEQQRK SKEIDKCLSR EKTYVKRLVK

 51 ILLLGAGESG KSTFLKQMRI IHGQDFDQRA REEFRPTIYS NVIKGMRVLV

101 DAREKLHIPW GDNKNQLHGD KLMAFDTRAP MAAQGMVETR VFLQYLPAIR

151 ALWEDSGIQN AYDRRREFQL GESVKYFLDN LDKLGVPDYI PSQQDILLAR

201 RPTKGIHEYD FEIKNVPFKM VDVGGQRSER KRWFECFDSV TSILFLVSSS

251 EFDQVLMEDR QTNRLTESLN IFETIVNNRV FSNVSIILFL NKTDLLEEKV

301 QVVSIKDYFL EFEGDPHCLR DVQKFLVECF RGKRRDQQQR PLYHHFTTAI

351 NTENIRLVFR DVKDTILHDN LKQLMLQ
```

Amino acids   377

Database accession numbers

	PIR	SWISSPROT	EMBL/GENBANK	REFERENCE
Mouse	B41095		M63660	1

References

1 Strathmann, M. P. and Simon, M. I. (1991)   Proc. Natl Acad. Sci. USA 88, 5582–5586.
2 Milligan, G. et al. (1992) FEBS Lett. 297, 186–188.

G-protein β-chains

β1

Molecular weights
Polypeptide	37 375
Reduced SDS PAGE	36 kDa [1]

Human gene location
Chromosome 1 [2].

Tissue Distribution

Ubiquitous, although highly expressed in retina and particularly rod cells where it represents major β-subunit of transducin [3–8].

Also expressed in cell lines eg. NIH 3T3 [7].

Function

Likely interaction with retinal $G_{\alpha t\text{-}1}$ [9–13] and γ1 or γ3 [5] to mediate stimulation of rod cell cyclic GMP-phosphodiesterase by photoactivated rhodopsin [14,15].

In GH3 cells β1 interacts specifically with $G_{\alpha oB}$ and γ3 to form a heterotrimeric G-protein mediating somatostatin receptor-dependent inhibition of calcium channel activity [16–18].

Wider specificities of interaction in heterotrimeric G-protein function unclear.

Comments
Forms dimers with either γ1, γ2, γ3 or γ5 G-protein subunits [19-21]. Both β1γ1, β1γ2 and β1γ5 dimers stimulate phosphoinositide hydrolysis by PLC-β2 [22].

Amino acid sequence

```
  1 MSELDQLRQE AEQLKNQIRD ARKACADATL SQITNNIDPV GRIQMRTRRT

 51 LRGHLAKIYA MHWGTDSRLL VSASQDGKLI IWDSYTTNKV HAIPLRSSWV

101 MTCAYAPSGN YVACGGLDNI CSIYNLKTRE GNVRVSRELA GHTGYLSCCR

151 FLDDNQIVTS SGDTTCALWD IETGQQTTTF TGHTGDVMSL SLAPDTRLFV

201 SGACDASAKL WDVREGMCRQ TFTGHESDIN AICFFPNGNA FATGSDDATC

251 RLFDLRADQE LMTYSHDNII CGITSVSFSK SGRLLLAGYD DFNCNVWDAL

301 KADRAGVLAG HDNRVSCLGV TDDGMAVATG SWDSFLKIWN
```

Amino acids 340
Cys25 is involved in interaction with G-protein γ-subunits [23].

Database accession numbers

	PIR	SWISSPROT	EMBL/GENBANK	REFERENCES
Human	A24853	P04901	J02956	24
		P04697	M16538	
			X04526	
Bovine	A24225	P04901	M13236	25,26
	A25457		X03073	

References

1 Amatruda, T. T. et al. (1988) J. Biol. Chem. 263, 5008–5011.
2 Blatt, C. et al. (1988) Proc. Natl Acad. Sci. USA 85, 7642–7646.
3 Fong, H. K. W. et al. (1987) Proc. Natl Acad. Sci. USA 84, 3792–3796.
4 Levine, M. A. et al. (1990) Proc. Natl Acad. Sci. USA 87, 2329–2333.
5 Peng, Y.-W. et al. (1992) Proc. Natl Acad. Sci. USA 89, 10882–10886.
6 Lee, R. H. et al. (1992) J. Biol. Chem. 267, 24776–24781.
7 Fung, B. K.-K. et al. (1992) J. Biol. Chem. 267, 24782–24788.
8 Von Weizsacker, E. et al. (1992) Biochem. Biophys. Res. Commun. 183, 350–356.
9 Lerea, C. L. et al. (1989) Neuron 3, 367–376.
10 Tanabe, T. et al. (1985) Nature 315, 242–245.
11 Medynski, D. et al. (1985) Proc. Natl Acad. Sci. USA 82, 4311–4315.
12 Yatsunami, K. and Gobind Khorana, H. (1985) Proc. Natl. Acad. Sci. USA 82, 4316–4320.
13 Lerea, C. L. et al. (1986) Science 234, 77–80.
14 **Hurley, J. B. (1987) Annu. Rev. Physiol. 49, 793–812.**
15 **Stryer, L. (1991) J. Biol. Chem. 266, 10711–10714.**
16 Kleuss, C. et al. (1991) Nature 353, 43–48.
17 Kleuss, C. et al. (1992) Nature 358, 424–426.
18 Kleuss, C. et al. (1993) Science 259, 832–834.
19 Schmidt, C. J. et al. (1992) J. Biol. Chem. 267, 13807–13810.
20 Pronin, A. N. and Gautam, N. (1992) Proc. Natl Acad. Sci. USA 89, 6220–6224.
21 Iniguez-Lluhi, J. A. (1992) J. Biol. Chem. 267, 23409–23417.
22 Katz, A. et al. (1992) Nature 360, 686–688.
23 Bubis, J. and Khorana, H. G. (1990) J. Biol. Chem. 12995–12999.
24 Codina, J. et al. (1986) FEBS Lett. 207, 187–192.
25 Sugimoto, K. et al. (1985) FEBS Lett. 191, 235–240.
26 Fong, H. K.W. et al. (1986) Proc. Natl Acad. Sci. USA 83, 2162–2166.
27 Hermonet, S. et al. (1993) FEBS Lett. 327, 183–188.

β2

Molecular weights

Polypeptide	37 329
Reduced SDS PAGE	35 kDa [1]

Human gene location

Chromosome 7 [2].

Tissue distribution

Wide distribution although less abundant than β1 and β3 [3,4].

Function

Component of heterotrimeric G-proteins although specific roles unclear.

Comments

Forms dimers with either γ2 or γ3 G-protein subunits although not γ1 [5–8]. β2γ5 dimer can activate PLCβ2 [10].

Amino acid sequence

```
  1 MSELEQLRQE AEQLRNQIRD ARKACGDSTL TQITAGLDPV GRIQMRTRRT

 51 LRGHLAKIYA MHWGTDSRLL VSASQDGKLI IWDSYTTNKV HAIPLRSSWV

101 MTCAYAPSGN FVACGGLDNI CSIYSLKTRE GNVRVSRELP GHTGYLSCCR
                                                      (N)
151 FLDDNQIITS SGDTTCALWD IETGQQTVGF AGHSGDVMSL SLAPDGRTFV

201 SGACDASIKL WDVRDSMCRQ TFIGHESDIN AVAFFPNGYA FTTGSDDATC

251 RLFDLRADQE LLMYSHDNII CGITSVAFSR SGRLLLAGYD DFNCNIWDAM

301 KGDRAGVLAG HDNRVSCLGV TDDGMAVATG SWDSFLKIWN
```

Amino acids   340

Independently cloned by two groups [3,9] with one nucleotide difference resulting in either Asn or Asp at residue 195 (bold type).

Database accession numbers

	PIR	SWISSPROT	EMBL/GENBANK	REFERENCES
Human	B26617	P11016	M16514	3,9
	B28040		M16538	
			M36429	
			J02956	
Bovine	A26617	P11017	M16480	3,9
	A28040		M16539	

References

[1] Amatruda, T. T. et al. (1988) J. Biol. Chem. 263, 5008–5011.

[2] Blatt, C. et al. (1988) Proc. Natl Acad. Sci. USA 85, 7642–7646.

[3] Fong, H. K. W. et al. (1987) Proc. Natl Acad. Sci. USA 84, 3792–3796.

[4] Von Weizsacker, E. et al. (1992) Biochem. Biophys. Res. Commun. 183, 350–356.

[5] Schmidt, C. J. et al. (1992) J. Biol. Chem. 267, 13807–13810.

[6] Pronin, A. N. and Gautam, N. (1992) Proc. Natl Acad. Sci. USA 89, 6220–6224.

[7] Iniguez-Lluhi, J. A. (1992) J. Biol. Chem. 267, 23409–23417.

[8] Robishaw, J. D. et al. (1992) Biochem. J. 286, 677–680.

[9] Gao, B. et al. (1987) Proc. Natl Acad. Sci. USA 84, 6122–6125.

[10] Wu, D. et al. (1993) Proc. Natl Acad Sci USA 90, 5297–5301.

β3

Molecular weight
Polypeptide   37 221
Reduced SDS PAGE   35 kDa

Human gene location
Chromosome 12 [1].

Tissue distribution

Widely expressed, although particularly high expression in retinal cone cells [1-4].

Function

Specific co-expression suggests interaction with retinal $G_{\alpha t-2}$ [5-9] and γ2 or γ3 [2] to mediate stimulation of cone cyclic GMP-phosphodiesterase by photoactivated opsins [10,11].

In GH$_3$ cells, β3 interacts specifically with $G_{\alpha oA}$ and γ4 to form a heterotrimeric G-protein mediating muscarinic M$_4$ receptor-dependent inhibition of calcium channel activity [12-14].

Wider specificities of interaction in heterotrimeric G-protein function unclear.

Comments
Does not form dimers with γ1 or γ2 [15,16].

Amino acid sequence

```
  1 MGEMEQLRQE AEQLKKQIAD ARKACADVTL AELVSGLEVV GRVQMRTRRT

 51 LRGHLAKIYA MHWATDSKLL VSASQDGKLI VWDSYTTNKV HAIPLRSSWV

101 MTCAYAPSGN FVACGGLDNM CSIYNLKSRE GNVKVSRELS AHTGYLSCCR

151 FLDDNNIVTS SGDTTCALWD IETGQQKTVF VGHTGDCMSL AVSPDFNLFI

201 SGACDASAKL WDVREGTCRQ TFTGHESDIN AICFFPNGEA ICTGSDDASC

251 RLFDLRADQE LICFSHESII CGITSVAFSL SGRLLFAGYD DFNCNVWDSM

301 KSERVGILSG HDNRVSCLGV TADGMAVATG SWDSFLKIWN
```

Amino acids   340

Database accession numbers

	PIR	SWISSPROT	EMBL/GENBANK	REFERENCE
Human	A35096	P16520	M31328	1

References

1 Levine, M. A. et al. (1990) Proc. Natl Acad. Sci. USA 87, 2329–2333.
2 Peng, Y.-W. et al. (1992) Proc. Natl Acad. Sci. USA 89, 10882–10886.
3 Lee, R. H. et al. (1992) J. Biol. Chem. 267, 24776–24781.
4 Fung, B. K.-K. et al. (1992) J. Biol. Chem. 267, 24782–24788.
5 Lerea, C. L. et al. (1989) Neuron 3, 367–376.
6 Tanabe, T. et al. (1985) Nature 315, 242–245.
7 Medynski, D. et al. (1985) Proc. Natl Acad. Sci. USA 82, 4311–4315.
8 Yatsunami, K. and Gobind Khorana, H. (1985) Proc. Natl Acad. Sci. USA 82, 4316–4320.
9 Lerea, C. L. et al. (1986) Science 234, 77–80.
10 Hurley, J. B. (1987) Annu. Rev. Physiol. 49, 793–812.
11 Stryer, L. (1991) J. Biol. Chem. 266, 10711–10714.
12 Kleuss, C. et al. (1991) Nature 353, 43–48.
13 Kleuss, C. et al. (1992) Nature 358, 424–426.
14 Kleuss, C. et al. (1993) Science 259, 832–834.
15 Schmidt, C. J. et al. (1992) J. Biol. Chem. 267, 13807–13810.
16 Pronin, A. N. and Gautam, N. (1992) Proc. Natl Acad. Sci. USA 89, 6220–6224.

β4

Molecular weight

Polypeptide	37 354
Reduced SDS PAGE	36 kDa

Tissue distribution

Wide distribution, with highest levels in brain, eye, lung, heart and testes [1].

Function

Component of heterotrimeric G-proteins although specific roles unclear.

Amino acid sequence (mouse)

```
  1 MSELEQLRQE AEQLRNQIQD ARKACNDATL VQITSNMDSV GRIQMRTRRT
 51 LRGHLAKIYA MHWGYDSRLL VSASQDGKLI IWDSYTTNKM HAIPLRSSWV
101 MTCAYAPSGN YVACGGLDNI CSIYNLKTRE GDVRVSRELA GHTGYLSCCR
151 FLDDGQIITS SGDTTCALWD IETGQQTTTF TGHSGDVMSL SLSPDLKTFV
201 SGACDASSKL WDIRDGMCRQ SFTGHISDIN AVSFFPSGYA FATGSDDATC
251 RLFDLRADQE LLLYSHDNII CGITSVAFSK SGRLLLAGYD DFNCSVWDAL
301 KGGRSGVLAG HDNRVSCLGV TDDGMAVATG SWDSFLRIWN
```

Amino acids   340

Database accession numbers

	PIR	SWISSPROT	EMBL/GENBANK	REFERENCE
Mouse	JS0669		M87286	1

Reference

1 Von Weizsacker, E. et al. (1992) Biochem. Biophys. Res. Commun. 183, 350–356.

G-protein γ-subunits

γ1

Molecular weights
Polypeptide   8413
Reduced SDS PAGE   6–8 kDa

Tissue distribution
Highly localized to rod cells of retina [1–3].

Function
Likely interaction with retinal $G_{\alpha t-1}$ [4–8] and β1 [3,9,10] to mediate stimulation of rod cell cyclic GMP-phosphodiesterase by photoactivated rhodopsin [11,12].

Comments
Farnesylation (C-15) and carboxymethylation at Cys71 (C-terminal consensus CAAX where A=aliphatic residue and X=unspecified) is important for membrane association and functional activity. β1γ1C71L is unable to stimulate PLC-β2 activity [13–17]. β1γ1 dimer stimulates phosphoinositide hydrolysis by PLC-β2 [17].

γ1 forms dimers with β1 although not β2 or β3 G-protein subunits [15,18,19].

Cys36 and/or Cys37 are important for interaction with β-subunits [20].

Amino acid sequence (bovine)

```
                                              β
  1 MPVINIEDLT EKDKLKMEVD QLKKEVTLER MLVSKCCEEF RDYVEERSGE
                              $
 51 DPLVKGIPED KNPFKELKGG CVIS
```

Amino acids   74
β   Important for interaction with β-subunits.
$ Site of carboxymethylation and farnesylation.

Database accession numbers

	PIR	SWISSPROT	EMBL/GENBANK	REFERENCES
Bovine	A03153	P02698	K02199	21,22
	A22944		K02436	
			K03255	

References
1 Gautam, N. et al. (1990) Proc. Natl Acad. Sci. USA 87, 7973–7977.
2 **Simon, M. I. et al. (1991) Science 252, 802–808.**
3 Peng, Y.-W. et al. (1992) Proc. Natl Acad. Sci. USA 89, 10882–10886.
4 Lerea, C. L. et al. (1989) Neuron 3, 367–376.
5 Tanabe, T. et al. (1985) Nature 315, 242–245.
6 Medynski, D. et al. (1985) Proc. Natl Acad. Sci. USA 82, 4311–4315.

7 Yatsunami, K. and Gobind Khorana, H. (1985) Proc. Natl Acad. Sci. USA 82, 4316–4320.
8 Lerea, C. L. etal. (1986) Science 234, 77–80.
9 Lee, R. H. et al. (1992) J. Biol. Chem. 267, 24776–24781.
10 Fung, B. K.-K. et al. (1992) J. Biol. Chem. 267, 24782–24788.
11 **Hurley, J. B. (1987) Annu. Rev. Physiol. 49, 793–812.**
12 **Stryer, L. (1991) J. Biol. Chem. 266, 10711–10714.**
13 Fukada, Y. et al. (1990) Nature 346, 658–660.
14 **Spiegel, A. M. et al. (1991) Trends Biochem. Sci. 16, 338–341.**
15 Iniguez-Lluhi, J. A. (1992) J. Biol. Chem. 267, 23409–23417.
16 Ohguro, H. et al. (1991) EMBO J. 10, 3669–3674.
17 Katz, K. et al. (1992) Nature 360, 686–689.
18 Pronin, A. N. and Gautam, N. (1992) Proc. Natl Acad. Sci. USA 89, 6220–6224.
19 Schmidt, C. J. et al. (1992) J. Biol. Chem. 267, 13807–13810.
20 Bubis, J. and Khorana, H. G. et al. (1990) J. Biol. Chem. 265, 12995–12999.
21 Hurley, J. B. et al. (1984) Proc. Natl Acad. Sci. USA 81, 6948–6952.
22 Yatsunami, K. et al. (1985) Proc. Natl Acad. Sci. USA 82, 1936–1940.

γ2

Molecular weights

Polypeptide	7850
Reduced SDS PAGE	6 kDa [1,2]

Tissue distribution

Wide expression, with highest levels in brain [3–5]. Immunoreactivity also detected in retinal cone cells [6].

Function

Component of heterotrimeric G-proteins although specific roles unclear.

Comments

γ2 was originally termed γ6 as a reflection of its apparent molecular mass [4,7].

Geranylgeranylation (C-20) and carboxymethylation at Cys68 (C-terminal consensus CAAX where A=aliphatic residue and X=unspecified) are crucial for membrane association.

β1γ2C68S is cytosolic and unable to regulate adenylyl cyclase types I and II [8–13].

γ2 forms dimers with β1 and β2 G-protein subunits although not with β3 [13–15].

β1γ2 dimer stimulates phosphoinositide hydrolysis by PLC-β2 [16].

Amino acid sequence (bovine)

```
 1 MASNNTASIA QARKLVEQLK MEANIDRIKV SKAAADLMAY CEAHAKEDPL
                 $
51 LTPVPASENP FREKKFFCAI L
```

Amino acids 71

$ Site of carboxymethylation and geranylgeranylation.

Database accession numbers

	PIR	SWISSPROT	EMBL/GENBANK	REFERENCES
Bovine	A41352	P16874	J05071	7,17
	B34228		M37183	

References

1 Graber, S. G. et al. (1992) J. Biol. Chem. 267, 13123–13126.
2 Robishaw, J. D. et al. (1992) Biochem. J. 286, 677–680.
3 Gautam, N. et al. (1990) Proc. Natl Acad. Sci. USA 87, 7973–7977.
4 Cali, J. J. et al. (1992) J. Biol. Chem. 267, 24023–24027.
5 **Simon, M. I. et al. (1991) Science 252, 802–808.**
6 Peng, Y.-W. et al. (1992) Proc. Natl Acad. Sci. USA 89, 10882–10886.
7 Robishaw, J. D. et al. (1989) J. Biol. Chem. 264, 15758–15761.
8 Yamane, H. K. et al. (1990) Proc. Natl Acad. Sci. USA 87, 5868–5872.
9 Mumby, S. M. et al. (1990) Proc. Natl Acad. Sci. USA 87, 5873–5877.
10 Simonds, W. F. et al. (1991) J. Biol. Chem. 266, 5363–5366.
11 Muntz, K. H. et al. (1992) Mol. Biol.Cell 3, 49–61.
12 **Spiegel, A. M. et al. (1991) Trends Biochem. Sci. 16, 338–341.**
13 Iniguez-Lluhi, J. A. (1992) J. Biol. Chem. 267, 23409–23417.
14 Schmidt, C. J. et al. (1992) J. Biol. Chem. 267, 13807–13810.
15 Pronin, A. N. and Gautam, N. (1992) Proc. Natl Acad. Sci. USA 89, 6220–6224.
16 Katz, K. et al. (1992) Nature 360, 686–689.
17 Gautam, N. et al. (1989) Science 244, 971–974.

γ3

Molecular weights

Polypeptide	8305
Reduced SDS PAGE	7 kDa [1,2]

Tissue distribution

Predominantly found in brain, although immunoreactivity also detected in retina [2–5].

Function

In GH3 cells, γ3 interacts specifically with $G_{\alpha oB}$ and β1 to form a heterotrimeric G-protein mediating somatostatin receptor-dependent inhibition of calcium channel activity [6–8]. Wider specificities of interaction in heterotrimeric G-protein function are unclear.

Comments

Forms dimers with β1 and β2 G-protein subunits [1,9].

Consensus (CAAX where A=aliphatic residue and X=unspecified) for carboxymethylation and isoprenylation at Cys72 [10].

Amino acid sequence (bovine)

```
 1 MKGETPVNST MSIGQARKMV EQLKIEASLC RIKVSKAAAD LMTYCDAHAC
                       $
51 EDPLITPVPT SENPFREKKF FCALL
```

Amino acids 75
$ Likely site for carboxymethylation and geranylgeranylation [10].

Database accession numbers

	PIR	SWISSPROT	EMBL/GENBANK	REFERENCE
Bovine			M37182	2
			M58349	

References
1 Robishaw, J. D. et al. (1992) Biochem. J. 286, 677–680.
2 Cali, J. J. et al. (1992) J. Biol. Chem. 267, 24023–24027.
3 Gautam, N. et al. (1990) Proc. Natl Acad. Sci. USA 87, 7973–7977.
4 **Simon, M. I. et al. (1991) Science 252, 802–808.**
5 Peng, Y.-W. et al. (1992) Proc. Natl Acad. Sci. USA 89, 10882–10886.
6 Kleuss, C. et al. (1991) Nature 353, 43–48.
7 Kleuss, C. et al. (1992) Nature 358, 424–426.
8 Kleuss, C. et al. (1993) Science 259, 832–834.
9 Iniguez-Lluhi, J. A. (1992) J. Biol. Chem. 267, 23409–23417.
10 **Spiegel, A. M. et al. (1991) Trends Biochem. Sci. 16, 338–341.**

γ4 (PARTIAL SEQUENCE)
Tissue distribution

Unknown.

Function

In GH3 cells, γ4 interacts specifically with $G_{\alpha oA}$ and β3 to form a heterotrimeric G-protein mediating muscarinic M4 receptor-dependent inhibition of calcium channel activity [1–3].
Wider specificities of interaction in heterotrimeric G-protein function unclear.

Amino acid sequence (mouse) (partial sequence)

```
 1 EACMDRVKVS QAASDLLAYC EAHVREDPLI IPV
```

Database accession numbers

	PIR	SWISSPROT	EMBL/GENBANK	REFERENCE
Mouse	C36204			4

References

[1] Kleuss, C. et al. (1991) Nature 353, 43–48.
[2] Kleuss, C. et al. (1992) Nature 358, 424–426.
[3] Kleuss, C. et al. (1993) Science 259, 832–834.
[4] Gautam, N. et al. (1990) Proc. Natl Acad. Sci. USA 87, 7973–7977.

γ5

Molecular weights

Polypeptide	7318
Reduced SDS PAGE	5–6 kDa [1]

Gene structure (bovine)

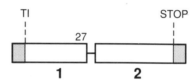

Organization of bovine γ5 subunit adapted from ref. 2. The gene spans approximately 7 kb and is divided into three exons, although the open-reading frame is distributed between exons 1 and 2. Exon 3 comprises the 3' untranslated sequence.

Tissue distribution

Widely expressed, including in kidney, heart, lung, liver and brain [1].

Function

Component of heterotrimeric G-proteins, although specific roles unclear.

Comments

The C-terminal of γ5 deviates from the CAAX (A=aliphatic residue and X=unspecified) motif believed important for polyisoprenylation [3], possessing an aromatic (Phe) and hydroxy (Ser) amino acid at -2 and -3 positions from the C-terminal.

$\beta_1\gamma_5$ and $\beta_2\gamma_5$ dimers stimulate phosphoinositide hydrolysis by PLCβ2 [4].

Amino acid sequence (bovine)

```
 1 MSGSSSVAAM KKVVQQLRLE AGLNRVKVSQ AAADLKQFCL QNAQHDPLLT

51 GVSSSTNPFR PQKVCSFL
```

Amino acids   68

Database accession numbers

	PIR	SWISSPROT	EMBL/GENBANK	REFERENCE
Bovine			M95779	2
Rat			M95780	2

References

1 Cali, J. J. et al. (1992) J. Biol. Chem. 267, 24023–24027.
2 Fisher, K. J. and Aronson, N. N. (1992) Mol. Cell. Biol. 12, 1585–1591.
3 Spiegel, A. M. et al. (1991) Trends Biochem. Sci. 16, 338–341.
4 Wu, O. et al. (1993) Proc. Natl Acad Sci USA 90, 5297–5301.

γ7

Molecular weights

Polypeptide	7553
Reduced SDS PAGE	5–6 kDa [1]

Tissue distribution

Widely expressed, including in brain, heart, spleen, kidney and lung [1].

Function

Component of heterotrimeric G-proteins, although specific roles unclear.

Comments

γ7 was originally termed γ5 as a reflection of its apparent molecular mass [2].

Consensus (CAAX where A=aliphatic residue and X=unspecified) for carboxymethylation and isoprenylation at Cys72 [3].

Amino acid sequence (bovine)

```
  1 MSATNNIAQA RKLVEQLRIE AGIERIKVSK ASSELMSYCE QHARNDPLLV
                    $
 51 GVPASENPFK DKKPCIIL
```

Amino acids 68
$ Likely site for carboxymethylation and geranylgeranylation [3].

Database accession numbers

	PIR	SWISSPROT	EMBL/GENBANK	REFERENCE
Bovine			M99393	1

References

1 Cali, J. J. et al. (1992) J. Biol. Chem. 267, 24023–24027.
2 Robishaw, J. D. et al. (1989) J. Biol. Chem. 264, 15758–15761.
3 Spiegel, A. M. et al. (1991) Trends Biochem. Sci. 16, 338–341.

G-PROTEIN LINKED EFFECTOR AND SECOND MESSENGER SYSTEMS

Introduction: G-protein linked effector and second messenger systems

Heterotrimeric G-protein linked effector and second messenger systems are structurally and functionally more diverse than their "upstream" signalling partners and include retinal cyclic GMP phosphodiesterases (GMP-PDE), ion channels (e.g. potassium, calcium), as well as several phospholipases and adenylyl cyclase subtypes. These control crucial cellular parameters such as membrane potential, cytosolic Ca^{2+} levels and activities associated with a multitude of kinases. In this introductory section we overview the range and characteristics of effector components currently identified molecularly.

PHOSPHOLIPASE C (PHOSPHOINOSITIDASE C)
Phospholipase C and signal transduction

Phosphoinositide-specific phospholipase C (PLC) plays an important role, mediating cellular actions of many hormones, neurotransmitters and some growth factors. Agonist-dependent activation of PLC leads to hydrolysis of membrane phosphatidylinositol 4,5-bisphosphate (PIP2) with consequent generation of the two second messengers inositol 1,4,5-trisphosphate (IP3) and diacylglycerol (DAG). IP3 binds specific intracellular receptors to trigger Ca^{2+} mobilization, while together with membrane phospholipids and Ca^{2+}, DAG mediates activation of a large family of protein kinase C (PKC) isoenzymes [1-3]. It has become clear that multiple PLC subtypes can catalyse this crucial hydrolytic reaction.

Phospholipase C diversity

Early purification of several distinct PLCs from different mammalian tissues [4-11] was followed by molecular cloning of some of these enzymes [12-16]. Based on molecular size, immunoreactivity and predicted amino acid sequences these were categorized into classes which included PLC-β, PLC-γ and PLC-δ [17]. Subsequent cloning of additional isoenzymes displaying structural similarities to each of these divisions [18-23] has led to the current classification of PLC subtypes: PLC-β1, PLC-β2, PLC-β3, PLC-γ1, PLC-γ2, PLC-δ1 and PLC-δ2 (Table 1) [24-26]. Although a PLC-δ3 has apparently been cloned from fibroblasts [24,25] this sequence is not currently

Table 1 *Overall percentage amino acid identities between PLC subtypes*

	PLC-β2	PLC-β3	PLC-γ1	PLC-γ2	PLC-δ1	PLC-δ2
PLC-β1	48	56	22	23	32	34
	PLC-β2	46	23	22	34	34
		PLC-β3	33	21	35	35
			PLC-γ1	51	35	36
				PLC-γ2	35	36
					PLC-δ1	49
						PLC-δ2

Percentage identities were calculated using the GAP routine of the Wisconsin GCG sequence analysis software package version 7. Note that percentage identities as calculated reflect matches over entire molecule and mask two more highly conserved regions (X and Y) identified within all PLC subtypes (see Table 2).

Cloned sequences are from rat (PLC-β1, PLC-β3, PLC-δ1), bovine (PLC-δ2) and human sources (PLC-γ1, PLC-γ2, PLC-β2).

available. It should also be noted that although the original classification included PLC-α purified from several sources [11,17,27], the authenticity of the cloned cDNA [16] is now in doubt as it displays no PLC activity, no homology with other PLC subtypes and is identical to thiol:protein-disulfide oxidoreductase [28-30].

PLC structural organization

Low overall sequence identity between PLC subtypes (21–56%; Table 1) masks two domains of higher homology identified in all isoforms. These stretch over approximately 170 (region X) and 260 residues (region Y) and display sequence identities of 47–77% and 35–80%, respectively (Table 2). Limited proteolysis and analysis of point and deletion mutants suggests that these domains are crucial for catalytic activity [19,31,32,151]. X and Y domains are separated in PLC-βs by a region of 70–120 amino acids rich in serine, threonine and acidic residues [23] and in PLC-γs by an insert of more than 400 residues containing two Src homology 2 (SH2) regions [33] and one SH3 region [34]. Other interesting structural features include a stretch of approximately 450 amino acids rich in basic amino acids following the Y region of PLC-βs [23]. Also, although all PLCs contain approximately 300 amino acids preceding their X domains, this region shows little homology between subtypes. These structural characteristics of PLC subtypes are shown schematically in Fig. 1.

Table 2 *Percentage identity between X and between Y regions (X/Y) of PLC isoenzymes*

	PLC-β2	PLC-β3	PLC-γ1	PLC-γ2	PLC-δ1	PLC-δ2
PLC-β1	76/62	77/80	49/38	52/39	50/36	49/41
	PLC-β2	70/63	52/39	49/38	50/37	51/40
		PLC-β3	47/42	48/40	52/38	52/40
			PLC-γ1	75/58	54/38	50/41
				PLC-γ2	56/35	54/39
					PLC-δ1	64/53
						PLC-δ2

Percentage identities between X and between Y regions (X/Y) were calculated using the GAP routine of the Wisconsin GCG sequence analysis software package version 7. Homology regions compared are PLC-β1 (X: 293–468; Y: 539–791), PLC-β2 (X: 289–464; Y: 541–793), PLC-β3 (partial sequence; X: 275–449; Y: 571–822), PLC-γ1 (X: 296–465; Y: 952–1207), PLC-γ2 (X: 288–457; Y: 929–1182), PLC-δ1 (X: 273–441; Y: 491-750) and PLC-δ2 (X: 267–436; Y: 489-751).

Sequences were taken from rat (PLC-β1, PLC-δ1), bovine (PLC-δ2) and human sources (PLC-γ1, PLC-γ2, PLC-β2) and are given in full in the section.

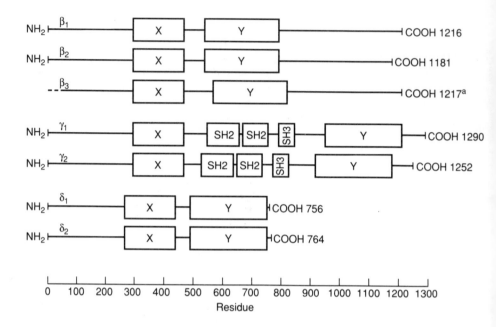

Figure 1 *Linear representation illustrating structural organization of PLC isoenzymes. Two regions of homology between PLC-β, -γ and -δ isoforms likely to play a crucial role in the catalytic activities of these enzymes are boxed and indicated as X and Y. In PLC-γ1 and -γ2 domains X and Y are separated by more than 400 amino acids displaying sequence similarities with regulatory regions of the Src proto-oncogene. Src-homology (SH) 2 and 3 regions are boxed and indicated SH2 and SH3. Note: PLC-β3 represents a partial sequence which, based on molecular size and homology, is likely to be missing approximately 10–20 residues at the N-terminus.*

Catalytic properties

All PLC subtypes display similar enzymatic properties and catalyse the hydrolysis of the three major inositol-containing phospholipids; phosphatidylinositol (PI), phosphatidylinositol 4-monophosphate (PIP) and phosphatidylinositol 4,5-bisphosphate (PIP$_2$) although not other phospholipids. Purified PLCs are highly sensitive to free Ca^{2+} concentrations and undergo powerful activation to hydrolyse PIP$_2$ at concentrations between 10^{-7} M and 10^{-5} M. PLC-dependent hydrolysis of PI generally requires higher Ca^{2+} levels and at low concentrations PIP and PIP$_2$ are preferentially hydrolysed [8–10,17,22-24,35,36].

Mechanisms of PLC activation

Distinct mechanisms of control have been identified for members of the PLC-β and PLC-γ classes while processes regulating PLC-δ isoforms are unknown. PLC-β1 and PLC-β3, and to a lesser extent PLC-β2 (although not PLC-γ1, PLC-γ2 or PLC-δ1), are activated by the pertussis-toxin-insensitive G-protein α-chains $G_{\alpha q}$ and $G_{\alpha 11}$ [20,23,37–43]. $G_{\alpha 14}$, $G_{\alpha 15}$ and $G_{\alpha 16}$ appear to share this activity [23,43,44] although α-chains of the G_i class are ineffective [37,38,41]. PLC-β isoforms (particularly β2 and β3) are also sensitive to activation by G-protein βγ dimers [45–49]. βγ dimers derived from $G_{\alpha i}$ or $G_{\alpha o}$ could therefore contribute to pertussis toxin-sensitive phosphoinositide hydrolysis observed in some cell types. A schematic representation of G-protein-dependent activation of PLC-β is shown in Fig. 2.

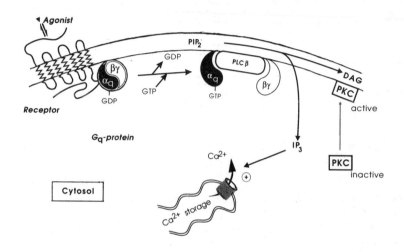

Figure 2 *Schematic representation of receptor-linked phosphoinositide hydrolysis by PLC-β isoforms. Agonist binding its receptor stimulates G-protein dissociation and α-chains of the G_q class mediate activation of PLC-β1 to hydrolyse membrane phosphatidylinositol 4,5-bisphosphate (PIP2). This degradation generates the second messengers inositol 1,4,5-trisphosphate (IP3) and diacylglycerol (DAG) which elevate cytosolic Ca^{2+} and activate protein kinase C (PKC), respectively. G-protein βγ dimers can also stimulate PLC-β (particularly PLC-β2 and PLC-β3 isoforms) and βγ derived from G-proteins of the G_i class may underly pertussis-toxin-sensitive receptor-linked phosphoinositide hydrolysis in some cell types.*

In contrast to PLC-β subtypes, PLC-γ1 activation follows phosphorylation of key tyrosine residues. Phosphorylation can be catalysed by receptors with intrinsic tyrosine kinase activity such as those for EGF, FGF, NGF (p140^{trk}) and PDGF, or by non-receptor tyrosine kinases of the *Src* family following activation of the T cell antigen receptor or immunoglobulin (Ig) receptors IgE (high affinity), IgM (membrane) and IgG [50–64]. PLC-γ2 undergoes similar tyrosine phosphorylation following activation of the PDGF receptor or B cell antigen receptor [31,65–67]. SH2 domains [33] within PLC-γ1 and PLC-γ2 play a key role facilitating catalytic activation through recognizing and binding tightly phosphotyrosines within tyrosine kinases and a model describing these processes has been proposed [25,26].

PHOSPHOLIPASE A$_2$

Phospholipase A$_2$ in signal transduction

Phospholipase A$_2$ (PLA$_2$) hydrolyses the 2-acyl ester bond of membrane phospholipids to generate free fatty acid and lysophospholipids. Arachidonic acid is one major product of PLA$_2$ activity and is the precursor of eicosanoids which have important autocrine and paracrine regulatory actions in many cells types [68,69]. In addition, *cis*-unsaturated fatty acids can facilitate protein kinase C activation [70] while arachidonic acid or its metabolites have also been implicated in control of NF1-GAP activity [71,72], ion channel conductance and neurotransmitter release [73–76]. Lysophospholipids formed in parallel by PLA$_2$ activity possess membrane-lytic activity [77] although diverse biological roles such as chemotaxis, smooth muscle contraction and mitogenesis have also been suggested [70]. Some forms of lysophospholipid can also be used for the synthesis platelet activating factor [78]. Together these observations illustrate the wide-ranging importance of PLA$_2$ activity in both long- and short-term control of cell function.

Stimulation of PLA$_2$ activity has been reported for several G-protein linked receptors as well as some growth factors [79–90] although the molecular steps underlying these responses remain unclear. Hitherto only one PLA$_2$ isoform regulated directly by cell surface receptors has been identified molecularly [91,92]. This enzyme is termed *cytosolic* or 85kD PLA$_2$ and hydrolyses selectively arachidonic acid-containing phospholipids and undergoes translocation to membrane fractions in response to physiologically relevant increases in free calcium [91,93]. Recombinantly expressed 85kD PLA$_2$ also mediates cellular arachidonic acid release by G-protein linked receptors, reflecting elevated cytosolic calcium as well as phosphorylation by MAP kinase [94–96]. 85kD PLA$_2$ also contributes to macrophage eicosanoid generation following bacterial challenge [152]. Direct interaction of cytosolic PLA$_2$ with G-proteins has not been demonstrated.

ADENYLYL CYCLASE
Adenylyl cyclase and signal transduction

Cyclic AMP synthesized by adenylyl cyclase from ATP represents a ubiquitous second messenger of critical importance and regulates functions as wide-ranging as gene transcription, mitogenesis, metabolism, smooth muscle contractile state and ion channel activity. Adenylyl cyclase activity reflects a family of several related but molecularly distinct subtypes under direct control by many hormones, neurotransmitters and growth factors through the intermediacy of heterotrimeric G-proteins. Stimulation follows interaction with G$_{\alpha s}$ subtypes while G$_{\alpha i}$ subtypes mediate enzyme inhibition. G-protein βγ dimers are also important regulators of some adenylyl cyclase subtypes [97–103]. In addition to direct control by G-proteins, some adenylyl cyclase subtypes are under regulation by calcium/calmodulin (Fig. 3) [103–109]. All adenylyl cyclases are sensitive to activation by the diterpene forskolin, or conversely, inhibition by 3'-AMP, adenosine and related analogues (P-site inhibitors) [103].

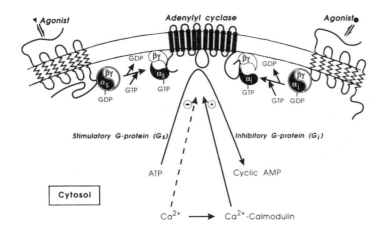

Figure 3 *Schematic representation of adenylyl cyclase regulation by G-protein linked receptors. Cyclic AMP generation follows receptor-linked activation of the heterotrimeric G-protein G_s and interaction of free $G_{\alpha s}$-chain with the adenylyl cyclase catalyst. Some enzyme subtypes are also sensitive to activation by G-protein $\beta\gamma$ (not necessarily derived from G_s) or by calmodulin in the presence of elevated intracellular calcium. Receptors that interact with members of the G_i family mediate adenylyl cyclase inhibition although molecular interactions underlying this remain unclear. Inhibitory mechanisms may involve both $G_{\alpha i}$-chains and $\beta\gamma$ dimers and could vary with different enzyme isoforms. High (submillimolar) concentrations of free calcium inhibit all adenylyl cyclases due to competition with magnesium for the catalytic site, but type VI (and possibly type V) are inhibited by physiologically relevent levels ($\sim 1\ \mu M$). All adenylyl cyclases are stimulated by the diterpene forskolin and inhibited by P-site adenosine analogues [68-80].*

Adenylyl cyclase diversity

Six distinct mammalian adenylyl cyclase isoenzymes currently identified molecularly (as well as an additional partial cDNA) display a primary amino acid sequence identity of 35–67% (Table 3). Adenylyl cyclase type I and III are divergent from each other and all other members hitherto identified. In contrast, type II displays high identity with type IV while type V is more homologous to type VI and these may form subclasses within the adenylyl cyclase family [110] (Fig. 4). Such diversity is likely to underlie distinct mechanisms of regulation by G-proteins, calmodulin and calcium. It is of note that an additional putative adenylyl cyclase cDNA cloned from bovine brain [111] appears to encode neural cell adhesion molecule [112,113] and is not discussed further here.

Adenylyl cyclase structural organization

All cloned adenylyl cyclase subtypes share a common topology consisting of two tandem repeats of six hydrophobic putative transmembrane domains separated by

Table 3 *Amino acid identity between adenylyl cyclase subtypes*

	AC-II	AC-III	AC-IV	AC-V	AC-VI	FRAG
AC-I	40	38	40	47	44	37
	AC-II	35	59	42	41	37
		AC-III	37	37	38	36
			AC-IV	42	42	37
				AC-V	68	40
					AC-VI	41
						FRAG

Percentage identities between adenylyl cyclase subtypes were calculated using the GAP routine of the Wisconsin GCG sequence analysis software package version 7. Note that percentage identities as calculated reflect matches over entire molecule and mask regions of higher conservation in major cytosolic loop located between putative sixth and seventh transmembrane region and in the carboxyl tail.

Sequences were taken from rat (AC-II, AC-III, AC-IV), bovine (AC-I) and canine sources (AC-V, AC-VI). FRAG corresponds to a partial cDNA encoding 675 amino acids of a human adenylyl cyclase with percentage identities calculated based on comparisons with second tandem repeats (C-terminal halves) of other isoenzymes.

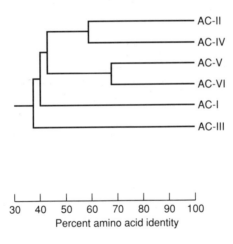

Figure 4 *Structural relationship between adenylyl cyclase subtypes which are grouped according to amino acid identity with branch junctions corresponding to approximate homologies. Percentage sequence identities were calculated using the GAP routine of the Wisconsin GCG sequence analysis software package version 7. Precise values for sequence identity comparisons between all cloned adenylyl cyclase isoforms are given in Table 3.*

a large cytoplasmic loop of approximately 40 kDa and terminating in an intracellular tail of similar size (Fig. 5). Both amino and carboxyl halves of adenylyl cyclase are required for catalytic activity [100,101,103,106]. Primary sequences of adenylyl cyclases differ substantially within the hydrophobic regions while the major intracellular loop between membrane spanning domains VI and VII (C1) and also the cytoplasmic tails (C2) are well conserved between subtypes (Table 4). These regions share homology with guanylyl cyclases and probably underlie the catalytic properties of adenylyl cyclases [110,114–118].

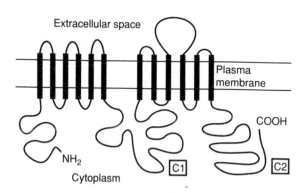

Figure 5 *Schematic representation of adenylyl cyclase topology showing two tandem repeats of six hydrophobic putative transmembrane domains separated by a large cytoplasmic loop (C1) and terminating in a large intracellular tail. (C2). C1 and C2 regions are believed to represent the catalytic core of adenylyl cyclases. Note: several adenylyl cyclase subtypes are believed to be highly glycosylated at sites within extracellular loops.*

Table 4 *Amino acid identity between major cytosolic regions of adenylyl cyclase subtypes*

	AC-II	AC-III	AC-IV	AC-V	AC-VI	FRAG
AC-I	64/55	57/54	61/56	72/59	73/58	/55
	AC-II	49/52	77/76	62/57	62/56	/56
		AC-III	50/52	58/52	58/51	/56
			AC-IV	57/56	57/59	/57
				AC-V	94/92	/58
					AC-VI	/57
						FRAG

Percentage identities between major cytosolic loops (C1) and between carboxyl tails (C2) (C1 and C2 are as indicated in Fig. 4) were calculated using the GAP routine of the Wisconsin GCG sequence analysis software package version 7 and are given above as C1/C2. Comparisons were made between the following residues of adenylyl cyclase type I, C1 248–432, C2 807–1070; type II, C1 212-416, C2 822–1088; type III, C1 248–446, C2 859–1132; type IV, C1 196–400, C2 806–1062; type V, C1 323–519, C2 930–1184; type VI, C1 309–504, C2 912–1165.

Sequences were taken from rat (types II, III and IV), bovine (type I) and canine sources (types V and VI) and are given in full in the section. FRAG indicates partial sequence for human adenylyl cyclase with comparisons made with C2 regions over residues 342–607.

RETINAL PHOTORECEPTOR 3',5'-CYCLIC-GMP PHOSPHODIESTERASE

cGMP-PDE oligomeric subunit components

Retinal 3',5'-cGMP phosphodiesterase (cGMP-PDE), located in photoreceptor outer segments, is light activated and plays a critical role in visual signal transduction. In retinal rod cells cGMP-PDE is oligomeric, consisting of an αPDE-, a βPDE-, and two identical inhibitory γPDE-subunits ($\alpha\beta(\gamma2)$). In retinal cones, cGMP-PDE is composed of a catalytic homodimer of two α'PDE-chains which are associated with several smaller subunits, one of which is similar to the rod γ-subunit. In addition to their catalytic activities to hydrolyse cGMP to 5'-GMP, both rod and cone enzymes bind cGMP with high affinity [119–124].

cGMP-PDE and rod visual signal transduction

In rod cells several crucial molecular events underlying the conversion of light information into electrical nerve activity have been identified (Fig. 6). In the dark, basal cyclicGMP levels maintain a state of rod photoreceptor membrane depolarization by allosterically increasing sodium and calcium conductance through cyclic nucleotide-gated cation channels. Upon light absorption, photoisomerization of the 11-*cis*-retinal chromophore covalently linked to

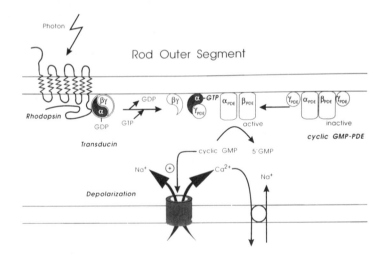

Figure 6 *Schematic representation of rhodopsin-linked activation of cGMP-phosphodiesterase (PDE) in rod outer segments. In the dark, basal cGMP levels maintain membrane depolarization by increasing cation channel conductance. Oligomeric cGMP-PDE is inactive due to interaction with inhibitory γPDE subunits. Photostimulated rhodopsin activates heterotrimeric transducin which dissociates with subsequent interaction between transducin α-chains and γPDE subunits liberating catalytically active $\alpha\beta$PDE. Cyclic GMP hydrolysis results in cation channel closure and hyperpolarization.*

rhodopsin leads to formation of an all-*trans* form of the chromophore and photoactivated receptor. This interacts with the heterotrimeric G-protein transducin ($G_{\alpha t-1}\beta\gamma$) which, following exchange of GDP for GTP, dissociates allowing free $G_{\alpha t-1}$ to activate cGMP-PDE. This then hydrolyses cGMP to 5'-GMP, leading to closure of the cGMP-gated cation channel, membrane hyperpolarization and consequently a change in nerve activity. Rather than a direct stimulatory action by transducin, it appears that cGMP-PDE activation follows interaction of $G_{\alpha t-1}$ with inhibitory γPDE-subunits (Fig. 6). Overall, the speed and degree of amplification wihin rod visual signal transduction is impressive, with photoexcitation of a single rhodopsin inducing hydrolysis of up to 10^5 cGMP molecules and closure of cation channels within a few hundred milli-seconds [125-134].

Recovery of dark state

Several mechanisms underly rapid recovery of the dark state after light excitation. These include rhodopsin phosphorylation (serine/threonine) by rhodopsin kinase, together with its association with the capping protein arrestin to block further interaction with transducin. Transducin is also deactivated by endogenous GTPase activity, a process which is regulated by interaction with cGMP-PDE and which leads to reassociation of $G_{\alpha t}$ with G-protein $\beta\gamma$ and liberation of the inhibitory γPDE-subunit. Also, under conditions of diminished cation influx during channel closure, calcium is continually exported and diminished intracellular levels both facilitate cGMP-dependent channel opening and promote guanylyl cyclase activation. Together these events elevate intracellular cGMP, re-open cation channels and re-establish a state of membrane depolarization [127,131,135-143].

Colour vision in cone cells

Retinal cone cells mediate vertebrate colour vision and signal transduction processes appear similar to rods. Three human light-sensitive pigments for red, blue or green light (opsins) also use 11-*cis*-retinal as a chromophore and, like rhodopsin, belong to the seven transmembrane receptor superfamily. Opsins interact with a cone-specific form of transducin ($G_{\alpha t-2}$) to mediate activation of cGMP-PDE itself comprising molecularly distinct subunits [131,144-148]. It is of note that despite many parallels between visual transduction in rods and cones there are clear functional differences in performance. Rods respond slowly but are extremely sensitive to light while cones are less sensitive although the response is terminated more rapidly and also adapts to high levels of light more readily than rods [149,150].

References.

[1] Berridge, M. J. (1987) Annu. Rev. Biochem. 56, 159–193.

[2] Nishizuka, Y. (1992) Science 258, 607–614.

[3] Michell, R. H. (1992) Trends Biochem. Sci. 17, 274–276.

[4] Hofmann, S. L. and Majerus, P. W. (1982) J. Biol. Chem. 257, 6461–6469.

[5] Ryu, S. H. et al. (1986) Biochem. Biophys. Res. Commun. 141, 137–144.

[6] Rebecchi, M. J. and Rosen, O. M. (1987) J. Biol. Chem. 262, 12526–12532.

7 Homma, Y. et al. (1988) J. Biol. Chem. 263, 6592–6598.

8 Ryu, S. H. et al. (1987) J. Biol. Chem. 262, 12511–12518.

9 Ryu, S. H. et al. (1987) Proc. Natl Acad. Sci. USA 84, 6649–6653.

10 Katan, M. and Parker, P. J. (1987) Eur. J. Biochem. 168, 413–418.

11 Bennett, C. F. and Crooke, S. T. (1987) J. Biol. Chem. 262, 13789–13797.

12 Stahl, M. L. et al. (1988) Nature 332, 269–272.

13 Suh, P. G. et al. (1988) Proc. Natl Acad. Sci. USA 85, 5419–5423.

14 Suh, P. G. et al. (1988) Cell 54, 161–169.

15 Katan, M. et al. (1988) Cell 54, 171–177.

16 Bennett, C. F. et al. (1988) Nature 334, 268–270.

17 **Rhee, S. G. et al. (1989) Science 244, 546–550.**

18 Ohta, S. et al. (1988) FEBS Lett. 242, 31–35.

19 Emori, Y. et al. (1989) J. Biol. Chem. 264, 21885–21890.

20 Park, D. et al. (1992) J. Biol. Chem. 267, 16048–16055.

21 Meldrum, E. et al. (1991) Eur. J. Biochem. 196, 159–165.

22 Carozzi, A. J. et al. (1992) Eur. J. Biochem. 210, 521–529.

23 Jhon, D.-Y. et al. (1993) J. Biol. Chem. 268, 6654–6661.

24 Kriz, R. et al. (1990) Ciba Found. Symp.150, 112–127.

25 **Rhee, S. G. (1991) Trends Biochem. Sci. 16, 297–301.**

26 **Rhee, S. G. and Choi, K. D. (1992) J. Biol. Chem. 267, 12393–12396.**

27 **Cockcroft, S. and Thomas, G. M. H. (1992) Biochem. J. 288, 1–14.**

28 Srivastava, S. P. et al. (1991) J. Biol. Chem. 266, 20337–20344.

29 Urade, R. et al. (1992) J. Biol. Chem. 267, 15152–15159.

30 Martin, J. L. et al. (1991) Biochem. Biophys. Res. Commun. 178, 679–685.

31 Sultzman, E. et al. (1991) Mol. Cell. Biol. 11, 2018–2025.

32 Wu, D. et al. (1993) J. Biol. Chem. 268, 3704–3709.

33 **Anne Koch, C et al. (1991) Science 252, 668–674.**

34 **Musacchio, A. et al. (1992) FEBS Lett. 307, 55–61.**

35 Meldrum, E. et al. (1989) Eur. J. Biochem. 182, 673–677.

36 Ginger, R. S. and Parker, P. J. (1992) Eur. J. Biochem. 210, 155–160

37 Taylor, S. J. et al. (1991) Nature 350, 516–518.

38 Smrcka, A. V. et al. (1991) Science 251, 804–807.

39 Taylor, S. J. and Exton, J. H. (1991) FEBS Lett. 286, 214–216.

40 Blank, J. L. et al. (1991) J. Biol. Chem. 266, 18206–18216.

41 Wu, D. et.al. (1992) J. Biol. Chem. 267, 1811–1817.

42 Bernstein, G. et al. (1992) J. Biol. Chem. 267, 8081–8088.

43 Wu, D. et al. (1992) J. Biol. Chem. 267, 25798–25802.

44 Lee, C. H. et al. (1992) J. Biol. Chem. 267, 16044–16047.

45 Camps, M. et al. (1992) Nature 360, 684–686.

46 Katz, A. et al. (1992) Nature 360, 686–689.

47 Blank, J. L. et al. (1992) J. Biol. Chem. 267, 23069–23075.

48 Carozzi, A. et al. (1993) FEBS Lett. 315, 340–342.

49 Park, D. et al. (1993) J. Biol. Chem. 268, 4573–4576.

50 Kim, H. K. et al. (1991) Cell 65, 435–441.

51 Wahl, M. I. et al. (1988) Science 241, 968–970.

52 Wahl, M. I. et al. (1992) J. Biol. Chem. 267, 10447–10456.

53 Goldschmidt-Clermont, P. J. et al. (1991) Science 251, 1231–1233.

54 Nishibe, S. et al. (1990) Science 250, 1253–1256.

[55] Vetter, M. L. et al. (1991) Proc. Natl Acad. Sci. USA 88, 5650–5654.
[56] Kim, U.-H. et al. (1991) J. Biol. Chem. 266, 1359–1362.
[57] Carter, R. H. et al. (1991) Proc. Natl Acad. Sci. USA 88, 2745–2749.
[58] Park, D. J. et al. (1991) Proc. Natl Acad. Sci. USA 88, 5453–5456.
[59] Weiss, A. et al. (1991) Proc. Natl Acad. Sci. USA 88, 5484–5488.
[60] Park, D. J. et al. (1991) J. Biol. Chem. 266, 24237–24240.
[61] Liao, F. et al. (1992) Proc. Natl Acad. Sci. USA 89, 3659–3663.
[62] Li, W. et al. (1992) Mol. Cell. Biol. 12, 3176–3182.
[63] **Schlessinger, J. and Ullrich, A. (1992) Neuron 9, 383–391.**
[64] Granja, C. et al. (1991) J. Biol. Chem. 266, 16277–16280.
[65] Coggeshall, K. M. et al. (1992) Proc. Natl Acad. Sci. USA 89, 5660–5664.
[66] Roifman, C. M. and Wang, G. (1992) Biochem. Biophys. Res. Commun. 183, 411–416.
[67] Homma, Y. et al. (1993) Biochem. J. 290, 649–653.
[68] **Smith, W. L. (1989) Biochem. J. 259, 315–324.**
[69] **Shimizu, T. and Wolfe, L. S. (1990) J. Neurochem. 55, 1–15.**
[70] **Nishizuka, Y. (1992) Science 258, 607–614.**
[71] Bollag, G. and McCormick, F. (1991) Nature 351, 576–579.
[72] Golubic, M. et al. (1991) EMBO J. 10, 2897–2903.
[73] Miller, B. et al. (1992) Nature 355, 722–725.
[74] Hwang, T.-C. et al. (1990) Proc. Natl Acad. Sci. USA 87, 5706–5709.
[75] Kim, D. and Clapham, D. E. (1989) Science 244, 1174–1176.
[76] Piomelli, D. and Greengard, P. (1990) Trends Pharmacol. Sci. 11, 367–373.
[77] Weltzien, H. U. (1979) Biochim. Biophys. Acta 559, 259–264.
[78] **Hanahan, D. J. (1986) Annu. Rev. Biochem. 55, 483–509.**
[79] Burch, R. M. et al.(1986) J. Biol. Chem. 261, 11236–11241.
[80] Ho, A. K. and Klein, D. (1987) J. Biol. Chem. 262, 11764–11770.
[81] **Axelrod, J. et al. (1988) Trends Neurosci. 11, 117–123.**
[82] Margolis, B. L. et al. (1988) Biochem. J. 265, 469–474.
[83] Felder, C. C. et.al. (1990) Proc. Natl Acad. Sci. USA 87, 2187–2191.
[84] Goldberg, H. J. et al. (1990) Biochem. J. 267, 461–465.
[85] Gupta, S. K. et al. (1990) Science 249, 662–666.
[86] Teitelbaum, I. et al. (1990).J. Biol. Chem. 265, 4218–4222.
[87] Bonventre, J. V. et al. (1990) J. Biol. Chem. 265, 4934–4938.
[88] Weiss, B. A. and Insel, P. A. (1991) J. Biol. Chem. 266, 2126–2133.
[89] Currie, S. et al. (1992) J. Biol. Chem. 267, 6056–6062.
[90] Spaargaren, M. et al. (1992) Biochem. J. 287, 37–43.
[91] Clark, J. D. et al. (1991) Cell 65, 1043–1051.
[92] Sharp, J. D. et al. (1991) J. Biol. Chem. 266, 14850–14853.
[93] **Mayer, R. J. and Marshall, L. A. (1993) FASEB J. 7, 339–348.**
[94] Lin, L.-L. et al. (1992) Proc. Natl Acad. Sci. USA 89, 6147–6151.
[95] Lin, L.-L. et al. (1993) Cell 72, 269–278.
[96] Nemenoff, R. A. et al. (1993) J. Biol. Chem. 268, 1960–1964.
[97] **Gilman, A. G. (1987) Annu. Rev. Biochem. 56, 615–649.**
[98] **Birnbaumer, L. et al. (1990) Biochim. Biophys. Acta 1031, 163–224.**
[99] **Levitzki, A. and Bar-Sinai, A. (1991) Pharma. Ther. 50, 271–283.**
[100] Tang, W.-J. et al. (1991) J. Biol. Chem. 266, 8595–8603.
[101] Tang, W.-J. and Gilman, A. G. (1991) Science 254, 1500–1503.

102 Federman, A. D. et al. (1992) Nature 356, 159–161.
103 **Tang, W.-J. and Gilman, A. G. (1992) Cell 70, 869–872.**
104 Harrison, J. K. et al. (1989) J. Biol. Chem. 264, 15880–15885.
105 Choi, E.-J. et al. (1992) J. Biol. Chem. 267, 12440–12442.
106 Tang, W.-J. et al. (1991) J. Biol. Chem. 266, 8595–8603.
107 Choi, E.-J. et al. (1992) Biochemistry 31, 6492–6498.
108 Caldwell, K. K. et al. (1992) Cell Calcium 13, 107–121.
109 Yoshimura, M. and Cooper, D. M. F. (1992) Proc. Natl Acad. Sci. USA 89, 6716–6720.
110 Katsushika, S. et al. (1992) Proc. Natl Acad. Sci. USA 89, 8774–8778.
111 Lipkin, V. M. et al. (1989) FEBS Lett. 254, 69–73.
112 Premont, R. T. (1992) FEBS Lett. 295, 230–231.
113 Lipkin, V. M. et al. (1992) FEBS Lett. 304, 9–11.
114 Chinkers, M. et al. (1989) Nature 338, 78–83.
115 **Chinkers, M. and Garbers, D. L. (1991) Annu. Rev. Biochem. 60, 553–576.**
116 Singh, S. et al. (1988) Nature 334, 708–712.
117 Krupinski, J. et al. (1989) Science 244, 1558–1564.
118 Gao, B. and Gilman, A. G. (1991) Proc. Natl Acad. Sci. USA 88, 10178–10182.
119 **Stryer, L. (1986) Annuu. Rev. Neurosci. 9, 87–119.**
120 **Hurley, J. B. (1987) Ann. Rev. Physiol. 49, 793–812.**
121 Deterre, P. et al. (1988) Proc. Natl Acad. Sci. USA 85, 2424–2428.
122 Gillespie, P. G. and Beavo, J. A. (1988) J. Biol. Chem. 263, 8133–8141.
123 Gillespie, P. G. and Beavo, J. A. (1989) Proc. Natl Acad. Sci. USA 86, 4311–4315.
124 Fung, B. K.-K. et al. (1990) Biochemistry 29, 2657–2664.
125 Hurley, J. B. and Stryer, L. (1982) J. Biol. Chem. 257, 11094–11099.
126 Brown, R. L. and Stryer, L. (1989) Proc. Natl Acad. Sci. USA 86, 4922–4926.
127 Fung, B. K.-K. and Griswold-Prenner, I. (1989) Biochemistry 28, 3133–3137.
128 Griswold-Prenner, I. et al. (1989) Biochemistry 28, 6145–6150.
129 Fung, B. K.-K. et al. (1990) Biochemistry 29, 2657–2664.
130 Wensel, T. G. and Stryer, L. (1990) Biochemistry 29, 2155–2161.
131 **Stryer, L. (1991) J. Biol. Chem. 266, 10711–10714.**
132 **Chabre, M. and Deterre, P. (1989) Eur. J. Biochem. 179, 255–266.**
133 Clerc, A. and Bennett, N. (1992) J. Biol. Chem. 267, 6620–6627.
134 Clerc, A. et al. (1992) J. Biol. Chem. 267, 19948–19953.
135 Wilden, U. et al. (1986) Proc. Natl Acad. Sci. USA 83, 1174–1178.
136 Miller, J. L. et al. (1986) Biochemistry 25, 4983–4988.
137 Yamazaki, A. et al. (1990) J. Biol. Chem. 265, 11539–11548.
138 Vuong, T. M. and Chabre, M. (1990) Nature 346, 71–74.
139 Dizhoor, A. M. et al. (1991) Science 251, 915–918.
140 Arshavsky, V. Y. and Bownds, M. D. (1992) Nature 357, 416–417.
141 Ray, S. et al. (1992) Proc. Natl Acad. Sci. USA 89, 5705–5709.
142 Pages, F. et al. (1992) J. Biol. Chem. 267, 22018–22021.
143 Hsu, Y.-T. and Molday, R. S. (1993) Nature 361, 76–79.
144 Nathans, J. et al. (1986) Science 232, 193–202.
145 Lerea, C. L. et al. (1986) Science 234, 77–79.
146 Hamilton, S. E. and Hurley, J. B. (1990) J. Biol. Chem. 265, 11259–11264.
147 Li, T. et al. (1990) Proc. Natl Acad. Sci. USA 87, 293–297.

148 Charbonneau, H. et al. (1990) Proc. Natl Acad. Sci. USA 87, 288–292.
149 Normann, R. A. and Werblin, F. S. (1974) J. Gen. Physiol. 63, 37–61.
150 Baylor, D. A. (1987) Invest. Ophthalmol. Vis. Sci. 28, 34–49.
151 Ellis, M. V. et al. (1993) Eur. J. Biochem. 213, 339–347.
152 Svensson, U. et al. (1993) Eur. J. Biochem. 213, 81–86.

PHOSPHOLIPASE C-β1

Molecular weights

Polypeptide	138 344
Reduced SDS PAGE	140–154 kDa

Gene structure and localization

Unknown.

Tissue Distribution

High levels in brain, particularly frontal and pyriform cortex, caudate-putamen, ventral pallidum, olfactory bulb, hippocampus, dendate gyrus, thalamus and granule layer of cerebellum. In the periphery, it is found in adrenal, lung, parotid, uterus, liver and ovary. Expressed in several cell lines including fibroblasts, colon carcinoma and osteosarcoma. Undetectable in HL60 promyelocytes and B cell lines [1–5].

Function

Hydrolysis of membrane phosphoinositides to generate second messengers IP_3 and DAG. Receptor-linked activation by pertussis-toxin-insensitive G-proteins of the Gq family including $G_{\alpha q}$, $G_{\alpha 11}$, $G_{\alpha 14}$, $G_{\alpha 15}$ and $G_{\alpha 16}$. Other G-proteins such as $G_{\alpha o}$, $G_{\alpha i}$ and $G_{\alpha z}$ are inactive [5–13]. Weak activation by G-protein βγ dimers also reported [14–16].

Comments

PLC-β1 has been detected in the nucleus of some cell types [17,18].

Phosphorylation by PKC at Ser887 may underlie feedback inhibition [19].

Regions X (residues 293–468) and Y (residues 539–791) conserved between PLC subtypes are believed important for catalytic activity [5,20,21].

Deletion mutants show residues 903–1030 (P box) and 1031–1142 (G box) are crucial for interaction with $G_{\alpha q}$. P box is also important for association with particulate fractions [20].

Amino acid sequence (rat)

```
  1 MAGAQPGVHA LQLKPVCVSD SLKKGTKFVK WDDDSTIVTP IILRTDPQGF

 51 FFYWTDQNKE TELLDLSLVK DARCGKHAKA PKDPKLRELL DVGNIGHLEQ

101 RMITVVYGPD LVNISHLNLV AFQEEVAKEW TNEVFSLATN LLAQNMSRDA

151 FLEKAYTKLK LQVTPEGRIP LKNIYRLFSA DRKRVETALE ACSLPSSRND

201 SIPQEDFTPD VYRVFLNNLC PRPEIDNIFS EFGAKSKPYL TVDQMMDFIN

                                                  x region
251 LKQRDPRLNE ILYPPLKQEQ VQVLIEKYEP NSSLAKKGQM SVDGFMRYLS
```

```
 301  GEENGVVSPE KLDLNEDMSQ PLSHYFINSS HNTYLTAGQL AGNSSVEMYR

 351  QVLLSGCRCV ELDCWKGRTA EEEPVITHGF TMTTEISFKE VIEAIAECAF

 401  KTSPFPILLS FENHVDSPKQ QAKMAEYCRL IFGDALLMEP LEKYPLESGV

 451  PLPSPMDLMY KILVKNKKKS HKSSEGSGKK KLSEQASNTY SDSSSVFEPS
                                                   y region
 501  SPGAGEADTE SDDDDDDDDC KKSSMDEGTA GSEAMATEEM SNLVNYIQPV

 551  KFESFETSKK RNKSFEMSSF VETKGLEQLT KSPVEFVEYN KMQLSRIYPK

 601  GTRVDSSNYM PQLFWNAGCQ MVALNFQTVD LAMQINMGMY EYNGKSGYRL

 651  KPEFMRRPDK HFDPFTEGIV DGIVANTLSV KIISGQFLSD KKVGTYVEVD

 701  MFGLPVDTRR KAFKTKTSQG NAVNPVWEEE PIVFKKVVLP SLACLRIAAY

 751  EEGGKFIGHR ILPVQAIRPG YHYICLRNER NQPLMLPAVF VYIEVKDYVP

 801  DTYADVIEAL SNPIRYVNLM EQRAKQLAAL TLEDEEEVKK EADPGETSSE

 851  APSETRTTPA ENGVNHTATL APKPPSQAPH SQPAPGSVKA PAKTEDLIQS

 901  VLTEVEAQTI EELKQQKSFV KLQKKHYKEM KDLVKRHHKK TTELIKEHTT

 951  KYNEIQNDYL RRRAALEKSA KKDSKKKSEP SSPDHGSSAI EQDLAALDAE

1001  MTQKLIDLKD KQQQQLLNLR QEQYYSEKYQ KREHIKLLIQ KLTDVAEECQ

1051  NNQLKKLKEI CEKEKKELKK KMDKKRQEKI TEAKSKDKSQ MEEEKTEMIR

1101  SYIQEVVQYI KRLEEAQSKR QEKLVEKHKE IRQQILDEKP KLQMELEQEY

1151  QDKFKRLPLE ILEFVQEAMK GKVSEDSNHG SAPPSLASDP AKVNLKSPSS

1201  EEVQGENAGR EFDTPL
```

Amino acids   1216
X and Y regions are boxed [5,6,21].
Ser887 is site of phosphorylation by PKC (in bold) [19].

Database accession numbers

	PIR	SWISSPROT	EMBL/GENBANK	REFERENCES
Rat	A28821	P10687	M20636	[22]
Bovine		P10894	J03137	[1]

References

1 Katan, M. et al. (1988) Cell 54, 171–177.
2 Gerfen, C. R. et al. (1988) Proc. Natl Acad. Sci. USA 85, 3208–3212.
3 Hempel, W. M. and DeFranco, A. L. (1991) J. Immunol. 146, 3713–3720.
4 Ross, C. A. et al. (1989) Proc. Natl Acad. Sci. USA 86, 2923–2927.
5 Jhon, D.-K. (1993) J. Biol. Chem. 268, 6654–6661.
6 Park, D. et al. (1992) J. Biol. Chem. 267, 16048–16055.
7 Smrcka, A. V. et al. (1991) Science 250, 804–807.
8 Taylor, S. J. et al. (1991) Nature 350, 516–518.
9 Wu, D. et al. (1992) J. Biol. Chem. 267, 1811–1817.
10 Bernstein, G. et al. (1992) J. Biol. Chem. 267, 8081–8088.
11 Blank, J. L. et al. (1991) J. Biol. Chem. 266, 18206–18216.
12 Lee, C. H. et al. (1992) J. Biol. Chem. 267, 16044–16047.
13 Wu, D. et al. (1992) J. Biol. Chem. 267, 25798–25802.
14 Park, D. et al. (1993) J. Biol. Chem. 268, 4573–4576.
15 Carozzi, A. et al. (1993) FEBS Lett. 315, 340–342.
16 Camps, M. et al. (1992) Nature 360, 684–686.
17 Martelli, A. M. et al. (1992) Nature 358, 242–245.
18 Divecha, N. et al. (1993) Biochem. J. 289, 617–620.
19 Ryu, S. H. et al. (1990) J. Biol. Chem. 265, 17941–17945.
20 Wu, D. et al. (1993) J. Biol. Chem. 268, 3704–3709.
21 **Rhee, S. G. and Choi, K. D. (1992) Adv. Sec. Mess. Phosph. Res. 26, 35–61.**
22 Suh, P.-G. et al. (1988) Cell 54, 161–169.

PHOSPHOLIPASE C-β2

Molecular weight

| Polypeptide | 133 700 |
| Reduced SDS PAGE | 140 kDa |

Gene structure and localization

Unknown.

Tissue distribution

Cloned from human promyelocytic cell line HL60 cDNA library [1]. Undetectable in many peripheral tissues including brain, liver, heart and kidney [2].

Function

Hydrolysis of membrane phosphoinositides to generate second messengers IP3 and DAG. Receptor-linked regulation by G-proteins although only weak activation by $G\alpha q$, $G\alpha 11$ and $G\alpha 16$ [2,3]. Powerful activation by βγ dimers indicates a potential role mediating pertussis-toxin-sensitive phosphoinositide hydrolysis by G-proteins of the G_i class [4-6].

Comments

Regions X (residues 289–464) and Y (residues 541–793) conserved between PLC subtypes are believed important for catalytic activity [1,2,7–9].

PLC-β2 activation by β1γl requires γ-subunit isoprenylation [4].

Amino acid sequence

```
   1  MSLLNPVLLP PKVKAYLSQG ERFIKWDDET TVASPVILRV DPKGYYLYWT

  51  YQSKEMEFLD ITSIRDTRFG KFAKMPKSQK LRDVFNMDFP DNSFLLKTLT

 101  VVSGPDMVDL TFHNFVSYKE NVGKAWAEDV LALVKHPLTA NASRSTFLDK

 151  ILVKLKMQLN SEGKIPVKNF FQMFPADRKR VEAALSACHL PKGKNDAINP

 201  EDFPEPVYKS FLMSLCPRPE IDEIFTSYHA KAKPYMTKEH LTKFINQKQR
                                                  X region
 251  DSRLNSLLFP PARPDQVQGL IDKYEPSGIN AQRGQLSP EG MVWFLCGPEN

 301  SVLAQDKLLL HHDMTQPLNH YFINSSHNTY LTAGQFSGLS SAEMYRQVLL

 351  SGCRCVELDC WKGKPPDEEP IITHGFTMTT DIFFKEAIEA IAESAFKTSP

 401  YPIILSFENH VDSPRQQAKM AEYCRTIFGD MLLTEPLEKF PLKPGVPLPS

 451  PEDLRGKILI KNKKNQFSGP TSSSKDTGGE AEGSSPPSAP AVWAGEEGTE
                                                       Y region
 501  LEEEEVEEEE EEESGNLDEE EIKKMQSDEG TAGLEVTAYE EMSSLVNYIQ

 551  PTKFVSFEFS AQKNRSYVIS SFTELKAYDL LSKASVQFVD YNKRQMSRIY

 601  PKGTRMDSSN YMPQMFWNAG CQMVALNFQT MDLPMQQNMA VFEFNGQSGY

 651  LLKHEFMRRP DKQFNPFSVD RIDVVATTL SITVISGQFL SERSVRTYVE

 701  VELFGLPGDP KRRYRTKLSP STNSINPVWK EEPFVFEKIL MPELASLRVA

 751  VMEEGNKFLG HRIIPINALN SGYHHLCLHS ESNMPLTMPA LFIF LEMKDY

 801  IPGAWADLTV ALANPIKFFS AHDTKSVKLK EAMGGLPEKP FPLASPVASQ

 851  VNGALAPTSN GSPAARAGAR EEAMKEAAEP RTASLEELRE LKGVVKLQRR

 901  HEKELRELER RGARRWEELL QRGAAQLAEL GPPGVGGVGA CKLGPGKGSR

 951  KKRSLPREES AGAAPGEGPE GVDGRVRELK DRLELELLRQ GEEQYECVLK

1001  RKEQHVAEQI SKMMELAREK QAAELKALKE TSENDTKEMK KKLETKRLER
```

```
1051  IQGMTKVTTD KMAQERLKRE INNSHIQEVV QVIKQMTENL ERHQEKLEEK

1101  QAACLEQIRE MEKQFQKEAL AEYEARMKGL EAEVKESVRA CLRTCFPSEA

1151  KDKPERACEC PPELCEQDPL IAKADAQESR L
```

Amino acids   1181
X and Y regions are boxed [1,2,7].

Database accession numbers

	PIR	SWISSPROT	EMBL/GENBANK	REFERENCE
Human			M95678	1

References
[1] Park, D. et al. (1992) J. Biol. Chem. 267, 16048–16055.
[2] Jhon, D.-J. et al. (1993) J. Biol. Chem. 268, 6654–6661.
[3] Lee, C. H. et al. (1992) J. Biol. Chem. 267, 16044–16047.
[4] Katz, A. et al. (1992) Nature 360, 686–688.
[5] Camps, M. et al. (1992) Nature 360, 684–686.
[6] Park, D. et al. (1993) J. Biol. Chem. 268, 4573–4576.
[7] **Rhee, S. G. and Choi, K. D. (1992) Adv. Sec. Mess. Phosph. Res. 26, 35–61.**
[8] **Rhee, S. G. and Choi, K. D. (1992) J. Biol. Chem. 267, 12393–12396.**
[9] Wu, D. et al. (1993) J. Biol. Chem. 268, 3704–3709.

PHOSPHOLIPASE C-β3

Molecular weight
Polypeptide	137 721 (incomplete sequence)
Reduced SDS PAGE	140–158 kDa (endogenous protein) [1,2]

Gene structure and localization

Unknown.

Tissue distribution

Detected in parotid gland, brain and testes. Less abundant in liver, lung and uterus with low levels in heart, adrenal gland and ovary [1]. Also expressed in many cell lines including NG108–15 neurogliohybridoma, 3T3 fibroblasts and HeLa S3 cells [2]. Cloned from FRTL thyroid cell cDNA library [1] and WI-38 fibroblasts [2].

Function

Hydrolysis of membrane phosphoinositides to generate second messengers IP$_3$ and DAG. Receptor-linked activation by heterotrimeric G-proteins of the Gq family including G$_{\alpha q}$, G$_{\alpha 11}$ and G$_{\alpha 16}$ [1]. Powerful activation also by G-protein βγ dimers [3,4]. PLC-β3 could underlie both pertussis-toxin-insensitive (Gq α-chains) and sensitive (βγ from G-proteins of G$_i$ class) receptor-linked phosphoinositide hydrolysis.

Comments

Comparison with PLC-β1 and PLC-β2 suggests rat sequence is incomplete and lacks 10–20 amino acids at the N-terminus [1]. Cloned human PLC-β3 sequence is also incomplete and lacks approximately 200 N-terminal residues [2].

Regions X (residues 275–449) and Y (residues 571–822) conserved between PLC subtypes are believed important for catalytic activity [1,2,5–7].

Amino acid sequence (rat) (incomplete sequence)

```
  1 VVETLRRGSK FIKWDEEASS RNLVTLRLDP NGFFLYWTGP NMEVDTLDIS

 51 SIRDTRTGRY ARLPKDPKIR EVLGFGGPDT RLEEKLMTVV AGPDPVNTTF

101 LNFMAVQDDT VKVWSEELFK LAMNILAQNA PEHVLRKAYT KLKLQVNQDG

151 RIPVKNILKM FSADKKRVET ALICGLNFNR SESIRPDEFS LEIFERFLNK

201 LLLRPDIDKI LLEIGAKGKP YLTLEQLMDF INQKQRDPRL NEVLYPPLRS
                                          X region
251 SQARLLIEKY EPNKQFLERD QMSMEGFSRY LGGEENGILP LEALDLSMDM

301 TQPLSAYFIN SSHNTYLTAG QLAGTSSVEM YRQALLWGCR CVELDVWKGR

351 PPEEEPFITH GFTMTTEVPL RDVLEAIAET AFKTSPYPVI LSFENHVDSA

401 KQQAKMAEYC RSIFGEALLI DPLDKYPLSA GTPLPSPQDL MGRILVKNKK

451 RHRPSTGVPD SSVAKRPLEQ SNSALSESSA ATEPSSPQLG SPSSDSCPGL

501 SNGEEVGLEK TSLEPQKSLG EEGLNRGPNV LMPDRDREDE EEDEEEETT
                                   Y region
551 DPKKPTTDEG TASSEVNATE EMSTLVNYVE PVKFKSFEAS RKRNKCFEMS

601 SFVETKAMEQ LTKSPMEFVE YNKQQLSRIY PKGTRVDSSN YMPQLFWNVG

651 CQLVALNFQT LDLPMQLNAG VFEYNGRSGY LLKPEFMRRP DKSFDPFTEV

701 IVDGIVANAL RVKVISGQFL SDRKVGIYVE VDMFGLPVDT RRKYRTRTSQ

751 GNSFNPVWDE EPFDFPKVVL PTLASLRIAA FEEGGRFVGH RILPVSAIRS

801 GYHYVCLRNE ANQPLCLPAL LIYTEASDYI PDDHQDYAEA LINPIKHVSL

851 MDQRAKQLAA LIGESEAQAS TEMCQETPSQ QQGSQLSSNP VPNPLDDSPR

901 WPPGPTTSPT STSLSSPGQR DDLIASILSE VTPTPLEELR SHKAMVKLRS

951 RQDRDLRELH KKHQRKAVAL TRRLLDGLAQ ARAEGKCRPS SSALSRATNV
```

```
1001  EDVKEEEKEA  ARQYREFQNR  QVQSLLELRE  AQADAETERR  LEHLKQAQQR

1051  LREVVLDAHT  TQFKRLKELN  EREKKELQKI  LDRKRNNSIS  EAKTREKHKK

1101  EVELTEINRR  HITESVNSIR  RLEEAQKQRH  ERLLAGQQQV  LQQLVEEEPK

1151  LVAQLTQECQ  EQRERLPQEI  RRCLLGETSE  GLGDGPLVAC  ASNGHAAGSG

1201  GHQSGADSES  QEENTQL
```

Amino acids   1217
X and Y regions are boxed.

Database accession numbers

	PIR	SWISSPROT	EMBL/GENBANK	REFERENCES
Rat			M99567	1
Human			Z16411	2

References

1 Jhon, D. Y. et al. (1993) J. Biol. Chem. 268, 6654–6661.
2 Carozzi, A. J. et al. (1992) Eur. J. Biochem. 210, 521–529.
3 Park, D. et al. (1993) J. Biol. Chem. 268, 4573–4576.
4 Carozzi, A. et al. (1993) FEBS Lett. 315, 340–342.
5 **Rhee, S. G. and Choi, K. D. (1992) Adv. Sec. Mess. Phosph. Res. 26, 35–61.**
6 **Rhee, S. G. and Choi, K. D. (1992) J. Biol. Chem. 267, 12393–12396.**
7 Wu, D. et al. (1993) J. Biol. Chem. 268, 3704–3709.

PHOSPHOLIPASE C-γ1

Molecular weight
Polypeptide   148 531
Reduced SDS PAGE   145–150 kDa

Gene structure and localization

Unknown.

Tissue distribution

Expressed in brain, particularly frontal and pyriform cortex, olfactory bulb, hippocampus, dendate gyrus, granule and Purkinje cell layers of cerebellum. In the periphery, it is also present in vascular endothelium, platelets and several cell lines including T cell, fibroblast, colon carcinoma and osteosarcoma. Low levels in some lymphocyte B cell lines. Undetectable in HL60 promyelocytes [1-6,38].

Function

Hydrolysis of membrane phosphoinositides to generate IP_3 and DAG following activation by some receptor tyrosine kinases (e.g. PDGF, EGF, FGF, NGF). T cell antigen receptor and some immunoglobulin receptors (mIgM on B cells, high-affinity IgE on mast cells and basophils, IgG on monocytic and natural killer cells) are also linked to PLC-γ1 through intracellular tyrosine kinases of the *Src* family. Phosphorylation of Tyr783 and Tyr1254 appears crucial for PLC-γ1 activation [7-27,37].

Comments

Two SH2 domains (SH2(N) and SH2(C)) play a key role facilitating PLC-γ1 phosphorylation and catalytic acivation through recognizing and binding tightly phosphotyrosine residues within specific regions of tyrosine kinases [10,11,26,28-30].

PLC-γ1 is substrate for *in vitro* phosphorylation by p56*lck*, p53/56*lyn*, p59*hck*, p59*fyn* and p60*src* [37].

Regions X (residues 296-465) and Y (residues 952-1207) conserved between PLC subtypes are believed important for catalytic activity [10,11,31].

Two-dimensional NMR analysis shows the SH3 domain of PLC-γ1 to be composed of two three-stranded β sheets and one two-stranded antiparallel β sheet [39].

The carboxyl end of SH2(C) forms a putative PLC inhibitor region including residues 747-754 (YRKMKLRY) suppressing strongly *in vitro* activities of PLC-β1, PLC-γ1, PLC-γ2 and PLC-δ1 [32].

Ser1258 represents site of phosphorylation by PKC and cAMP-dependent PKA and suppresses T cell receptor-dependent activation [33].

Subcellular localization suggests association with fibroblast cytoskeleton [34].

Increased levels reported in human breast carcinomas [35].

Amino acid sequence

```
  1 MAGAASPCAN GCGPGAPSDA EVLHLCRSLE VGTVMTLFYS KKSQRPERKT

 51 FQVKLETRQI TWSRGADKIE GAIDIREIKE IRPGKTSRDF DRYQEDPAFR

101 PDQSHCFVIL YGMEFRLKTL SLQATSEDEV NMWIKGLTWL MEDTLQAPTP

151 LQIERWLRKQ FYSVDRNRED RISAKDLKNM LSQVNYRVPN MRFLRERLTD

201 LEQRSGDITY GQFAQLYRSL MYSAQKTMDL PFLEASTLRA GERPELCRVS

                                                     X region
251 LPEFQQFLLD YQGELWAVDR LQVQEFMLSF LRDPLREIEE PYFFLDEFVT

301 FLFSKENSVW NSQLDAVCPD TMNNPLSHYW ISSSHNTYLT GDQFSSESSL

351 EAYARCLRMG CRCIELDCWD GPDGMPVIYH GHTLTTKIKF SDVLHTIKEH

401 AFVASEYPVI LSIEDHCSIA QQRNMAQYFK KVLGDTLLTK PVEISADGLP

451 SPNQLKRKIL IKHKKLAEGS AYEEVPTSMM YSENDISNSI KNGILYLEDP

501 VNHEWYPHYF VLTSSKIYYS EETSSDQGNE DEEEPKEVSS STELHSNEKW
```

SH2(N)

551 FHGKLGAGRD GRHIAERLLT EYCIETGAPD GSFLVRESET FVGDYTLSFW

601 RNGKVQHCRI HSRQDAGTPK FFLTDNLVFD SLYDLITHYQ QVPLRCNEFE

SH2(C)

651 MRLSEPVPQT NAHESKEWYH ASLTRAQAEH MLMRVPRDGA FLVRKRNEPN

701 SYAISFRAEG KIKHCRVQQE GQTVMLGNSE FDSLVDLISY YEKHPLYRKM

SH3

751 KLRYPINEEA LEKIGTAEPD YGALYEGRNP GFYVEANPMP TFKCAVKALF

801 DYKAQREDEL TFIKSAIIQN VEKQEGGWWR GDYGGKKQLW FPSNYVEEMV

851 NPVALEPERE HLDENSPLGD LLRGVLDVPA CQIAIRPEGK NNRLFVFSIS

901 MASVAHWSLD VAADSQEELQ DWVKKIREVA QTADARLTEG KIMERRKKIA

Y region

951 LELSELVVYC RPVPFDEEKI GTERACYRDM SSFPETKAEK YVNKAKGKKF

1001 LQYNRLQLSR IYPKGQRLDS SNYDPLPMWI CGSQLVALNF QTPDKPMQMN

1051 QALFMTGRHC GYVLQPSTMR DEAFDPFDKS SLRGLEPCAI SIEVLGARHL

1101 PKNGRGIVCP FVEIEVAGAE YDSTKQKTEF VVDNGLNPVW PAKPFHFQIS

1151 NPEFAFLRFV VYEEDMFSDQ NFLAQATFPV KGLKTGYRAV PLKNNYSEDL

1201 ELASLLIKID IFPAKENGDL SPFSGTSLRE RGSDASGQLF HGRAREGSFE

1251 SRYQQPFEDF RISQEHLADH FDSRERRAPR RTRVNGDNRL

Amino acids   1290

X and Y regions are boxed [31].

Two SH2 regions (SH2(N) and SH2(C)) and one SH3 domain are indicated by shaded boxes [29,36].

Tyr771, Tyr783 and Tyr1254 of rat PLC-γ1 identified as major phosphorylation sites [9].

Database accession numbers

	PIR	SWISSPROT	EMBL/GENBANK	REFERENCES
Human	A36466	P19174	M34667	[1]
Rat	A31317	P10686	J03806	[2]
Bovine	S00666	P08487	Y00301	[3]

References

[1] Burgess, W. H. et al. (1990) Mol. Cell. Biol. 10, 4770–4777.

[2] Suh, P.-G. et al. (1988) Proc. Natl Acad. Sci. USA 85, 5419–5423.

[3] Stahl, M. L. et al. (1988) Nature 332, 269–272.

[4] Hempel, W. M. and DeFranco, A. L. (1991) J. Immunol. 146, 3713–3720.
[5] Ross, C. A. et al. (1989) Proc. Natl Acad. Sci. USA 86, 2923–2927.
[6] Gerfen, C. R. et al. (1988) 85, 3208–3212.
[7] Meisenhelder, J. et al. (1989) Cell 57, 1109–1122.
[8] Morrison, D. K. et al. (1990) Mol. Cell. Biol. 10, 2359–2366.
[9] Kim, H. K. et al. (1991) Cell 65, 435–441.
[10] **Rhee, S. G. (1991) Trends Biochem. Sci. 16, 297–301.**
[11] **Rhee, S. G. and Choi, K. D. (1992) J. Biol. Chem. 267, 12393–12396.**
[12] Wokim, J. et al. (1990) J. Biol. Chem. 265, 3940–3943.
[13] Wahl, M. I. et al. (1990) J. Biol. Chem. 265, 3944–3948.
[14] Wahl, M. I. et al. (1988) Science 241, 968–970.
[15] Wahl, M. I. et al. (1992) J. Biol. Chem. 267, 10447–10456.
[16] Goldschmidt-Clermont, P. J. et al. (1991) Science 251, 1231–1233.
[17] Nishibe, S. et al. (1990) Science 250, 1253–1256.
[18] Vetter, M. L. et al. (1991) Proc. Natl Acad. Sci. USA 88, 5650–5654.
[19] Kim, U.-H. et al. (1991) J. Biol. Chem. 266, 1359–1362.
[20] Carter, R. H. et al. (1991) Proc. Natl Acad. Sci. USA 88, 2745–2749.
[21] Park, D. J. et al. (1991) Proc. Natl Acad. Sci. USA 88, 5453–5456.
[22] Weiss, A. et al. (1991) Proc. Natl Acad. Sci. USA 88, 5484–5488.
[23] Park, D. J. et al. (1991) J. Biol. Chem. 266, 24237–24240.
[24] Liao, F. et al. (1992) Proc. Natl Acad. Sci. USA 89, 3659–3663.
[25] Li, W. et al. (1992) Mol. Cell. Biol. 12, 3176–3182.
[26] **Schlessinger, J. and Ullrich, A. (1992) Neuron 9, 383–391.**
[27] Granja, C. et al. (1991) J. Biol. Chem. 266, 16277–16280.
[28] Cantley, L. C. et al. (1991) Cell 64, 281–302.
[29] **Koch, C. A. et al. (1991) Science 252, 668–674.**
[30] Songyang, Z. et al. (1993) Cell 72, 767–778.
[31] **Rhee, S. G. and Choi, K. D. (1992) Adv. Sec. Mess. Phosph. Res. 26, 35–61.**
[32] Homma, Y and Takenawa, T. (1992) J. Biol. Chem. 267, 21884–21849.
[33] Park, D. J. et al. (1992) J. Biol. Chem. 267, 1496–1501.
[34] McBride, K. et al. (1991) Proc. Natl Acad. Sci. USA 88, 7111–7115.
[35] Arteaga, C. L. et al. (1991) Proc. Natl Acad. Sci. USA 88, 10435–10439.
[36] **Musacchio, A. et al. (1992) FEBS Lett. 307, 55–61.**
[37] Liao, F. et al. (1993) Biochem. Biophys. Res. Commun. 191, 1028–1033.
[38] Baldassare, J. J. et al. (1993) Biochem. J. 291, 235–240.
[39] Kohda et al. (1993) Cell 72, 953–960.

PHOSPHOLIPASE C-γ2

Molecular weight

Polypeptide	146 119
Reduced SDS PAGE	140–150 kDa

Gene structure and localization

Unknown.

Tissue distribution

High levels in lung, spleen, thymus, skeletal muscle and lymphocyte B cell lines. Also expressed in platelets as well as cultured aortic vascular smooth muscle cells and HL60 promyelocytes [1-7,17].

Function

Hydrolysis of membrane phosphoinositides to produce second messengers IP$_3$ and DAG. Activation follows phosphorylation by receptor tyrosine kinases or members of *Src* kinase family. Regulatory receptors include B-lymphocyte antigen and PDGF. Tyr753 and Tyr759 identified as phosphorylation sites [4,7-9].

Comments

Two SH2 domains (SH2(N) and SH2(C)) believed important for PLC-γ2 phosphorylation and activation through recognizing and binding tightly phosphotyrosine residues within specific regions of tyrosine kinases [8-13].

PLC-γ2 is modified by tyrosine phosphorylation [4,7]. Tyr759 is cognate to Tyr783 of PLC-γ1 which is crucial for catalytic activation.

PLC-γ2 is substrate for *in vitro* phosphorylation by p56[lck], p53/56[lyn], p59[hck], p59[fyn] and p60[src] [16].

Regions X (residues 288-457) and Y (residues 929-1182) conserved between PLC subtypes are believed important for catalytic activity [2,8,9].

C-terminal end of SH2(C) forms putative PLC inhibitor region including residues 726-733 (YRKMRLRY) suppressing strongly *in vitro* activities of PLC-β1, PLC-γ1, PLC-γ2 and PLC-δ1 [14].

Primary sequences of cloned human and rat PLC-γ2 are 93.7% identical with final 12 amino acids of human enzyme replaced by unrelated 25 residue sequence in rat enzyme [1,2].

Amino acid sequence

```
  1 MSTTVNVDSL AEYEKSQIKR ALELGTVMTV FSFRKSTPER RTVQVIMETR

 51 QVAWSKTADK IEGFLDIMEI KEIRPGKNSK DFERAKAVRQ KEDCCFTILY

101 GTQFVLSTLS LAADSKEDAV NWLSGLKILH QEAMNASTPT IIESWLRKQI

151 YSVDQTRRNS ISLRELKTIL PLINFKVSSA KFLKDKFVEI GAHKDELSFE

201 QFHLFYKKLM FEQQKSILDE FKKDSSVFIL GNTDRPDASA VYLHDFQRFL
                                                    X region
251 IHEQQEHWAQ DLNKVRERMT KFIDDTMRET AEPFLFV|DEF LTYLFSRENS|

301 |IWDEKYDAVD MQDMNNPLSH YWISSSHNTY LTGDQLRSES SPEAYIRCLR|

351 |MGCRCIELDC WDGPDGKPVI YHGWTRTTKI KFDDVVQAIK DHAFVTSSFP|
```

```
 401   VILSIEEHCS VEQQRHMAKA FKEVFGDLLL TKPTEASADQ LPSPSQLREK

 451   IIIKHKKLGP RGDVDVNMED KKDEHKQQGE LYMWDSIDQK WTRHYCAIAD
                                                  SH2 (N)
 501   AKLSFSDDIE QTMEEEVPQD IPPTELHFGE KWFHKKVEKR TSAEKLLQEY

 551   CMETGGKDGT FLVRESETFP NDYTLSFWRS GRVQHCRIRS TMEGGTLKYY
                                                          SH2 (C)
 601   LTDNLRFRRM YALIQHYRET HLPCAEFELR LTDPVPNPNP HESKPWYYDS

 651   LSRGEAEDML MRIPRDGAFL IRKREGSDSY AITFRARGKV KHCRINRDGR

 701   HFVLGTSAYF ESLVELVSYY EKHSLYRKMR LRYPVTPELL ERYNTERDIN
                                    SH3
 751   SLYDVSRMYV DPSEINPSMP QRTVKALYDY KAKRSDELSF CRGALIHNVS

 801   KEPGGWWKGD YGTRIQQYFP SNYVEDISTA DFEELEKQII EDNPLGSLCR

 851   GILDLNTYNV VKAPQGKNQK SFVFILEPKE QGDPPVEFAT DRVEELFEWF
                                             X region
 901   QSIREITWKI DSKENNMKYW EKNQSIAIEL SDLVVYCKPT SKTKDNLENP

 951   DFREIRSFVE TKADSIIRQK PVDLLKYNQK GLTRVYPKGQ RVDSSNYDPF

1001   RLWLCGSQMV ALNFQTADKY MQMNHALFSL NGRTGYVLQP ESMRTEKYDP

1051   MPPESQRKIL MTLTVKVLGA RHLPKLGRSI ACPFVEVEIC GAEYGNNKFK

1101   TTVVNDNGLS PIWAPTQEKV TFEIYDPNLA FLRFVVYEED MFSDPNFLAH

1151   ATYPIKAVKS GFRSVPLKNG YSEDIELASL LVFCEMRPVL ESEEELYSSC

1201   RQLRRRQEEL NNQLFLYDTH QNLRNANRDA LVKEFSVNEN HSSCTRRNAT

1251   RG
```

Amino acids   1252

X and Y regions are boxed and based on sequence alignment with homologous regions indicated in PLC-γ1 [8]. Two SH2 domains SH2(N) and SH2(C) and one SH3 region are indicated by shaded boxes [11,15]. Tyr753 and Tyr759 are sites of phosphorylation in PLC-γ2 [8,9].

Database accession numbers

	PIR	SWISSPROT	EMBL/GENBANK	REFERENCES
Human	S02004	P16885	M37238 X14034	[1]
Rat	A34163	P24135	J05155	[2]

References

1 Ohta, S. et al. (1988) FEBS Lett. 242, 31–35.
2 Emori, Y. et al. (1989) J. Biol. Chem. 264; 21885–21890.
3 Hempel, W. M. and De Franco, A. L. (1991) J. Immunol. 11, 3713–3720.
4 Coggeshall, K. M. et al. (1992) Proc. Natl Acad. Sci. USA 89, 5660–5664.
5 Banno, Y. et al. (1990) Biochem. Biophys. Res. Commun. 167, 396–401.
6 Roifman, C. M. and Wang, G. (1992) Biochem. Biophys. Res. Commun. 183, 411–416.
7 Homma, Y. et al. (1993) Biochem. J. 290, 649–653.
8 **Rhee, S. G. and Choi, K. D. (1992) Adv. Sec. Mess. Phosph. Res. 26, 35–61.**
9 **Rhee, S. G. and Choi, K. D. (1992) J. Biol. Chem. 267, 12393–12396.**
10 **Schlessinger, J. and Ullrich, A. (1992) Neuron 9, 383–391.**
11 **Koch, C. A. et al. (1991) Science 252, 668–674.**
12 Songyang, Z. et al. (1993) Cell 72, 767–778.
13 Homma, Y. et al. (1992) J. Biol. Chem. 267, 3778–3782.
14 Homma, Y and Takenawa, T. (1992) J. Biol. Chem. 267, 21884–21849.
15 **Musacchio, A. et al. (1992) FEBS Lett. 307, 55–61.**
16 Liao, F. et al. (1993) Biochem. Biophys. Res. Commun. 191, 1028–1033.
17 Baldassare, J. J. et al. (1993) Biochem. J. 291, 235–240.

PHOSPHOLIPASE C-δ1

Molecular weights

Polypeptide	85 962
Reduced SDS PAGE	85 kDa

Gene structure and localization

Unknown.

Tissue distribution

Expressed in adrenal, C6Bu-1 glioma cell line and brain. Higher levels in hippocampus. Undetectable in lymphocyte B cell lines [1-3].

Function

Hydrolysis of membrane phosphoinositides to produce second messengers IP_3 and DAG. Cellular mechanism for control unknown [4-6].

Comments

Regions X (residues 273–441) and Y (residues 491–750) conserved between PLC subtypes are believed important for catalytic activity [4,5]. Some residues on the N-terminal side of region X also appear important for catalytic activity [7].

PLC-δ1 binds vesicles containing PIP_2 with high affinity and specificity. Calcium is not required for this interaction although essential for catalytic activity [8-10].

Increased PLC-δ1 levels and activity have been detected in aortas from spontaneously hypertensive rats [11].

Amino acid sequence (rat)

```
  1 MDSGRDFLTL HGLQDDPDLQ ALLKGSQLLK VKSSSWRRER FYKLQEDCKT

 51 IWQESRKVMR SPESQLFSIE DIQEVRMGHR TEGLEKFARD IPEDRCFSIV

101 FKDQRNTLDL IAPSPADAQH WVQGLRKIIH HSGSMDQRQK LQHWIHSCLR

151 KADKNKDNKM NFKELKDFLK ELNIQVDDGY ARKIFRECDH SQTDSLEDEE

201 IETFYKMLTQ RAEIDRAFEE AAGSAETLSV ERLVTFLQHQ QREEEAGPAL
                                     X region
251 ALSLIERYEP SETAKAQRQM TKDGFLMYLL SADGNAFSLA HRRVYQDMDQ

301 PLSHYLVSSS HNTYLLEDQL TGPSSTEAYI RALCKGCRCL ELDCWDGPNQ

351 EPIIYHGYTF TSKILFCDVL RAIRDYAFKA SPYPVILSLE NHCSLEQQRV

401 MARHLRAILG PILLDQPLDG VTTSLPSPEQ LKGKILLKGK KLGGLLPAGG
                                                 Y region
451 ENGSEATDVS DEVEAAEMED EAVRSQVQHK PKEDKLKLVP ELSDMIIYCK

501 SVHFGGFSSP GTSGQAFYEM ASFSESRALR LLQESGNGFV RHNVSCLSRI

551 YPAGWRTDSS NYSPVEMWNG GCQIVALNFQ TPGPEMDVYL GCFQDNGGCG

601 YVLKPAFLRD PNTTFNSRAL TQGPWWRPER LRVRIISGQQ LPKVNKNKNS

651 IVDPKVIVEI HGVGRDTGSR QTAVITNNGF NPRWDMEFEF EVTVPDLALV

701 RFMVEDYDSS SKNDFIGQST IPWNSLKQGY RHVHLLSKNG DQHPSATLFV

751 KISIQD
```

Amino acids 756
X and Y regions are boxed [4].

Database accession numbers

	PIR	SWISSPROT	EMBL/GENBANK	REFERENCES
Rat	B28821	P10688	M20637	[1]
Bovine	C28821	P10895	M20638	[1]

References
[1] Suh, P.-G. et al. (1988) Cell 54, 161–169.
[2] Hempel, W. M. and DeFranco, A. L. (1991) J. Immunol. 146, 3713–3720.
[3] Ross, C. A. et al. (1989) Proc. Natl Acad. Sci. USA 86, 2923–2927.
[4] **Rhee, S. G. and Choi, K. D. (1992) Adv. Sec. Mess. Phosph. Res. 26, 35–61.**
[5] **Rhee, S. G. and Choi, K. D. (1992) J. Biol. Chem. 267, 12393–12396.**

6 Ginger, R. S. and Parker, P. J. (1992) Eur. J. Biochem. 210, 155–160.

7 Ellis, M. V. et al. (1993) Eur. J. Biochem. 213, 339–347.

8 Rebecchi, M. J. et al. (1992) Biochemistry 31, 12742–12747.

9 Rebecchi, M. J. et al. (1992) Biochemistry 31, 12748–12753.

10 Rebecchi, M. J. et al. (1993) J. Biol. Chem. 268, 1735–1741.

11 Kato, H. et al. (1992) J. Biol. Chem. 267, 6483–6487.

PHOSPHOLIPASE C-δ2

Molecular weights

Polypeptide	86 941
Reduced SDS PAGE	85kD

Gene structure (bovine)

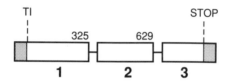

Organization of bovine PLC-δ2 gene adapted from ref. 1.

Tissue distribution

Unknown but purified from bovine brain [2].

Function

Hydrolysis of membrane phosphoinositides to produce second messengers IP$_3$ and DAG. Cellular regulatory mechanism unknown [2-4].

Comments

Regions X (residues 267–436) and Y (residues 489–751) conserved between PLC subtypes are believed important for catalytic activity [3,4].

C-terminal 95 residues not crucial for catalytic activity [1].

Amino acid sequence (bovine)

```
  1 MAYLLQGRLP INQDLLLMQK GTMMRKVRSK SWKKLRFFRL QDDGMTVWHA

 51 RQAGGRAKPS FSISDVDTVR EGHESELLRN LAEEFPLEQG FTIVFHGRRS

101 NLDLVANSVQ EAQTWMQGLQ LLVGFVTNMD QQERLDQWLS DWFQRGDKNQ
```

```
151 DGRMSFGEVQ RLLHLMNVEM DQEYAFQLFQ TADTSQSGTL EGEEFVEFYK

201 SLTQRPEVQE LFEKFSSDGQ KLTLLEFVDF LQEEQKEGER ASDLALELID
                   X region
251 RYEPSESGKL RHVLSMDGFL GYLCSKDGDI FNPTCHPLYQ DMTQPLNHYY

301 INSSHNTYLV GDQLCGQSSV EGYIRALKRG CRCVEVDIWD GPSGEPIVYH

351 GHTLTSRIPF KDVVAAIGQY AFQTSDYPVI LSLENHCSWE QQEIIVRHLT

401 EILGDQLLTT ALDGQPPTQL PSPEDLRGKI LVKGKKLMLE EEEEEPEAEL
                                               Y region
451 EAEQEARLDL EAQLESEPQD LSPRSEDKKK KPKAILCPAL SALVVYLKAV

501 TFYSFTHSRE HYHFYETSSF SETKAKSLIK EAGDEFVQHN AWQLSRVYPS

551 GLRTDSSNYN PQEFWNAGCQ MVAMNMQTAG LEMDLCDGLF RQNAGCGYVL

601 KPDFLRDAQS SFHPERPISP FKAQTLIIQE PWLQVISGQQ LPKVDNTKEQ

651 SIVDPLVRVE IFGVRPDTTR QETSYVENNG FNPYWGQTLC FRILVPELAL

701 LRFVVKDYDW KSRNDFIGQY TLPWSCMQQG YRHIHLLSKD GLSLHPASIF

751 VHICTQEVSE EAES
```

Amino acids 764
X and Y regions are boxed and based on sequence alignment with homologous region indicated in PLC-δ1 [3].

Database accession numbers

	PIR	SWISSPROT	EMBL/GENBANK	REFERENCES
Bovine	S14113			1

References
[1] Meldrum, E. et al. (1991) Eur. J. Biochem. 196, 159–165.
[2] Meldrum, E. et al. (1989) Eur. J. Biochem. 182, 673–677.
[3] **Rhee, S. G. and Choi, K. D. (1992) Adv. Sec. Mess. Phosph. Res. 26, 35–61.**
[4] **Rhee, S. G. and Choi, K. D. (1992) J. Biol. Chem. 267, 12393–12396.**

Phospholipase A₂

[phospholipase sn-2 acylhydrolase, EC 3.1.1.4]

85kD PLA₂ (cPLA₂)

Molecular weights

Polypeptide	85 210
Reduced SDS PAGE	95–110 kDa

Gene structure and localization

Unknown.

Tissue distribution

Neutrophils, platelets, brain, lung, liver, kidney, spleen, fibroblasts, monocytes, U937 monoblasts as well as RAW264.7 and J774 macrophages [1-15].

Function

85kD PLA₂ displays high-selectivity for sn-2-arachidonoyl-containing phospholipids. It has an important role in generating bioactive and proinflammatory lipid mediators such as eicosanoids (e.g. prostaglandins, leukotrienes) and platelet activating factor is likely [1,3,7,8,10,11,15-17].

Comments

Cellular mechanism for activation is unclear although 85kD PLA₂ undergoes translocation to membrane fractions in the presence of submicromolar calcium concentrations. An N-terminal 140 amino acid fragment shares this activity [1,10,16]. Phosphorylation at Ser505 by MAP kinase also activates 85kD PLA₂ [19,20]. MAP kinase may mediate 85kD PLA₂ activation by PKC as well as by both G-protein linked and tyrosine kinase receptors [13,18-20].

85kD PLA₂ expression can be regulated by tumour necrosis factor, interleukin 1, transforming growth factor β, macrophage colony stimulating factor and glucocorticoid [12,13,21]. 85kD PLA₂ levels are also developmentally regulated in embryonic rat brain [6].

Note: 85kD PLA₂ displays no detectable sequence homology to mammalian secretory PLA₂s which are of small molecular size (~14 kDa), are inhibited by sulfhydryl reducing agents, display no fatty acid specificity at the sn-2 position, require calcium as catalytic cofactor, and are active following extracellular release [15].

Amino acid sequence

```
  1 MSFIDPYQHI IVEHQYSHKF TVVVLRATKV TKGAFGDMLD TPDPYVELFI

 51 STTPDSRKRT RHFNNDINPV WNETFEFILD PNQENVLEIT LMDANYVMDE

101 TLGTATFTVS SMKVGEKKEV PFIFNQVTEM VLEMSLEVCS CPDLRFSMAL

151 CDQEKTFRQQ RKEHIRESMK KLLGPKNSEG LHSARDVPVV AILGSGGGFR
```

```
201 AMVGFSGVMK ALYESGILDC ATYVAGLSGS TWYMSTLYSH PDFPEKGPEE

251 INEELMKNVS HNPLLLLTPQ KVKRYVESLW KKKSSGQPVT FTDIFGMLIG

301 ETLIHNRMNT TLSSLKEKVN TAQCPLPLFT CLHVKPDVSE LMFADWVEFS

351 PYEIGMAKYG TFMAPDLFGS KFFMGTVVKK YEENPLHFLM GVVWGSAFSIL

401 FNRVLGVSGS QSRGSTMEEE LENITTKHIV SNDSSDSDDE SHEPKGTENE

451 DAGSDYQSDN QASWIHRMIM ALVSDSALFN TREGRAGKVH NFMLGLNLNT

501 SYPLSPLSDF ATQDSFDDDE LDAAVADPDE FERIYEPLDV KSKKIHVVDS

551 GLTFNLPYPL ILRPQRGVDL IISFDFSARP SDSSPPFKEL LLAEKWAKMN

601 KLPFPKIDPY VFDREGLKEC YVFKPKNPDM EKDCPTIIHF VLANINFRKY

651 KAPGVPRETE EEKEIADFDI FDDPESPFST FNFQYPNQAF KRLHDLMHFN

701 TLNNIDVIKE AMVESIEYRR QNPSRCSVSL SNVEARRFFN KEFLSKPKA
```

Amino acids 749
Ser505 is phosphorylation site for catalytic activation by MAPkinase [19,20].

Database accession numbers

	PIR	SWISSPROT	EMBL/GENBANK	REFERENCES
Human			M68874	1,2
Mouse				1

References

[1] Clark, J. D. et al.(1991) Cell 65, 1043–1051.
[2] Sharp, J. D. et al. (1991) J. Biol. Chem. 266, 14850–14853.
[3] Clark, J. D. et al. (1990) Proc. Natl Acad. Sci. USA 87, 7708–7712.
[4] Kramer, R. M. et al. (1991) J. Biol. Chem. 5268–5272.
[5] Yoshihara, Y. and Watanabe, Y. (1990) Biochem. Biophys. Res. Commun. 170, 484–490
[6] Yoshihara, Y. et al. (1992) Biochem. Biophys. Res. Commun. 185, 350–355.
[7] Gronich, J. H. et al. (1990) Biochem. J. 271, 37–43.
[8] Wijkander, J. and Sundler, R. (1991) Eur. J. Biochem. 202, 873–880.
[9] Kim, D. K. et al. (1991) Biochem. Biophys. Res. Commun. 174, 189–196.
[10] Channon, J. Y. and Leslie, C. C. (1990) J. Biol. Chem. 265, 5409–5413.
[11] Takayama, K. et al. (1991) FEBS Lett. 282, 326–330.
[12] Lin, L.-L. et al. (1992) J. Biol. Chem. 267, 23451–23454.
[13] Nakamura, T. et al. (1992) EMBO J. 11, 4917–4922.
[14] Fujimori, Y. et al. (1992) J. Biochem. (Tokyo) 111, 54–60.
[15] **Mayer, R. J. and Marshall, L. A. (1993) FASEB 7, 339–348.**
[16] Diez, E. and Mong, S. (1990) J. Biol. Chem. 265, 14654–14661.

17 Diez, E. et al. (1992) J. Biol. Chem. 267, 18342–18348.
18 Lin, L.-L. et al. (1992) Proc. Natl Acad. Sci. USA 89, 6147–6151.
19 Nemenoff, R. A. et al. (1993) J. Biol. Chem. 268, 1960–1964.
20 Lin, L.-L. et al. (1993) Cell 72, 269–278.
21 Schalkwijk, C. G. et al. (1992) Eur. J. Biochem. 210, 169–176.

ATP pyrophosphate-lyase
(cyclizing), EC 4.6.1.1

ADENYLYL CYCLASE TYPE I

Molecular weights
Polypeptide	123 978
Reduced SDS PAGE	110–120 kDa

Structure

Two tandem repeats of six hydrophobic putatively transmembrane domains separated by a large cytosolic loop and terminating in a large intracellular tail.

Carbohydrate
N-linked sites: Asn706.

Gene structure and localization

Unknown.

Tissue distribution

Predominantly neuronal. Expressed in adrenal medulla, retina and brain, particularly neocortex, entorhinal cortex, hippocampus, dendate gyrus, cerebellum and olfactory system. Not detected in caudate-putamen, globus pallidus, corpus collosum, substantia nigra, brainstem or spinal cord. Also not expressed in heart, aorta, skeletal muscle, lung, liver, kidney, thyroid, adipose, spleen, uterus or lymph node. Cell lines N1E-115 neuroblastoma, NG-108 neurogliohybridoma, 36B-10 glioma and PC-12 phaeochromocytoma cells also negative [1–3].

Function

Generation of cyclic AMP. Receptor-linked activation of adenylyl cyclase type I by $G_{\alpha s}$ or by calcium/calmodulin following calcium mobilization. This activity is inhibited by $G_{\alpha i}$ subtypes and $\beta\gamma$ dimer. Dimers of $\beta 1\gamma 2$, $\beta 1\gamma 3$, $\beta 2\gamma 2$ and $\beta 2\gamma 3$ mediate similarly potent inhibition in vitro while $\beta 1\gamma 1$ is markedly less active [4–10].

Comments
This subtype could mediate calcium/calmodulin-dependent cyclic AMP generation in brain and could also be important in learning and memory processes [1,2,11,12].

Inhibition by G-protein $\beta 1\gamma 2$ requires γ-subunit isoprenylation [9].

Sensitivity to forskolin and calcium/calmodulin regulated by PKC [13,14].

Amino acid sequence (bovine)

```
  1 MAGAPRGRGG GGGGGGAGES GGAERAAGPG GRRGLRACDE EFACPELEAL
                  TM1                            TM2
 51 FRGYTLRLEQ AATLKALAVL SLLAGALALA ELLGAPGPAP GLAKGSHPVH
                               TM3
101 CVLFLALLVV TNVRSLQVPQ LQQVGQLALL FSLTFALLCC PFALGGPAGA
```

```
              TM                                    TM5
    151  HAGAAAVPAT ADQGVWQLLL VTFVSYALLP VRSLLAIGFG LVVAASHLLV
                 TM6
    201  TATLVPAKRP RLWRTLGANA LLFLGVNVYG IFVRILAERA QRKAFLQARN

    251  CIEDRLRLED ENEKQERLLM SLLPRNVAME MKEDFLKPPE RIFHKIYIQR

    301  HDNVSILFAD IVGFTGLASQ CTAQELVKLL NELFGKFDEL ATENHCRRIK

    351  ILGDCYYCVS GLTQPKTDHA HCCVEMGLDM IDTITSVAEA TEVDLNMRVG

    401  LHTGRVLCGV LGLRKWQYDV WSNDVTLANV MEAAGLPGKV HITKTTLACL

    451  NGDYEVEPGH GHERNSFLKT HNIETFFIVP SHRRKIFPGL ILSDIKPAKR

    501  MKFKTVCYLL VQLMHCRKMF KAEIPFSNVM TCEDDDKRRA LRTASEKLRN

    551  RSSFSTNVVQ TTPGTRVNRY IGRLLEARQM ELEMADLNFF TLKYKQAERE
                       TM7                         TM8
    601  RKYHQLQDEY FTSAVVLALI LAALFGLVYL LIIPQSVAVL LLLVFCICFL
                              TM9
    651  VACVLYLHIT RVQCFPGCLT IQIRTVLCIF IVVLIYSVAQ GCVVGCLPWS
                                                      TM10
    701  WSSSPNGSLV VLSSGGRDPV LPVPPCESAP HALLCGLVGT LPLAIFLRVS
             TM11                          TM12
    751  SLPKMILLAV LTTSYILVLE LSGYTKAMGA GAISGRSFEP IMAILLFSCT

    801  LALHARQVDV KLRLDYLWAA QAEEERDDME KVKLDNKRIL FNLLPAHVAQ

    851  HFLMSNPRNM DLYYQSYSQV GVMFASIPNF NDFYIELDGN NMGVECLRLL

    901  NEIIADFDEL MDKDFYKDLE KIKTIGSTYM AAVGLAPTAG TKAKKCISSH

    951  LSTLADFAIE MFDVLDEINY QSYNDFVLRV GINVGPVVAG VIGARRPQYD

   1001  IWGNTVNVAS RMDSTGVQGR IQVTEEVHRL LRRGSYRFVC RGKVSVKGKG

   1051  EMLTYFLEGR TDGNGSQTRS LNSERKMYPF GRAGLQTRLA AGHPPVPPAA

   1101  GLPVGAGPGA LQGSGLAPGP PGQHLPPGAS GKEA
```

Amino acids   1134

Asn706 is potential site for *N*-linked glycosylation located within putative extracellular loop between TM9 and TM10.

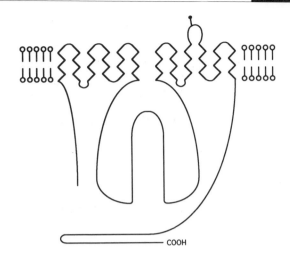

Database accession numbers

	PIR	SWISSPROT	EMBL/GENBANK	REFERENCE
Bovine	A41350	P19754	M25579	15

References

1 Xia, Z. et al. (1991) Neuron 6, 431–443.
2 Xia, Z. et al. (1993) J. Neurochem. 60, 305–311.
3 Glatt, C. E. and Snyder, S. H. (1993) Nature 361, 536–538.
4 Tang, W.-J. and Gilman, A. G. (1991) Science. 254, 1500–1503.
5 Tang, W.-J. et al. (1991) J. Biol. Chem. 266, 8595–8603.
6 Yoshimura, M. and Cooper, D. M. F. (1992) Proc. Natl Acad. Sci. USA 89, 6716–6720.
7 Feinstein, P. G. et al. (1991) Proc. Natl Acad. Sci. USA 88, 10173–10177.
8 Choi, E.-J. et al. (1992) Biochemistry 31, 6492–6498.
9 Iniguez, J. A. et al. (1992) J. Biol. Chem. 267, 23409–23417.
10 Taussig, R. et al. (1993) J. Biol. Chem. 268, 9–12.
11 Choi, E.-J. et al. (1992) J. Biol. Chem. 267, 12440–12442.
12 Levin, L. R. et al. (1992) Cell 68, 479–489.
13 Jacobowitz, O. et al. (1993) J. Biol. Chem. 268, 3829–3832.
14 Choi, E.-J. et al. (1993) Biochemistry 32, 1891–1894.
15 Krupinski, J. et al. (1989) Science 244, 1558–1564.

ADENYLYL CYCLASE TYPE II

Molecular weight

Polypeptide   123 315

Structure

Two tandem repeats of six hydrophobic putatively transmembrane domains separated by a large cytosolic loop and terminating in a large intracellular tail.

Carbohydrate

N-linked sites: Asn712 and Asn717.

Gene structure and localization

Chromosome 5 (5p15).

Tissue distribution

Expressed in brain, particularly granule cells of cerebellum, CA-1 and CA-4 of hippocampus, the piriform cortex, olfactory bulb granule cells of dendate gyrus some thalamic nuclei, mammillary nucleus dorsal raphe nucleus, parabrachial nucleus and locus coeruleus. Lower levels in olfactory epithelium and lung. Undetectable in kidney, liver, intestine and heart [1,2].

Function

Generation of cyclic AMP. Receptor-linked activation by $G_{\alpha s}$ and this stimulation is enhanced by G-protein βγ. Dimers of β1γ2, β1γ3, β2γ2 and β2γ3 are similarly effective *in vitro* while β1γ1 is markedly less active. Adenylyl cyclase type II is inhibited by mutationally activated $G_{\alpha 12}$ although is insensitive to calcium/calmodulin. Adenyl cyclase type II activation is increased following PKC activation by phorbol ester [1,3-8].

Comments

Potentiation of $G_{\alpha s}$ activation by G-protein β1γ2 requires γ-subunit iso-prenylation [5]. Adenyl cyclase control by several regulatory inputs suggests role integrating signals from multiple receptor subtypes.

Amino acid sequence (rat)

```
                                                              TM1
    1  MRRRRYLRDR AEAAAAAAG GGEGLQRSRD WLYESYYCMS QQHPLIVFLL
                              TM2
   51  LIVMGACLAL LAVFFALGLE VEDHVAFLIT VPTALAIFFA IFILVCIESV
         TM3                                TM4
  101  FKKLLRVFSL VIWICLVAMG YLFMCFGGTV SAWDQVSFFL FIIFVVYTML
         TM5                                TM6
  151  PFNMRDAIIA SILTSSSHTI VLSVYLSATP GAKEHLFWQI LANVIIFICG

  201  NLAGAYHKHL MELALQQTYR DTCNCIKSRI KLEFEKRQQE RLLLSLLPAH

  251  IAMEMKAEII QRLQGPKAGQ MENTNNFHNL YVKRHTNVSI LYADIVGFTR

  301  LASDCSPGEL VHMLNELFGK FDQIAKENEC MRIKILGDCY YCVSGLPISL

  351  PNHAKNCVKM GLDMCEAIKK VRDATGVDIN MRVGVHSGNV LCGVIGLQKW
```

```
401   QYDVWSHDVT  LANHMEAGGV  PGRVHISSVT  LEHLNGAYKV  EEGDGEIRDP

451   YLKQHLVKTY  FVINPKGERR  SPQHLFRPRH  TLDGAKMRAS  VRMTRYLESW

501   GAAKPFAHLH  HRDSMTTENG  KISTTDVPMG  QHNFQNRTLR  TKSQKKRFEE

551   ELNERMIQAI  DGINAQKQWL  KSEDIQRISL  LFYNKNIEKE  YRATALPAFK
                  TM7                                TM8
601   YYVTCACLIF  LCIFIVQILV  LPKTSILGFS  FGAAFLSLIF  ILFVCFAGQL
                                                      TM9
651   LQCSKKASTS  LMWLLKSSGI  IANRPWPRIS  LTIVTTAIIL  TMAVFNMFFL
                                                      TM10
701   SNSEETTLPT  ANTSNANVSV  PDNQASILHA  RNLFFLPYFI  YSCILGLISC
                  TM11
751   SVFLRVNYEL  KMLIMMVALV  GYNTILLHTH  AHVLDAYSQV  LFQRPGIWKD
                  TM12
801   LKTMGSVSLS  IFFITLLVLG  RQSEYYCRLD  FLWKNKFKKE  REEIETMENL

851   NRVLLENVLP  AHVAEHFLAR  SLKNEELYHQ  SYDCVCVMFA  SIPDFKEFYT

901   ESDVNKEGLE  CLRLLNEIIA  DFDDLLSKPK  FSGVEKIKTI  GSTYMAATGL

951   SAIPSQEHAQ  EPERQYMHIG  TMVEFAYALV  GKLDAINKHS  FNDFKLRVGI

1001  NHGPVIAGVI  GAQKPQYDIW  GNTVNVASRM  DSTGVLDKIQ  VTEETSLILQ

1051  TLGYTCTCRG  IINVKGKGDL  KTYFVNTEMS  RSLSQSNLAS
```

Amino acids   1090

Asn712 and Asn717 represent potential sites for *N*-linked glycosylation located within the putative extracellular loop between TM9 and TM10.

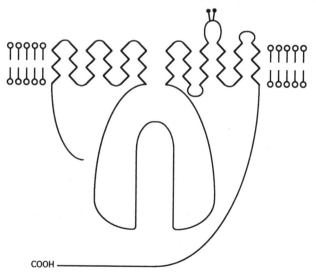

COOH

Database accession numbers

	PIR	SWISSPROT	EMBL/GENBANK	REFERENCES
Rat			M80550	*1*

References

1 Feinstein, P. G. et al. (1991) Proc. Natl Acad. Sci. USA 88, 10173–10177.
2 Glatt, C. E. and Snyder, S. H. (1993) Nature 361, 536–538.
3 Tang, W.-J. and Gilman, A. G. (1991) Science. 254, 1500–1503.
4 Federman, A. D. et al. (1992) Nature 356; 159–161.
5 Iniguez, J. A. et al. (1992) J. Biol. Chem. 267, 23409–23417.
6 Taussig, R. et al. (1993) J. Biol. Chem. 268, 9–12.
7 Jacobowitz, O. et al. (1993) J. Biol. Chem. 268, 3829–3832.
8 Yoshimura, M. and Cooper, D. M. F. (1993) J. Biol. Chem. 268, 4604–4607.

ADENYLYL CYCLASE TYPE III

Molecular weights
Polypeptide	128 935
Reduced SDS PAGE	180–200 kDa

Structure

Two tandem repeats of six hydrophobic putatively transmembrane domains separated by a large cytosolic loop and terminating in a large intracellular tail.

Carbohydrate
N-linked sites: Asn158, Asn734 and Asn827.

Gene structure and localization

Unknown.

Tissue Distribution

High levels in olfactory sensory neuronal cilia. Also detected in brain, spinal cord, heart, aorta, lung, retina, adrenal medulla and cortex. In brain higher levels expressed in granule cells of cerebellum, hippocampus and olfactory bulb. Also detected in 293 kidney cells, PC-12 phaeochromocytoma, human embryonic kidney and enythroleukemia cells. Not expressed in neuroblastoma N1E-115 and neural glial hybridoma NG-108 cells [1-5].

Function

Generation of cyclic AMP. Receptor-linked activation by $G_{\alpha s}$ and $G_{\alpha olf}$ and also by calcium/calmodulin. Adenylyl cyclase type III is insensitive to G-protein $\beta\gamma$ [1,6,7].

Comments

Co-expressed with G$_{\alpha olf}$ in olfactory neurons, suggesting a role in odorant signal transduction [1,4,5].

Less sensitive to activation by calcium/calmodulin than adenylyl cyclase type I, although this regulated by PKC [5,8,9].

Amino acid sequence (rat)

```
   1 MTEDQGFSDP EYSAEYSAEY SVSLPSDPDR GVGRTHEISV RNSGSCLCLP
                                                   TM1
  51 RFMRLTFVPE SLENLYQTYF KRQRHETLLV LVVFAALFDC YVVVMCAVVF
          TM2                        TM3
 101 SSDKLAPLMV AGVGLVLDII LFVLCKKGLL PDRVSRKVVP YLLWLLITAQ
                TM4                        TM5
 151 IFSYLGLNFS RAHAASDTVG WQAFFVFSFF ITLPLSLSPI VIISVVSCVV
                                  TM6
 201 HTLVLGVTVA QQQQDELEGM QLLREILANV FLYLCAIIVG IMSYYMADRK

 251 HRKAFLEARQ SLEVKMNLEE QSQQQENLML SILPKHVADE MLKDMKKDES

 301 QKDQQQFNTM YMYRHENVSI LFADIVGFTQ LSSACSAQEL VKLLNELFAR

 351 FDKLAAKYHQ LRIKILGDCY YCICGLPDYR EDHAVCSILM GLAMVEAISY

 401 VREKTKTGVD MRVGVHTGTV LGGVLGQKRW QYDVWSTDVT VANKMEAGGI

 451 PGRVHISQST MDCLKGEFDV EPGDGGSRCD YLDEKGIETY LIIASKPEVK

 501 KTAQNGLNGS ALPNGAPASK PSSPALIETK EPNGSAHASG STSEEAEEQE

 551 AQADNPSFPN PRRRLRLQDL ADRVVDASED EHELNQLLNE ALLERESAQV
                                                   TM7
 601 VKKRNTFLLT MRFMDPEMET RYSVEKEKQS GAAFSCSCVV LFCTAMVEIL
                TM8
 651 IDPWLMTNYV TFVVGEVLLL ILTICSMAAI FPRAFPKKLV AFSSWIDRTR
                TM9
 701 WARNTWAMLA IFILVMANVV DMLSCLQYYM GPYNVTTGIE LDGGCMENPK
             TM10                    TM11
 751 YYNYVAVLSL IATIMLVQVS HMVKLTLMLL VTGAVTAINL YAWCPVFDEY
                                                   TM12
 801 DHKRFQEKDS PMVALEKMQV LSTPGLNGTD SRLPLVPSKY SMTVMMFVMM

 851 LSFYYFSRHV EKLARTLFLW KIEVHDQKER VYEMRRWNEA LVTNMLPEHV

 901 ARHFLGSKKR DEELYSQSYD EIGVMFASLP NFADFYTEES INNGGIECLR

 951 FLNEIISDFD SLLDNPKFRV ITKIKTIGST YMAASGVTPD VNTNGFTSSS

1001 KEEKSDKERW QHLADLADFA LAMKDTLTNI NNQSFNNFML RIGMNKGGVL
```

```
1051  AGVIGARKPH  YDIWGNTVNV  ASRMESTGVM  GNIQVVEETQ  VILREYGFRF

1101  VRRGPIFVKG  KGELLTFFLK  GRDRPAAFPN  GSSVTLPHQV  VDNP
```

Amino acids 1144

Asn158, Asn734 and Asn 827 represent potential *N*-linked glycosylation sites located in presumed extracellular loops between TM3 and TM4, TM9. and TM10, and TM11 and TM12 respectively.

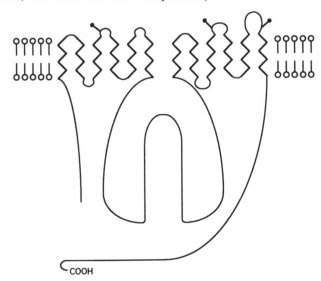

Database accession numbers

	PIR	SWISSPROT	EMBL/GENBANK	REFERENCE
Rat	A39833	P21932	M55075	1

References

1 Bakalyar, H. A. and Reed, R. R. (1990) Science 250, 1403–1406.
2 Glatt, C. E. and Snyder, S. H. (1993) Nature 361, 536–538.
3 Jones, D. T. and Reed, R. R. (1989) Science 244, 790–795.
4 Menco, B. P. M. et al. (1992) Neuron 8, 441–453.
5 Xia, Z. et al. (1992) Neurosci. Lett. 144, 169–173.
6 Choi, E.-J. et al. (1992) Biochemistry 31, 6492–6498.
7 Tang, W.-J. and Gilman, A. G. (1991) Science 254, 1500–1503.
8 Choi, E.-J. et al. (1993) Biochemistry 32, 1892–1894.
9 Jacobowitz, O. et.al. (1993) J. Biol. Chem. 268, 3829–3832.

ADENYLYL CYCLASE TYPE IV

Molecular weight

Polypeptide	118 798
Reduced SDS PAGE	110 kDa

Structure

Two tandem repeats of six hydrophobic putatively transmembrane domains separated by a large cytosolic loop and terminating in a large intracellular tail.

Carbohydrate
N-linked sites: Asn694 and Asn701.

Gene structure and localization

Unknown.

Tissue distribution

Detected in brain, heart, kidney, liver, lung and intestine [1].

Function

Generation of cyclic AMP. Receptor-linked activation by $G_{\alpha s}$ and this stimulation is enhanced by G-protein βγ. Adenylyl cyclase type IV is insensitive to calcium/calmodulin [1,2].

Amino acid sequence (rat)

```
                                              TM1
   1  MARLFSPRPP PSEDLFYETY YSLSQQYPLL ILLLVIVLCA IVALPAVAWA
                           TM2                            TM3
  51  SGRELTSDPS FLTTVLCALG GFSLLLGLAS REQQLQRWTR PLSGLIWAAL
                           TM4                            TM5
 101  LALGYGFLFT GGVVSAWDQV SFFLFIIFTV YAMLPLGMRD AAAAGVISSL
                           TM6
 151  SHLLVLGLYL GWRPESQRDL LPQLAANAVL FLCGNVVGAY HKALMERALR
 201  ATFREALSSL HSRRRLDTEK KHQEHLLLSI LPAYLAREMK AEIMARLQAG
 251  QSSRPENTNN FHSLYVKRHQ GVSVLYADIV GFTRLASECS PKELVLMLNE
 301  LFGKFDQIAK EHECMRIKIL GDCYYCVSGL PLSLPDHAIN CVRMGLDMCR
 351  AIRKLRVATG VDINMRVGVH SGSVLCGVIG LQKWQYDVWS HDVTLANHME
 401  AGGVPGRVHI TGATLALLAG AYAVERADME HRDPYLRELG EPTYLVIDPW
 451  AEEEDEKGTE RGLLSSLEGH TMRPSLLMTR YLESWGACKP FAHLSHVDSP
 501  ASTSTPLPEK AFSPQWSLDR SRTPRGLHDE LDTGDAKFFQ VIEQLNSQKQ
                                                     TM7
 551  WKQSKDFNLL TLYFREKEME KQYRLSALPA FKYYAACTFL VFLSNFTIQM
```

```
                       TM8
601  LVTTRPPALA TTYSITFLLF LLLLFVCFSE HLTKCVQKGP KMLHWLPALS
                       TM9
651  VLVATRPGLR VALGTATILL VFTMAVVSLL FLPVSSDCPF LAPNVSSVAF
                   TM10                      TM11
701  NTSWELPASL PLISIPYSMH CCVLGFLSCS LFLHMSFELK LLLLLLWLVA
                                                      TM12
751  SCSLFLHSHA WLSDCLIARL YQGSLGSRPG VLKEPKLMGA IYFFIFFFTL

801  LVLARQNEYY CRLDFLWKKK LRQEREETET MENVLPAHVA PQLIGQNRRN

851  EDLYHQSYEC VCVLFASIPD FKEFYSESNI NHEGLECLRL LNEIIADFDE

901  LLSKPKFSGV EKIKTIGSTY MAATGLNATP GQDTQQDAER SCSHLGTMVE

951  FAVALGSKLG VINKHSFNNF RLRVGLNHGP VVAGVIGAQK PQYDIWGNTV

1001 NVASRMESTG VLGKIQVTEE TARALQSLGY TCYSRGVIKV KGKGQLCTYF

1051 LNTDLTRTGS PSAS
```

Amino acids 1064
Asn694 and Asn701 represent potential N-linked glycosylation sites located in presumed extracellular loop between TM9 and TM10.

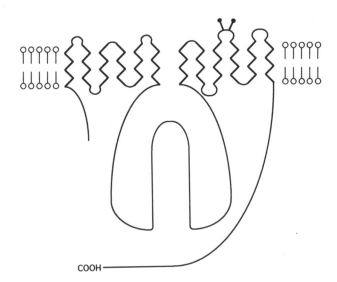

Database accession numbers

	PIR	SWISSPROT	EMBL/GENBANK	REFERENCES
Rat			M80633	1

References
[1] Gao, B. and Gilman, A. G. (1991) Proc. Natl Acad. Sci. USA 88, 10178–10182.
[2] Tang, W.-J. and Gilman, A. G. (1991) Science 254, 1500–1503.

ADENYLYL CYCLASE TYPE V

Molecular weight
Polypeptide   130 880

Structure

Two tandem repeats of six hydrophobic putatively transmembrane domains separated by a large cytosolic loop and terminating in a large intracellular tail.

Carbohydrate
N-linked sites: Asn792, Asn809 and Asn895.

Gene structure and localization
Unknown.

Tissue distribution

Expressed in heart, kidney, pancreas, gut, adrenal gland, lung, liver, and brain, where it appears enriched in corpus striatum. Also detected in cell lines including GH4 (pituitary) and PC12 (pheochromocytoma) but not S49 (lymphoma) or BAEC (vascular endothelial). Undetected in lung, skeletal muscle and testes [1-3].

Function

Generation of cyclic AMP. Receptor-linked activation by $G_{\alpha s}$ and potentially $G_{\alpha olf}$ (see Comments). Adenylyl cyclase type V is not regulateed by $\beta\gamma$ dimer and is calcium/calmodulin insensitive although free calcium inhibits at low concentrations. [1-3].

Adenylyl cyclase type V is inhibited by $G_{\alpha i}$ subtypes, but not by $G_{\alpha o}$.

Comments

Co-localization of adenylyl cyclase type V and $G_{\alpha olf}$ in corpus striatum of brain could reflect important role mediating dopaminergic cyclic AMP generation [3,4].

Truncated form of adenylyl cyclase type V (V-α) constituting N-terminal half of molecule and diverging at end of the first cytoplasmic loop has been identified [5].

Amino acid sequence (canine)

```
   1 MCSSSSAWPS AGAATTTPRW AATTPWPGAS ASASAPGRPG RSAAATTAGA

  51 AAGGGGGARR AGAAPGRPCG RRRRPGGGGR GGGAPPLGGA GPGRAAGPGP

 101 RRARARGRGR RPRGRPRGAG RRRPAGPAAC CRALLQIFRS KKFPSDKLER
                                           TM1                    TM2
 151 LYQRYFFRLN QSSLTMLMAV LVLVCLVMLA FHAARPPLRL PHLAVLAAAV
                       TM3
 201 GVILVMAVLC NRAAFHQDHM GLACYALIAV VLAVQVVGLL LPQPRSASEG
        TM4                        TM5
 251 IWWTVFFIYT IYTLLPVRMR AAVLSGVLLS ALHLAIALRA NAQDRFLLKQ
        TM6
 301 LVSNVLIFSC TNIVGVCTHY PAEVSQRQAF QETRECIQAR LHSQRENQQQ

 351 ERLLLSVLPR HVAMEMKADI NAKQEDMMFH KIYIQKHDNV SILFADIEGF

 401 TSLASQCTAQ ELVMTLNELF ARFDKLAAEN HCLRIKILGD CYYCVSGLPE

 451 ARADHAHCCV EMGMDMIEAI SLVREVTGVN VNMRVGIHSG RVHCGVLGLR

 501 KWQFDVWSND VTLANHMEAG GKAGRIHITK ATLSYLNGDY EVEPGCGGER

 551 NAYLKEHSIE TFLILRCTQK RKEEKAMIAK MNRQRTNSIG HNPPHWGAER

 601 PFYNHLGGNQ VSKEMKRMGF EDPKDKNAQE SANPEDEVDE FLGRAIDARS
                                                        TM7
 651 IDRLRSEHVR KFLLTFREPD LEKKYSKQVD DRFGAYVACA SLVFLFICFV
                            TM8
 701 QITIVPHSVF MLSFYLTCFL LLTLVVFVSV IYSCVKLFPG PLQSLSRKIV
                            TM9
 751 RSKTNSTLVG VFTITLVFLS AFVNMFMCNS EDLLGCLADE HNISTSRVNA
                                                        TM10
 801 CHVAASAANL SLGDEQGFCG TPWPSCNFPE YFTYSVLLSL LACSVFLQIS
                TM11
 851 CIGKLVLMLA IELIYVLVVE VPRVTLFDNA DLLVTANAID FNNNNGTSQC
              TM12
 901 PEHATKVALK VVTPIIISVF VLALYLHAQQ VESTARLDFL WKLQATEEKE

 951 EMEELQAYNR RLLHNILPKD VAAHFLARER RNDELYYQSC ECVAVMFASI

1001 ANFSEFYVEL EANNEGVECL RVLNEIIADF DEIISEDRFR QLEKIKTIGS

1051 TYMAASGLND STYDKVGKTH IKALADFAMK LMDQMKYINE HSFNNFQMKI

1101 GLNIGPVVAG VIGARKPQYD IWGNTVNVAS RMDSTGVPDR IQVTTDMYQV

1151 LAANTYQLEC RGVVKVKGKG EMMTYFLNGG PPLS
```

Amino acids 1184
Asn792, Asn809 and Asn895 represent potential sites for *N*-linked glycosylation and are located in presumed extracellular loops between TM9 and TM10, and TM11 and TM12.

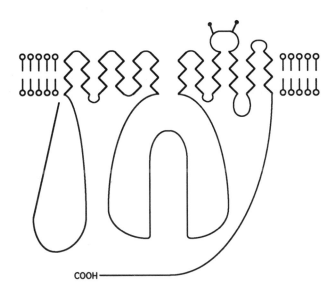

Database accession numbers

	PIR	SWISSPROT	EMBL/GENBANK	REFERENCES
Canine			M88649	1
Canine (V-α)			M97886	5
Rat			M96159	2
Rat				3

References

1 Ishikawa, Y. et al. (1992) J. Biol. Chem. 267, 13553–13557.
2 Premont, R. T. et al. (1992) Proc. Natl Acad. Sci. USA 89, 9809–9813.
3 Glatt, C. E. and Snyder, S. H. (1993) Nature 361, 536–538.
4 Drinnan, S. L. et al. (1991) Mol. Cell. Neurosci. 2, 66–70.
5 Katsushika, S. et al. (1993) J. Biol. Chem. 268, 2273–2276.

ADENYLYL CYCLASE TYPE VI

Molecular weight
Polypeptide   130 300

Structure

Two tandem repeats of six hydrophobic putatively transmembrane domains separated by a large cytosolic loop and terminating in a large intracellular tail.

Carbohydrate
N-linked sites: Asn762, Asn775, Asn790 and Asn875.

Gene structure and localization
Unknown.

Tissue distribution

Wide distribution including heart, brain, intestine, lung, spleen, kidney as well as cell lines S49 (lymphoma), A10 (smooth muscle), GH4 (pituitary), C6–2B (glioma) and NCB-20 (neuroblastoma-embryonic brain explant hybrid) [1–3,5].

Function

Generation of cyclic AMP. Receptor-linked activation by $G_{\alpha s}$ inhibited by mutationally activated $G_{\alpha i2}$. Adenylyl cyclase type VI is insensitive to calcium/calmodulin and G-protein βγ. Inhibition by submicromolar free calcium suggests a potential regulatory mechanism [1–3].

Comments
Type VI adenylyl cyclase was originally referred to as type V by Yoshimura and Cooper [2].

Amino acid sequence (canine)

```
  1 MSWFSGLLVP KVDERKTAWG ERNGQKRPRR GTRTSGFCTP RYMSCLRDAQ

 51 PPSPTPAAPP RCPWQDEAFI RRGGPGKGTE LGLRAVALGF EDTEAMSAVG

101 AAGGGPDVTP GSRRSCWRRL AQVFQSKQFR SAKLERLYQR YFFQMNQSSL
              TM1                            TM2
151 TLLMAVLVLL TAVLLAFHAA PARPQPAYVA LLACAATLFV ALMVVCNRHS
                TM3                             TM4
201 FRQDSMWVVS YVVLGILAAV QVGGALAANP RSPSVGLWCP VFFVYITYTL
              TM5                             TM6
251 LPIRMRAAVF SGLGLSTLHL ILAWQLNRGD AFLWKQLGAN MLLFLCTNVI

301 GICTNYPAEV SQRQAFQETR GYIQARLHLP DENRQQERLL LSVLPQHVAM

351 EMKEDINTKK EDMMFHKIYI QKHDNVSILF ADIEGFTSLA SQCTAQELVM

401 TLNELFARFD KLAAENHCLR IKILGDCYYC VSGLPEARAD HAHCCVEMGV

451 DMIEAISLVR EVTGVNVNMR VGIHSGRVHC GVLGLRKWQF DVWSNDVTLA

501 NHMEAARAGR IHITRATLQY LNGDYEVEPG RGGERNAYLK EQHIETFLIL
```

```
 551 GASQKRKEEK  AMLAKLQRTR  ANSMEGLMPR  WVPDRAFSRT  KDSKAFRQMG

 601 IDDSSKDNRG  AQDALNPEDE  VDEFLGRAID  ARSIDQLRKD  HVRRFLLTFQ
                                    TM7
 651 REDLEKKYSR  KVDPRFGAYV  ACALLVFCFI  CFIQLLVFPH  STVMLGIYAS
           TM8                                            TM9
 701 IFVLLLITVL  TCAVYSCGSL  FPKALRRLSR  SIVRSRAHST  VVGIFSVLLV

 751 FTSAIANMFT  CNHTPIRTCA  ARMLNVTPAD  ITACHLQQLN  YSLGLDAPLC
                               TM10                TM11
 801 EGTAPTCSFP  EYFVGNMLLS  LLASSVFLHI  SSIGKLAMIF  VLGLIYLVLL
                               TM12
 851 LLGPPSTIFD  NYDLLLGVHG  LASSNDTFDG  LDCPAAGRVA  LKYMTPVILL

 901 VFALALYLHA  QQVESTARLD  FLWKLQATGE  KEEMEELQAY  NRRLLHNILP

 951 KDVAAHFLAR  ERRNDELYYQ  SCECVAVMFA  SIANFSEFYV  ELEANNEGVE

1001 CLRLLNEIIA  DFDEIISEER  FRQLEKIKTI  GSTYMAASGL  NASTYDQAGR

1051 SHITALADYA  MRLMEQMKHI  NEHSFNNFQM  KIGLNMGPVV  AGVIGARKPQ

1101 YDIWGNTVNV  SSRMDSTGVP  DRIQVTTDLY  QVLAAKRYQL  ECRGVVKVKG

1151 KGEMTTYFLN  GGPPS
```

Amino acids 1165

Asn762, Asn775, Asn790 and Asn875 represent potential sites for *N*-linked glycosylation located within presumed extracellular loops between TM9 and TM10, and TM11 and TM12.

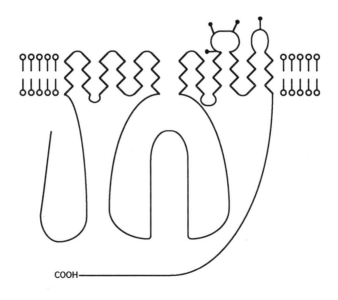

COOH

Database accession numbers

	PIR	SWISSPROT	EMBL/GENBANK	REFERENCES
Canine			M94968	1
Murine			M93422	2,4
			M96653	
Rat			M96160	3

References

1 Katsushika, S. et al. (1992) Proc. Natl Acad. Sci. USA 89, 8774–8778.
2 Yoshimura, M. and Cooper, D. M. F. (1992) Proc. Natl Acad. Sci. USA 89, 6716–6720.
3 Premont, R. T. et al. (1992) Proc. Natl Acad. Sci. USA 89, 9809–9813.
4 Premont, R. T. et al. (1992) Endocrinology 131, 2774–2784.
5 Krupinski, J. Et al. (1993) J. Biol. Chem. 267, 24858–24862.

ADENYLYL CYCLASE PARTIAL SEQUENCE (TYPE VIII)

Molecular weight

Polypeptide   76 422 (partial sequence)

Structure

Partial sequence probably corresponds to the second tandem repeat of six hydrophobic putatively transmembrane domains terminating in large intracellular tail.

Carbohydrate

N-linked sites: Asn241 and Asn312.

Gene structure and localization

Unknown.

Tissue distribution

Distributed widely in brain, particularly cerebellum, hippocampus and olfactory system [1]. Distribution in peripheral tissues unknown.

Function

Full length enzyme presumed active in synthesis of cyclic AMP.

Amino acid sequence (partial)

```
  1 LRKHNIETYL IKQPEDSLLS LPEDIVKESV SSSDRRNSGA TFTEGSWSPE

 51 LPFDNIVGKQ NTLAALTRNS INLLPNHLAQ ALHVQSGPEE INKRIEHTID
                                                    TM7
101 LRSGDKLRRE HIKPFSLMFK DSSLEHKYSQ MRDEVFKSNL VCAFIVLLFI
```

```
                        TM8
151 TAIQSLLPSS RVMPMTIQFS ILIMLHSALV LITTAEDYKC LPLILRKTCC
         TM9
201 WINETYLARN VIIFASILIN FLGAILNILW CDFDKSIPLK NLTFNSSAVF
         TM10                                   TM11
251 TDICSYPEYF VFTGVLAMVT CAVFLRLNSV LKLAVLLIMI AIYALLTETV
                        TM12
301 YAGLFLRYDN LNHSGEDFLG TKEVSLLLMA MFLLAVFYHG QQLEYTARLD

351 FLWRVQAKEE INEMKELREH NENMLRNILP SHVARHFLEK DRDNEELYSQ

401 SYDAVGVMFA SIPGFADFYS QTEMNNQGVE CLRLLNEIIA DFDELLGEDR

451 FQDIEKIKTI GSTYMAVSGL SPEKQQCEDK WGHLCALADF SLALTESIQE

501 INKHSFNNFE LRIGISHGSV VAGVIGAKKP QYDIWGKTVN LASRMDSTGV

551 SGRIQVPEET YLILKDQGFA FDYRGEIYVK GISEQEGKIK TYFLLGRVQP

601 NPFILPPRRL PGQYSLAAVV LGLVQSLNRQ RQKQLLNENN NTGIIKGHYN

651 RRTLLSPSGT EPGAQAEGTD KSDLP
```

Amino acids 675 (partial sequence)

Underlined are regions of hydrophobicity indicating potential trans-membrane domains as predicted using PEPPLOT routine (based on hydrophobicity measures of Kyte and Doolittle and of Goldman, Engelman and Steiz) of the Wisconsin GCG sequence analysis software package version 7. Regions are denoted TM7–TM12, as partial sequence shows highest homology with second tandem repeat (c-terminal half) of other adenylyl cyclase subtypes.

Asn241 and Asn312 represent potential sites for N-linked glycosylation located within presumed extracellular loops located between TM9 and TM10, and TM11 and TM12 of full-length molecule.

Database accession numbers

	PIR	SWISSPROT	EMBL/GENBANK	REFERENCE
Human	PQ0227		M83533	2

References

1 Matsuoka, I. et al. (1992) J. Neurosci. 12, 3350–3360.
2 Parma, J. et al. (1991) Biochem. Biophys. Res. Commun. 179, 455–462.

Retinal 3',5'-cGMP phosphodiesterase

EC 3.1.4.17

ROD cGMP-PDE α-SUBUNIT (αPDE)

Molecular weights
Polypeptide   99 592
Reduced SDS PAGE   88–90 kDa

Gene structure
Unknown.

Human gene location
Chromsome 5 (5q31.2–5q34) [1].

Tissue Distribution
Retinal rod cells.

Function

Visual signal transduction under conditions of low illumination. Rod αPDE associates with βPDE- and γPDE-subunits to form inactive oligomeric αβ(γ)2 holoenzyme. Activation is linked to photoactivated rhodopsin by transducin (Gαt-1) with rapid cGMP hydrolysis leading to closure of cyclic nucleotide-gated channels and rod cell hyperpolarization. Catalytically active cGMP-PDE probably reflects an αβPDE heterodimer formed following Gαt association with inhibitory γPDE subunits [2–9].

Amino acid sequence

```
  1   MGEVTAEEVE KFLDSNIGFA KQYYNLHYRA KLISDLLGAK EAAVDFSNYH

 51   SPSSMEESEI IFDLLRDFQE NLQTEKCIFN VMKKLCFLLQ ADRMSLFMYR

101   TRNGIAELAT RLFNVHKDAV LEDCLVMPDQ EIVFPLDMGI VGHVAHSKKI

151   ANVPNTEEDE HFCDFVDILT EYKTKNILAS PIMNGKDVVA IIMAVNKVDG

201   SHFTKRDEEI LLKYLNFANL IMKWYHLSYL HNCETRRGQI LLWSGSKVFE

251   ELTDIERQFH KALYTVRAFL NCDRYSVGLL DMTKQKEFFD VWPVLMGEVP

301   PYSGPRTPDG REINFYKVID YILHGKEDIK VIPNPPPDHW ALVSGLPTYV

351   AQNGLICNIM NAPSEDFFAF QKEPLDESGW MIKNVLSMPI VNKKEEIVGV

401   ATFYNRKDGK PFDEMDETLM ESLTQFLGWS VLNPDTYESM NKLENRKDIF

451   QDIVKYHVKC DNEEIQKILK TREVYGKEPW ECEEEELAEI LQAELPDADK
```

```
501 YEINKFHFSD LPLTEI ELVK CGIQMYYELK VVDKFHIPQE ALVRFMYSLS

551 KGYRKITYHN WRHGFNVGQT MFSLLVTGKL KRYFTDLEAL AMVTAAFCHD

601 IDHRGTNNLY QMKSQNPLAK LHGSSILERH HLEFGKTLLR DESLNIFQNL

651 NRRQHEHAIH MMDIAIIATD LALYFKKRTM FQKIVDQSKT YESEQEWTQY

701 MMLEQTRKEI VMAMMMTACD LSAITKPWEV QSQVALLVAA EFWEQGDLER

751 TVLQQNPIPM MDRNKADELP KLQVGFIDFV CTFVYKEFSR FHEEITPMLD

801 GITNNRKEWK ALADEYDAKM KVQEEKKQKQ QSAKSAAAGN QPGGNQPRGA

851 TTSKSCCIQ
```

Amino acids 859

Cys856 is potential site of isoprenylation and carboxymethylation [15].

Boxed region represents predicted catalytic core based on similarities with other PDEs [12,13].

Shaded boxed regions represent potential noncatalytic cGMP-binding sites [12–14].

Database accession numbers

	PIR	SWISSPROT	EMBL/GENBANK	REFERENCES
Human	B34611	P16499	M26061	1
Bovine	A34611	P11541	M26043	1,10
	S00161		M27541	
	S06418		M36683	
	S08516		X12756	
Murine	S13030			11

References

1 Pittler, S. J. et al. (1990) Genomics 6, 272–283.
2 **Stryer, L. (1986) Annu. Rev. Neurosci. 9, 87–119.**
3 **Hurley, J. B. (1987) Annu. Rev. Physiol. 49, 793–812.**
4 **Liebman, P. A. et al. (1987) Ann. Rev. Physiol. 49, 765–791.**
5 **Chabre, M. and Deterre, P. (1989) Eur. J. Biochem. 179, 255–266**
6 **Stryer, L. (1991) J. Biol. Chem. 266, 10711–10714.**
7 Hurley, J. B. and Stryer, L. (1982) J. Biol. Chem. 257, 11094–11099.
8 Deterre, P. et al. (1988) Proc. Natl Acad. Sci. USA 85, 2424–2428.
9 Fung, B. K-K. et al. (1990) Biochemistry 29, 2657–2664.
10 Ovchinnikov, Y. A. et al. (1987) FEBS Lett. 223, 169–173.
11 Baehr, W. et al. (1991) FEBS Lett. 278, 107–114.
12 Charbonneau, H. et al. (1990) Proc. Natl Acad. Sci. USA 87, 288–292.
13 Li, T. et al. (1990) Proc. Natl Acad. Sci. USA 87, 293–297.
14 Gillespie, P. G. and Beavo, J. A. (1989) Proc. Natl Acad. Sci. USA 86, 4311–4315.
15 Ong, O. C. et al. (1989) Proc. Natl Acad. Sci. USA 86, 9238–9242.

CONE cGMP-PDE α'-SUBUNIT (α'PDE)

Molecular weights

Polypeptide   98 797
Reduced SDS PAGE   90–94 kDa

Gene structure and localization

Unknown.

Tissue distribution

Retinal cone cells.

Function

Colour vision signal transduction. α'PDE interacts with additional small subunits (11 kDa, 13 kDa, 15 kDa) to form inactive oligomeric holoenzyme. Activation is linked to photoactivated opsins by transducin ($G_{\alpha t-2}$) with rapid cGMP hydrolysis leading to closure of cyclic nucleotide-gated channels and cone cell hyperpolarization. Catalytically active cone cGMP-PDE probably reflects an α'α'PDE homodimer [1-5].

Comments

Cone cGMP-PDE is substantially more sensitive to activation by transducin than rod enzyme [3].

Amino acid sequence (bovine)

```
  1   MGEISQETVE KYLEANPQFA KEYFNRKLQV EVPSGGAQAP ASASFPGRTL

 51   AEEAALYLEL LEVLLEEAGS VELAAHRALQ RLAQLLQADR CSMFLCRARN

101   GTPEVASKLL DVTPTSKFED NLVVPDREAV FPLDVGIVGW VAHTKKTFNV

151   PDVKKNSHFS DFMDKQTGYV TRNLLATPIV MGKEVLAVFM AVNKVDASEF

201   SKQDEEVFSK YLSFVSIILK LHHTNYLYNI ESRRSQILMW SANKVFEELT

251   DVERQFHKAL YTVRTYLNCE RYSIGLLDMT KEKEFYDEWP VKLGEVEPYK

301   GPKTPDGREV IFYKIIDYIL HGKEEIKVIP TPPMDHWTLI SGLPTYVAEN

351   GFICNMLNAP ADEYFTFQKG PVDETGWVIK NVLSLPIVNK KEDIVGVATF

401   YNRKDGKPFD EYDEHIAETL TQFLGWSLLN TDTYEKMNKL ENRKDIAQEM

451   LMNHTKATPD EIKSILKFKE KLNIDVIEDC EEKQLVTILK EDLPDPRTAD
```

```
501 LYEFRFRHLP ITEHELIKCG LRLFFEINVV EKFKVPVEVL TRWMYTVRKG

551 YRAVTYHNWR HGFNVGQTMF TLLMTGRLKK YYTDLEAFAM LAAAFCHDID

601 HRGTNNLYQM KSTSPLARLH GSSILERHHL EYSKTLLQDE SLNIFQNLNK

651 RQYETVIHLF EVAIIATDLA LYFKKRTMFQ KIVDACEKME TEEEAIKYVT

701 IDPTKKEIIM AMMMTACDLS AITKPWEVQS QVALLVANEF WEQGDLERTV

751 LQQQPIPMMD RNKKDELPKL QVGFIDFVCT FVYKEFSRFH KEITPMLNGL

801 QNNRVEWKSL ADEYDEKMKV IEEMKKQEEG NTTEKAVEDS GGGGDDKKSK

851 TCLML
```

Amino acids 855
Cys852 is potential site of carboxymethylation and isoprenylation.
Boxed region represents predicted catalytic core based on similarities with other PDEs [4,5].
Shaded boxes represent potential noncatalytic cGMP-binding sites [3–5].

Database accession numbers

	PIR	SWISSPROT	EMBL/GENBANK	REFERENCES
Bovine	A34809	P16586	M29465	[4,5]
	A34810		M37838	

References

[1] **Hurley, J. B. (1987) Annu. Rev. Physiol. 49, 793–812.**
[2] Nathans, J. et al. (1986) Science 232,193–202.
[3] Gillespie, P. G. and Beavo, J. A. (1988) J. Biol. Chem. 263, 8133–8141.
[4] Charbonneau, H. et al. (1990) Proc. Natl Acad. Sci. USA 87, 288–292.
[5] Li, T. et al. (1990) Proc. Natl Acad. Sci. USA 87, 293–297.

ROD cGMP-PDE β-SUBUNIT (βPDE)

Molecular weight
Polypeptide   98 441
Reduced SDS PAGE   84–86 kDa

Gene structure (human)

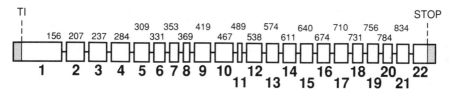

Organization of human cGMP-PDE β-subunit gene adapted from ref. 1.

Human gene location and size
Chromosome 4 (4p16.3); ~43 kb [1].

Tissue distribution

Retinal rod cells. Undetectable in adrenal, lung, oesophagus, skeletal muscle, kidney and liver [1].

Function

Visual signal transduction under conditions of low illumination. Rod βPDE associates with αPDE- and γPDE-subunits to form inactive oligomeric αβ(γ)$_2$ holoenzyme. Activation is linked to photoactivated rhodopsin by transducin (G$_{\alpha t\text{-}1}$) with rapid cGMP hydrolysis leading to closure of cyclic nucleotide-gated channels and rod cell hyperpolarization. Catalytically active cGMP-PDE probably reflects an αβPDE heterodimer formed following G$_{\alpha t}$ association with inhibitory γPDE subunits [2–9].

Comments

Mutations within βPDE could underlie lesions in autosomal recessive retinal degeneration (*rd*) mouse and rod/cone dysplasia in Irish setter dogs which are models for human retinitis pigmentosa) associated with elevated retinal cGMP [10].

An additional β-subunit generated by alternative splicing of the βPDE gene has been reported in mouse retina. This variant is truncated at the C-terminus, although it contains most of the predicted catalytic core [11].

Amino acid sequence

```
  1   MSLSEEQARS FLDQNPDFAR QYFGKKLSPE NVGRGCEDGC PPDCDSLRDL

 51   CQVEESTALL ELVQDMQESI NMERVVFKVL RRLCTLLQAD RCSLFMYRQR

101   NGVAELATRL FSVQPDSVLE DCLVPPDSEI VFPLDIGVVG HVAQTKKMVN

151   VEDVAECPHF SSFADELTDY KTKNMLATPI MNGKDVVAVI MAVNKLNGPF

201   FTSEDEDVFL KYLNFATLYL KIYHLSYLHN CETRRGQVLL WSANKVFEEL

251   TDIERQFHKA FYTVRAYLNC ERYSVGLLDM TKEKEFFDVW SVLMGESQPY

301   SGPRTPDGRE IVFYQVIDYL LHGKEEIKVI PTPSADHWAL ASGLPSYVAE

351   SGFICNIMNR SADEMFKFQE GALDDSGWLI KNVLSMPIVN KKEEIVGVAT

401   FYNRKDGKPF DEQDEVLMES LTQFLGWSVM NTDTYDKMNK LENRKDIAQD
```

```
451 MVLYHVKCDR DEIQLILPTR ARLGKEPADC DEDELGEILK EELPGPTTFD

501 IYEFHFSDLE CTELDLVKCG IQMYYELGVV RKFQIPQEVL VRFLFSISKG

551 YRRITYHNWR HGFNVAQTMF TLLMTGKLKS YYTDLEAFAM VTAGLCHDID

601 HRGTNNLYQM KSQNPLAKLH GSSILERHHL EFGKFLLSEE TLNIYQNLNR

651 RQHEHVIHLM DIAIIATDLA LYFKKRAMFQ KIVDESKNYQ DKKSWVEILS

701 LETTRKEIVM AMMMTACDLS AITKPWEVQS KVALLVAAEF WEQGDLERTV

751 LDQQPIPMMD RNKAAELPKL QVGFIDFVCT FVYKEFSRFH EEILPMFDRL

801 QNNRKEWKAL ADEYEAKVKA LEEKEEEERV AAKKVGTEIC NGGPAPKSST

851 CCIL
```

Amino acids 854
Cys851 is potential site for carboxymethylation and isoprenylation.
Boxed region represents predicted catalytic core.
Shaded boxes indicate potential noncatalytic cGMP-binding sites based on
sequence alignment with human cone α'PDE subunit [13].

Database accession numbers

	PIR	SWISSPROT	EMBL/GENBANK	REFERENCES
Human	S18715		X62692-X62695	1
Bovine	A36617	P23439	J05553	12
			X57146	
Murine	S13031	P23440	X55968	10,11
	S13121			

References

1 Weber, B. et al. (1991) Nucleic Acids Res. 19, 6263–6268.
2 **Stryer, L. (1986) Annu. Rev. Neurosci. 9, 87–119.**
3 **Hurley, J. B. (1987) Annu. Rev. Physiol. 49, 793–812.**
4 **Liebman, P. A. et al. (1987) Annu. Rev. Physiol. 49, 765–791.**
5 **Chabre, M. and Deterre, P. (1989) Eur. J. Biochem. 179, 255–266.**
6 **Stryer, L. (1991) J. Biol. Chem. 266, 10711–10714.**
7 Hurley, J. B. and Stryer, L. (1982) J. Biol. Chem. 257, 11094–11099.
8 Deterre, P. et al. (1988) Proc. Natl Acad. Sci. USA 85, 2424–2428.
9 Fung, B. K-K. et al. (1990) Biochemistry 29, 2657–2664.
10 Bowes, C. et al. (1990) Nature 347, 677–680.
11 Baehr, W. et al. (1991) FEBS Lett. 278, 107–114.
12 Lipkin, V. M. et al. (1990) J. Biol. Chem. 265, 12955–12959.
13 Li, T. et al. (1990) Proc. Natl Acad. Sci. USA 87, 293–297.

ROD cGMP-PDE γ-SUBUNIT (γPDE)

Molecular weights

Polypeptide	9643
Reduced SDS PAGE	9–11 kDa

Gene structure

Unknown.

Human gene location

Chromosome 17 (17q21.1) [1].

Tissue distribution

Retinal rod cells [2].

Function

Visual signal transduction under conditions of low illumination. Rod γPDE associates with αPDE- and βPDE-subunits to form oligomeric αβ(γ)$_2$ holoenzyme. γPDE is inhibitory on cGMP-PDE activity. Light-dependent activation is mediated by transducin (G$_{αt-1}$) which associates directly with γPDE to remove its inhibitory constraint. Full activation requires two molecules of activated transducin to interact with both γPDES and leads to rapid cGMP hydrolysis, closure of cyclic nucleotide-gated channels and rod cell hyperpolarization [3–13].

Comments

Enzymatic digestion of C-terminal 7 residues or a deletion mutant ending at Asn74 both lead to a loss of cGMP-PDE inhibitory activity [14,15].

Residues 46–87 involved in interaction with C-terminal region of transducin [16].

γPDE N-terminal is acetylated [17].

Amino acid sequence

```
 1 MNLEPPKAEF RSATRVAGGP VTPRKGPPKF KQRQTRQFKS KPPKKGVQGF

51 GDDIPGMEGL GTDITVICPW EAFNHLELHE LAQYGII
```

Amino acids 87

Database accession numbers

	PIR	SWISSPROT	EMBL/GENBANK	REFERENCES
Human	JH0142	P18545	M36476	[1]
Bovine	A25268	P04972	X04823	[17]
	A29580		X04270	
Murine	S00612	P09174	Y00746	[18]

References

1 Tuteja, N. et al. (1990) Gene 88, 227–232.
2 Hamilton, S. E. and Hurley, J. B. (1990) J. Biol. Chem. 265, 11259–11264.
3 **Stryer, L. (1986) Annu. Rev. Neurosci. 9, 87–119.**
4 **Hurley, J. B. (1987) Annu. Rev. Physiol. 49, 793–812.**
5 **Liebman, P. A. et al. (1987) Annu. Rev. Physiol. 49, 765–791.**
6 **Chabre, M. and Deterre, P. (1989) Eur. J. Biochem. 179, 255–266**
7 **Stryer, L. (1991) J. Biol. Chem. 266, 10711–10714.**
8 Hurley, J. B. and Stryer, L. (1982) J. Biol. Chem. 257, 11094–11099.
9 Deterre, P. et al. (1988) Proc. Natl Acad. Sci. USA 85, 2424–2428.
10 Fung, B. K.-K. et al. (1989) Biochemistry 28, 3133–3137.
11 Griswold-Prenner, I. et al. (1989) Biochemistry 28, 6145–6150.
12 Fung, B. K.-K. et al. (1990) Biochemistry 29, 2657–2664.
13 Wensel, T. G. and Stryer, L. (1990) Biochemistry 29, 2155–2161.
14 Lipkin, V. M. et al. (1988) FEBS Lett. 234; 287–290.
15 Brown, R. L. and Stryer, L. (1989) Proc. Natl Acad. Sci. USA 86, 4922–4926.
16 Artemyev, N. O. et al. (1992) J. Biol. Chem. 267, 25067–25072.
17 Ovchinnikov, Y. A. et al. (1986) FEBS Lett. 204, 288–292.
18 Tuteja, N. and Farber, D. B. (1988) FEBS Lett. 232, 182–186.

CONE cGMP-PDE γ'-SUBUNIT (γ'PDE)

Molecular weight
Polypeptide   9102
Reduced SDS PAGE   13 kDa

Gene structure and localization

Unknown.

Tissue Distribution

Expressed in a subset of retinal cone cells [1].

Function

Colour vision signal transduction. Oligomeric cone cGMP-PDE includes a catalytic α'α' homodimer isolated as a complex with three other subunits (11 kDa, 13 kDa and 15 kDa) [2]. Cone cell γPDE may be a 13 kDa inhibitory subunit with a role analogous to rod γPDE [1].

Amino acid sequence (bovine)

```
  1 MSDNTVLAPP TSNQGPTTPR KGPPKFKQRQ TRQFKSKPPK KGVKGFGDDI

 51 PGMEGLGTDI TVICPWEAFS HLELHELAQF GII
```

Amino acids 83

Database accession numbers

	PIR	SWISSPROT	EMBL/GENBANK	REFERENCE
Bovine	A42229	P22571	J05578	1

References

[1] Hamilton, S. E. and Hurley, J. B. (1990) J. Biol. Chem. 265, 11259–11264.
[2] Gillespie, P. G. and Beavo, J. A. (1988) J. Biol. Chem. 263, 8133–8141.

Index